From Catchment Management to Managing River Basins

From Catchment Management to Managing River Basins

Editor: Gloria Shaw

Callisto Reference
www.callistoreference.com

Callisto Reference,
118-35 Queens Blvd., Suite 400,
Forest Hills, NY 11375, USA

Visit us on the World Wide Web at:
www.callistoreference.com

ISBN: 978-1-64116-824-3 (Hardback)

Cataloging-in-Publication Data

From catchment management to managing river basins / edited by Gloria Shaw.
p. cm.
Includes bibliographical references and index.
ISBN 978-1-64116-824-3
1. Watershed management. 2. Water resources development. 3. Watersheds. I. Shaw, Gloria.
TC413 .F76 2023
333.91--dc23

Table of Contents

Preface

A catchment refers to an area of land commonly surrounded by hills or mountains in which the water flows and gets collected. Catchment management is a sustainable and potentially low-carbon impact method of altering the quality of raw water at its source by managing land use practices on a catchment scale. It can also aid in the enhancement of resilience and the reduction of flood risk. Rivers supply water for power generation, irrigation, industry, domestic use and for a variety of other ecosystem services, and intrinsic and biodiversity benefits. River basin refers to an area of land which is drained by a river and its tributaries. It is considered to be the basic hydrological unit for managing water resources. Challenges related to sustainable water management on a river basin scale include changes in land use, increasing water usage and climate change. This book elucidates the concepts and innovative models around prospective developments with respect to catchment and river basin management. It will also provide interesting topics for research, which interested readers can take up. A number of latest researches have been included to keep the readers up-to-date with the global concepts in this area of study.

This book unites the global concepts and researches in an organized manner for a comprehensive understanding of the subject. It is a ripe text for all researchers, students, scientists or anyone else who is interested in acquiring a better knowledge of this dynamic field.

I extend my sincere thanks to the contributors for such eloquent research chapters. Finally, I thank my family for being a source of support and help.

Editor

1

Current and Future Ecological Status Assessment: A New Holistic Approach for Watershed Management

André R. Fonseca [1,*], João A. Santos [1,2], Simone G.P. Varandas [1,3], Sandra M. Monteiro [1,4], José L. Martinho [5,6], Rui M.V. Cortes [1,3] and Edna Cabecinha [1,4]

1 Centre for the Research and Technology of Agro-Environmental and Biological Sciences, CITAB, 5000-801 Vila Real, Portugal; jsantos@utad.pt (J.A.S.); simonev@utad.pt (S.G.P.V.); smonteir@utad.pt (S.M.M.); rcortes@utad.pt (R.M.V.C.); edna@utad.pt (E.C.)

2 Departamento de Física, Universidade de Trás-os-Montes e Alto Douro, UTAD, 5000-801 Vila Real, Portugal

3 Departamento de Ciências Florestais e Arquitetura Paisagista, Universidade de Trás-os-Montes e Alto Douro, UTAD, 5000-801 Vila Real, Portugal

4 Departamento de Biologia e Ambiente, Universidade de Trás-os-Montes e Alto Douro, UTAD, 5000-801 Vila Real, Portugal

5 Departamento de Geologia, Universidade de Trás-os-Montes e Alto Douro, UTAD, 5000-801 Vila Real, Portugal; martinho@utad.pt

6 Geosciences Center, University of Coimbra, 3030-790 Coimbra, Portugal

* Correspondence: andre.fonseca@utad.pt.

Abstract: The Paiva River catchment, located in Portugal, integrates the Natura 2000 network of European Union nature protection areas. Resorting to topography, climate and land-use data, a semi-distributed hydrological model (Hydrological Simulation Program–FORTRAN) was run in order to simulate the hydrological cycle of the river and its tributaries. The model was calibrated over a 25-year period and validated within a 31-year period. Its performance was verified by comparing the recorded and simulated daily flows. The values of the Nash–Sutcliffe coefficient of efficiency of 0.95 and 0.76, and coefficient of determination of 0.95 and 0.82, were achieved for calibration and validation, respectively, thus showing a quite satisfactory model performance. Subsequently, the climate change impacts on temperature and precipitation, as well as their extremes, and on the flowrates were also assessed for a future period (2041–2070) under two anthropogenic forcing scenarios (representative concentration pathways 4.5 and 8.5). A procedure for selecting the most relevant metrics for assessing the ecological condition of the Paiva River was developed based upon a set of 52 invertebrate families sampled. Correspondence analyses were carried out for biological datasets (traits/metrics) with physicochemical and land use/land cover matrices separately. Out of all variables, water quality and flow and agriculture land use explained most of the variance observed. The integrated analysis undertaken in the present study is an important advance when compared to previous studies and it provides key information to stakeholders and decision-makers, particularly when planning suitable adaptation measures to cope with changing climates in the forthcoming decades.

Keywords: hydrological modelling; climate change; water quality; biological functional traits; macroinvertebrates; HSPF

1. Introduction

Natural integrity of river basins and the potential impact of anthropogenic pressures on water resources is raising social awareness worldwide, particularly regarding their socioeconomic prominence [1–3]. Hydrological modelling and simulation aim to quantitatively answer questions regarding the behaviour of water within a given catchment basin and assess its ecological status.

Hydrological models are effective tools for assessing and predicting surface runoff, nutrient transport and fate, soil erosion, and impacts of land-use changes [4], among others. Modelling has indeed become one of the most powerful tools for catchment management [5,6]. It can be used to simulate the response of a given catchment to different inputs (e.g., meteorological or land-use changes) and management scenarios (e.g., best management practices, blue-green infrastructure implementation) [5]. They are frequently used for generating time series of potential discharges and water quality under different scenarios. Nevertheless, model simulations encompass important uncertainties that are due not only to approximations to the real-world, but also because many model parameters are difficult, if not impossible, to measure and need to be estimated [7,8]. However, the application of hydrological models is well established in predicting both an acceptable parameter set for calibration and the uncertainty that is associated with the final predictions [7,9–21].

Accurate climate information, at the basin-scale, is demanded by hydrological models so as to be useful decision support systems to policymakers and other stakeholders, such as for water resource management [22] and ecosystem modelling [23]. Exploring water basin resources with semi-distributed and fully distributed hydrological models commonly requires extensive and detailed climate data, which is not always accessible or available. Although global datasets with coarser resolutions are available (e.g., E-OBS gridded dataset) [24], which can be an alternative when suitable local network stations are lacking [25], hydrologically-relevant regional or local climate gradients may not be properly resolved. In Portugal, many river catchments have been targeted in hydrological modelling studies, although meteorological data are frequently obtained from low-density weather station networks [7,16,17,26–29]. Nonetheless, as more data have been archived into large digital databases and coupled with Geographical Information Systems (GIS), data availability and spatial resolution has been significantly improved [30].

Besides flowrate, water quality parameters can also be simulated by hydrological models, but they require an appropriate calibration and validation to local conditions, while using observed data. Water quality is usually determined based on physical, chemical and biological indicators, taking into account the guidelines provided by national and international agencies, such as the World Health Organization—WHO [31]. According to Stanford et al. [32], hydrologic processes, food resources, nutrient dynamics, riparian vegetation, and many other factors alter the structure and functioning of aquatic ecosystems along the longitudinal profile, and these factors may be changed by various anthropogenic influences.

Because climate change is a fact (changing patterns of precipitation and temperature), approaches are needed for the early detection of effects on ecosystems. The Mediterranean basin is recognised as a biodiversity hotspot and freshwater communities in these areas may be particularly vulnerable to the expected scenarios reported by the International Panel on Climate Change [33]. The climatic scenarios for this region over the next 50–100 years predict an increase in the mean air temperature (between 1–5 °C) and extreme events frequency accompanied by a decrease in the annual precipitation. According to this Report, the regional risks include: (1) "increasing water restrictions"; (2) "significant reduction in water availability from river abstraction and from groundwater resources, combined with increased water demand (e.g., for irrigation, energy and industry, domestic use) and with reduced water drainage and runoff as a result of increased evaporative demand"; and, (3) " ... increasing risk of wildfires".

Common bioindicators that are used to assess river ecosystem quality usually comprise benthic macroinvertebrate analysis and their taxonomical groups. They play an important role in the food webs of streams and rivers [34,35] and respond to various river stressors [36–38], such as organic pollution and sediments [36,39–42], heavy metals and other contaminants, thermal pollution [43], flow disruption [43,44], acid rain [45], acid mine drainage [46], and climate change [47–49], thus depicting information on the overall ecosystem status. Brown and Kroll [48] highlight that the ability of macroinvertebrate taxa to survive climate changes will depend on its resistance, resilience, and dispersal capabilities, as well as available habitat.

According to Poff [50], "traits presumably represent functional relationships with important environmental selective forces, such as stream flooding or drying, local shear stress, temperature extremes, and human pollution". This idea was emphasised by Bonada et al. [51], which found that studies using biological functional traits (at larger spatial scales and higher biological organizational levels) may anticipate the impacts of climate change. Also, Charvet et al. [52], Phillips [53], and Vieira et al. [54] stressed that, unlike taxonomic composition, biological traits are relatively stable across large environmental gradients (e.g., geology, altitude, geographical coordinates, stream order, and slope), allowing for their use for biomonitoring over large geographical areas. Accordingly, the evaluation protocols must be based on both conventional attributes and trait-based variables once together are sufficiently sensitive to distinguish impairment. Undeniably, indices that are only based on traits or together with traditional taxonomic methods are the future of the assessment of causal relationships between specific stressors and macroinvertebrate community response [55,56].

This study assesses the current water and ecological status of the Paiva River catchment, in Portugal, and the corresponding potential modifications under climate change projections. In order to achieve this goal, the objectives of this study are three-fold: (1) to calibrate and validate flowrate and water quality in the Paiva River catchment for the historical period 1950–2005; (2) collect climatic datasets from a five-member ensemble of GCM-RCM chain simulations in order to assess the impact of climate projections in flowrate and water quality under two different anthropogenic forcing scenarios (Representative Concentration Pathways, RCP), RCP4.5, and RCP8.5; and, (3) to demonstrate the importance of the macroinvertebrate traits in discriminating the different levels of water ecological status and to identify which of them can be more effectively and accurately used in future monitoring of the Paiva River catchment under a climate change context. Lastly, some preventive measures that are devoted to the conservation of this sensible aquatic ecosystem will be presented contributing future management measures.

2. Materials and Methods

2.1. Study Area

The Paiva River (Portugal) is part of Natura 2000 network (code PTCON0059), of nature protection areas in the territory of the European Union. It flows between two other Sites of Community Importance: "Serra de Montemuro" (to the north) and "Serras da Freita e Arada" (to the south) (Figure 1). It is one of the Douro River main tributaries of the left margin and covers an area of approximately 796 km^2, with a total extension of nearly 115 km [57] (Figure 1). The Paiva River catchment shows a very complex orography, with the spring at an elevation of approximately 1000 m and the river mouth at 10 m altitude, flowing in a deep canyon, and nearly halfway between very steep mountains. The land-use is mainly comprised of shrubs (38%), plantations of *Pinus Pinaster* (17%) and *Eucalyptus* (16%), followed by agriculture (14%). In its final section, the slopes show high coverage and good plant density despite the increase in the area planted with eucalyptus, already reflecting humid maritime conditions [58]. Temperate Mediterranean climate characterizes the catchment, with an average annual temperature of 13 °C and average annual precipitation greater than 1000 mm [45,46,59].

2.2. Climate Data

Very high resolution (~1 km) gridded daily mean, minimum, and maximum temperatures over mainland Portugal (PT.TG.HRES, PT.TN.HRES and PT.TX.HRES) over the baseline period of 1950–2015 were retrieved for the grid points covering the Paiva River catchment. The methodology for the development of these datasets is based on the E-OBS observational gridded dataset [24], as the underlying station data were already subjected to quality control procedures and homogeneity analysis. A two-step approach was delineated: (1) geostatistical interpolation of temperatures to capture daily mean spatial variability and (2) bilinear interpolation of daily temperature anomalies to capture temporal variability. The complete methodology followed to obtain these gridded datasets, their validation and characterization can be found in Fonseca and Santos [59]. Gridded daily precipitation totals at ~1 km spatial resolution (PT.P.HRES) over the same baseline period were also retrieved for the Paiva River catchment. The PT.PHRES was developed based on gridded daily precipitation totals at ~20 km spatial resolution (PT02 database) [60], supplied by the Portuguese Weather Service (www.ipma.pt). The steps that were followed for the calculation of the gridded dataset PT.P.HRES can be found in Fonseca and Santos [21]. These very high-resolution datasets provide detailed information on temperature and precipitation spatial and temporal variability in the Paiva River catchment, which will be an important added value for the subsequent hydrological simulations.

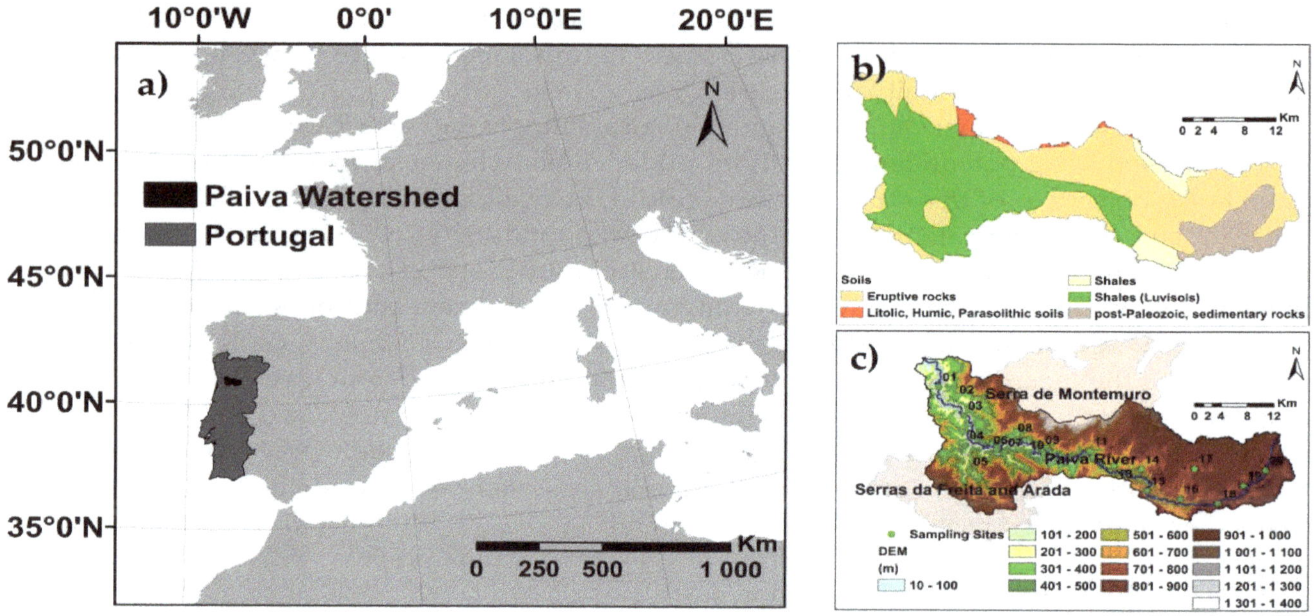

Figure 1. The Paiva River catchment: (**a**) geographical location in southwestern Europe and Portugal; (**b**) soil types (cf. legend for details on classes); and, (**c**) hypsometric chart showing the selected Digital Elevation Model, DEM (elevation in meters).

Gridded climatic datasets from a five-member ensemble of GCM-RCM chain simulations (see Table S1, Supplementary Material), developed within the framework of the EURO-CORDEX project, were retrieved for the development of climate change projections (http://www.euro-cordex.net/). The data were also subject to a bias correction performed under the previous project (SMHI-DBS45-MESAN, 1989–2010). The original spatial resolution is of about 12 km (EUR-11, 0.11° latitude × longitude) on a Gaussian mesh. The gridded daily precipitation totals (RR), minimum (TN), and maximum (TX) 2-meter air temperatures were extracted for the future period of 2041–2070 over a geographical sector covering the Paiva River catchment and under two different scenarios, RCP4.5 and RCP8.5. The data were then bilinearly interpolated to a latitude-longitude regular grid.

Several steps were undertaken for downscaling the climate model data to ~1 km spatial resolution: (1) for each model, scenario, and variable, data were bilinearly interpolated from ~12 km to ~1 km grid; (2) the gridded differences of monthly means (monthly sum) of temperature (precipitation) between model and observed data (PT.TG.HRES, PT.TN.HRES, and PT.TX.HRES) were calculated for a common baseline (1981–2010); (3) the corresponding long-term monthly differences at each grid point were then linearly interpolated to the daily timescale and smoothed out by a low-pass Lanczos filter, with a cut-off frequency at 30 days and 500 coefficients, to remove high frequency noise; and, (4) the gridded long-term daily differences were subsequently added (multiplied) to the simulated temperature (precipitation) for the period 2041–2070, under RCP4.5 and RCP8.5.

2.3. Hydrological Simulations

The climatic data were subsequently used as input in the Hydrological Simulation Program FORTRAN (HSPF) model in order to assess daily flowrates in the Paiva River. A graphical user interface of HSPF is included in the Better Assessment Science Integrating Point and Non-point Sources (BASINS), which is designed to perform water quantity and quality-based studies and was developed by the United States Environmental Protection Agency (USEPA) (https://www.epa.gov/ceam/better-assessment-science-integrating-point-and-non-point-sources-basins).

HSPF is a semi-distributed model that simulates water and contaminant transport over spatially distributed areas within a given catchment [61]. It is the combination of several previous models [62–66] and it is able to simulate continuous dynamic events, of both hydrologic and water quality processes, through surface and groundwater regimes in the catchment. The application of HSPF to model water processes and contaminant transport has been reported in many previous studies and for a wide range of types of catchments and environmental conditions [7,16,17,19,21,67,68].

The calibration process of water quantity and quality was carried out by coupling HSPF with GIS in the delineation of the river catchment data. The calibration and validation parameter values were estimated until good agreement was achieved between observed and simulated flowrates and water quality, which was then compared to the Portuguese legislation for surface freshwater.

Daily and monthly flowrates and quality constituents were calibrated and validated at the hydrometric station "Fragas da Torre", hereafter FT (Figure 1b), for the periods of 1950–1974 and 1980–2011, respectively. This is the only hydrometric station with enough observed data in the catchment to perform adequate model calibration and validation. The daily flowrates were obtained from the Sistema Nacional de Informação de Recursos Hídricos (SNIRH, www.snirh.pt). The verification of the model performance was assessed by three statistical metrics (performance measures): Coefficient of Determination (R^2), Percent Bias (PBIAS) and Nash–Sutcliffe efficiency coefficient (E) [69]. The Nash–Sutcliffe model efficiency coefficient can range from $-\infty$ to 1, where 1 is a perfect fit of modelled discharge to the observed data, whereas zero indicates that the model prediction is as accurate as the observed mean.

Regarding the water quality constituents, owing to the lack of timeliness data and accuracy, as well as the low number of records, a different approach was undertaken for calibration. It is not reasonable to expect the model to simulate a daily average concentration equal to an instant observed value, as the recorded values are typically sampled infrequently (monthly at best) and only represent an instant in time. Therefore, observed long-term means were compared against simulated long-term means to reach an agreement for calibration (best-fit parameter set). The monthly variation is also shown as the final simulation result. The calibrated parameters were water temperature, biochemical oxygen demand, dissolved oxygen, nitrates, orthophosphates, chlorophyll-a, faecal coliforms, and total suspended solids. In HSPF, nitrogen and phosphorus are modelled based on nutrient cycle empirical relationships, under the modules of PQUAL (quality constituents for pervious land, PERLND), IQUAL (quality constituents for impervious land, IMPLND), and QGUAL (quality constituents for reaches, RCHRES). Phosphorus load is usually correlated to the sediment load while using a power factor approach, while nitrogen, which is not normally deposited into the sediments, uses a method of accumulation and depletion based on the first-order wash off rate. Biological, chemical, and physical processes in streams were also simulated.

2.4. Environmental Variables, Biological and Ecological Indicators

Land-use/Land Cover (LU/LC) data (Table 1) were obtained by GIS from COS 2018 according to DGT (2019) (see Table S2, Supplementary Material). A proximity method was used to overcome the absence of strong associations between environmental descriptors at higher spatial scales and local aquatic communities, defining a circle with a radius of 1 km around each site based on Cortes et al. [70] (Figure 2). This buffer was imposed on LU/LC maps in order to obtain the percentages of each LU for each site. These variables were calculated from catchment to the river line. Temperature, flow, volume, and river size variables were derived from the above described models (see Table S2, Supplementary Material).

Table 1. Sampling points, respective river or tributary, Land-Use Land Cover (LULC) classes, elevation and typology. M—Northern mountain rivers; N1 ≤ 100—small northern rivers; N1 > 100—medium-large northern rivers [71]. Main rivers are in bold. Eu—Eucalyptus, BW—Broadleaf Woodland, CW–Coniferous Woodland, SH—Scrub & Shrubs, UR—Urban Area, and Ag—Agriculture.

		% of Land-Use Land Cover (LULC)							
	River	**Eu**	**BW**	**CW**	**SH**	**UR**	**Ag**	**Elevation (m)**	**Typology**
P1	River Paiva	52	26	12	1	0	5	118	N1 > 100
P2	River Ardena	86	7	2	0	1	4	222	N1 ≤ 100
P3	River Paiva	71	12	3	4	1	9	146	N1 >100
P4	River Paivô	56	4	21	18	0	1	256	N1 ≤ 100
P5	River de Frades	68	6	12	13	0	1	278	M
P6	River Paiva	26	5	19	40	0	8	199	N1 > 100
P7	Deilão Creek	29	8	15	46	0	1	228	N1 ≤ 100
P8	River Tenente	32	11	34	7	2	14	302	M
P9	River Sonso	37	7	29	14	4	6	300	M
P10	River Paiva	41	18	26	8	0	4	240	M
P11	Carvalhosa Creek	1	31	48	7	1	10	458	M
P12	River Paiva	2	17	46	0	6	28	437	N1 > 100
P13	River Paivó	0	12	34	1	24	29	555	N1 ≤ 100
P14	River Mau	0	6	39	35	4	16	563	N1 ≤ 100
P15	River Paiva	0	9	13	61	3	14	535	N1 > 100
P16	River Covo	0	10	51	34	0	4	640	N1 ≤ 100
P17	River Covo/Touro	11	11	10	48	0	20	764	N1 ≤ 100
P18	River Paiva	0	11	44	17	12	15	749	N1 ≤ 100
P19	River Paiva tributary	3	8	31	24	3	31	780	N1 ≤ 100
P20	River Paiva	1	16	43	22	0	16	783	N1 ≤ 100

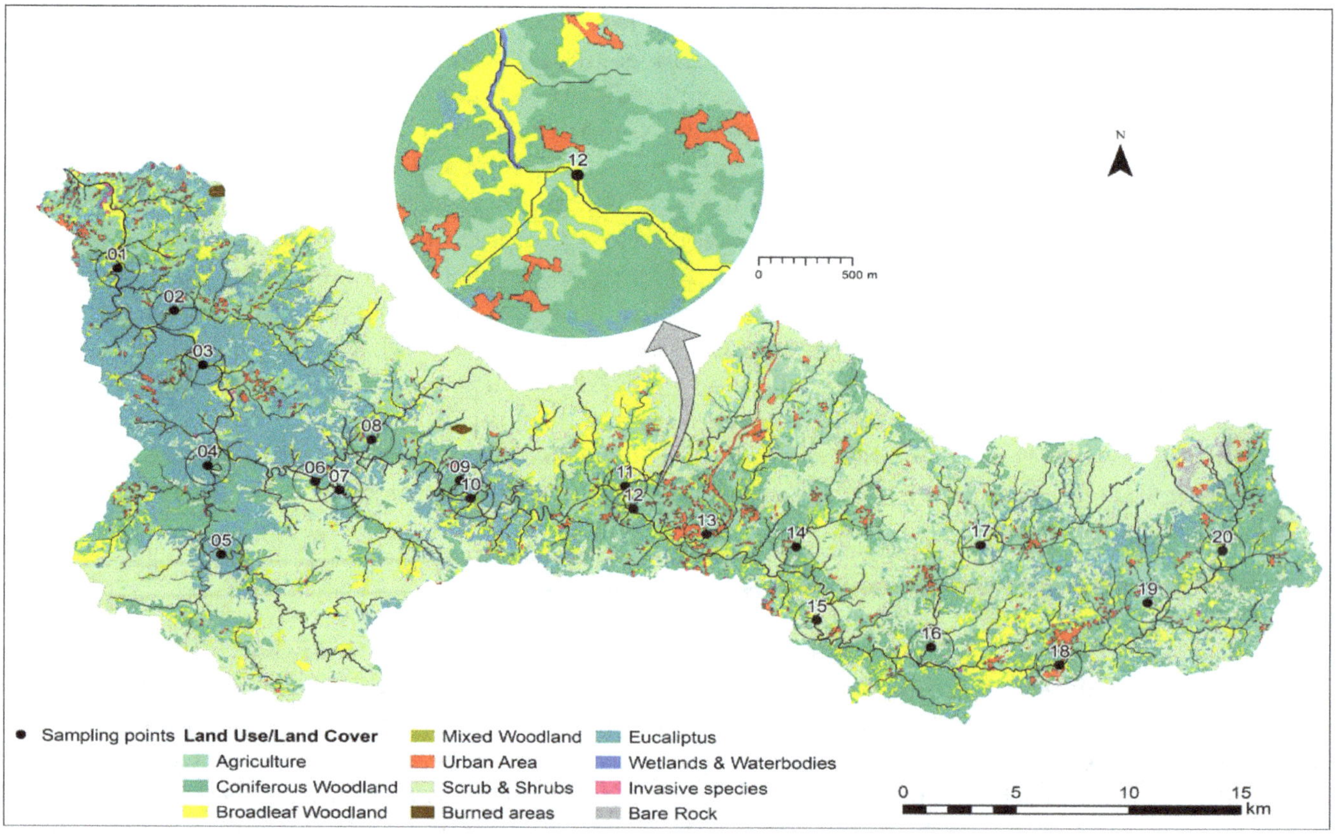

Figure 2. Location of the sampling sites and the circular buffers in their surroundings.

Physicochemical in situ parameters were measured in 20 sampling sites, which were distributed along the Paiva River catchment (Figure 2). Water samples were also collected, preserved in cooling conditions (4 °C), and then transported to the laboratory for analyses using standard methods. The analysed parameters were alkalinity (mg HCO_3^-·L^{-1}), chlorides (mg Cl^-·L^{-1}), nitrates (mg NO_3^-·L^{-1}), phosphate (mg PO_4^{3-}·L^{-1}), Total Suspended Solids (mg TSS·L^{-1}), sulphates (mg SO_4^{2-}·L^{-1}), hardness (mg $CaCO_3$·L^{-1}), calcium (mg Ca·L^{-1}), iron (μg Fe·L^{-1}), magnesium (mg Mg·L^{-1}), cadmium (μg Cd·L^{-1}), potassium (mg K·L^{-1}), sodium (mg Na·L^{-1}), total aluminium (μg Al·L^{-1}), arsenic (μg As·L^{-1}), lead (μg Pb·L^{-1}), copper (μg Cu·L^{-1}), manganese (μg Mn·L^{-1}), zinc (μg Zn·L^{-1}), nickel (μg Ni·L^{-1}), and chromium (μg Cr·L^{-1}). The values were classified according to the criteria of the ex-Portuguese Water Institute (INAG). This classification comprises five different classes and a site is considered to be unimpaired if it belongs to class A and impaired if it belongs to classes D or E. The environmental variables were standardized prior to all analyses in order to preserve the original scale.

Invertebrate assemblages were simultaneously collected with physicochemical parameters for each site with a hand-net following a WFD compliant protocol (INAG I.P., 2008). The samples were sorted and preserved with an ethanol concentration of ca. 70%. Subsequently, they were keyed to the family level, except for individuals belonging to the Oligochaeta and Hydracarina classes. The work was carried out during the autumn of 2018.

The macroinvertebrate data were organized into two data groups: (a) relative abundance of taxa, resulting in a matrix comprising 52 invertebrate families (see Table S3, Supplementary Material). The data of taxa were log(x+1) transformed before subsequent analyses; (b) biological traits (food and feeding habits), ecological traits (longitudinal and transversal distribution—zonation, current and microhabitat preferences, locomotion type), and conventional metrics (richness, composition and dominance measures, descriptors, and biotic indices) reflecting the life history of taxa and adaptive responses to environmental factors (see Table S4 [72], Supplementary Material). To obtain this second matrix, macroinvertebrate data were imported into the ASTERICS 4.03 software (AQUEM/STAR Ecological River Classification System) in order to calculate traits and metrics. The data were log(x + 1) transformed, except for metrics expressed in percentage. These metrics were not transformed, since they were normally distributed (see Table S4, Supplementary Material).

The Northern Portugal Invertebrate Index ($IPtI_N$) was used in order to assess the biological status of freshwater [73–77]. This index depicts information on how to differentiate the communities that live in the river sections with good ecological integrity. Its calculation is based on the presence of benthic invertebrates in freshwater due to their sensitivity to a myriad of contamination sources [1,78]. The ecological status was determined with the software Amiib@—"Application for the calculation of Benthic Invertebrate Metrics and Indices" developed by the Portuguese Water Institute (Lisbon, Portugal). It allows for the determination of several Portuguese official indexes for assessing the ecological status of a river, based on the benthic invertebrate biological element, including $IPtI_N$.

2.5. Correspondence Analysis

Ordinations of sampling sites according to macroinvertebrate relative abundance and traits/metrics were configured in order to explore the ecological patterns (i.e., whether or not the axes of ordinations could characterize gradients) by using correspondence analysis (CA), starting with 25 environmental variables. CAs were accomplished for both biological datasets, with the environmental matrices separately (physicochemical and LU/LC) providing bi-plots of all datasets. This multivariate analysis was applied in order to reduce the dimensionality and extract the dominant factors. CA can be viewed as a method projecting data into a multidimensional space, where the chi-square distance metric is preserved, but also as a method that computes latent predictor variables for (approximated) models of unimodal response. The CA was carried out using CANOCO 5 (version 5.12, Microcomputer Power, Ithaca, NY, USA) [79].

Relating species traits/metrics to physicochemical and LU/LC can provide important insights into the structure and functioning of stream communities. The Spearman correlation test was used to avoid the collinearity within traits/metrics. Hence, a non-redundant matrix of 13 traits/metrics from an initial matrix with 77 was built. The remaining traits belong to current preferences, food and feeding habits, locomotion type, and taxonomic groups (number of taxa) (Table S2, Electronic Supplementary Material). The main objectives are to demonstrate the importance of the different macroinvertebrate traits that are related to perturbation patterns and identify which of them allow more effective and accurate use in future monitoring works.

3. Results and Discussion

3.1. Climatic Characterization

3.1.1. Current Conditions

The mean annual temperature for the period 1950–2015 (Figure 3a) shows higher temperatures along the river line, reaching values as high as 15.5 °C, while much lower temperatures can be found in the mountains north and south of the river (~11 °C), thus showing a clear footprint of elevation (negative vertical temperature gradient). For mean temperature (see Figure S1, Supplementary Material), the observed variation is between 20 °C in summer and 5.5 °C in winter. Hence, according to the Köppen-Geiger climate classification, the Paiva River catchment presents a Csb climate (Mediterranean climate with cool summers), or predominant mesomediterranean climate with supramediterranean climate at the highest elevations [80,81]. Regarding the mean annual precipitation for the period 1950–2015 (Figure 3d), it increases as we move upstream the river (from 1050 to 1200 mm), but it shows higher values in the mountains (1400 mm). The precipitation pattern is also a clear footprint of the orography, as the condensation barrier effect that is imposed by the mountains to the Atlantic westerly and moist winds underlies the observed precipitation pattern, which is further enhanced by the Atlantic-facing deep river canyon, with a tunnelling and uplift effect on the air masses. During winter (DJF), the mean precipitation in the Paiva River catchment shows values of 620 mm, whereas much lower values (ca. 85 mm) are recorded in summer (JJA) (see Figure S1, Supplementary Material). In terms of aridity, the climate can be considered to be humid everywhere, as the annual mean precipitation is higher than annual mean potential evapotranspiration [82]. Mid-seasons (MAM and SON) show similar spatial patterns, with a variation of temperature between 10.5 and 15.5 °C and precipitation between 275 and 425 mm.

Although the Paiva River catchment depicts a typical Mediterranean climate (see Figure S2, Supplementary Material), the maritime influence from the North Atlantic is apparent, with a commonly short dry season (mid-June to mid-August) and considerably wet winters. While, in winter, frequent synoptic-scale disturbances travel from the North Atlantic to the region, thus generating high precipitation amounts, the prevalence of strong and northerly displaced Azores high-pressure systems in summer, accompanied by air sinking, is largely unfavourable to rain-generating conditions and it explains the summer minimum [83]. Therefore, the strong seasonality in the precipitation regime has significant effects on river flow, as well as on water resources and water quality. The irregularity in the monthly precipitation amounts is also noteworthy (see Figure S2, Supplementary Material).

3.1.2. Climate Change Projections

The projections for both RCP scenarios (for the period 2041–2070) (Figure 3b,c) show an overall increase of temperature for the whole of the region, where scenario RCP8.5 shows an increase for the entire catchment up to 2 °C. Concerning the precipitation scenarios for the period 2041–2070 (Figure 3e,f), there is a projected decreased in precipitation for scenario RCP4.5 of approximately 40 mm (maximum), while, in RCP8.5, there is an observed increase of precipitation in the eastern part of the catchment, varying from −40 to 30 mm.

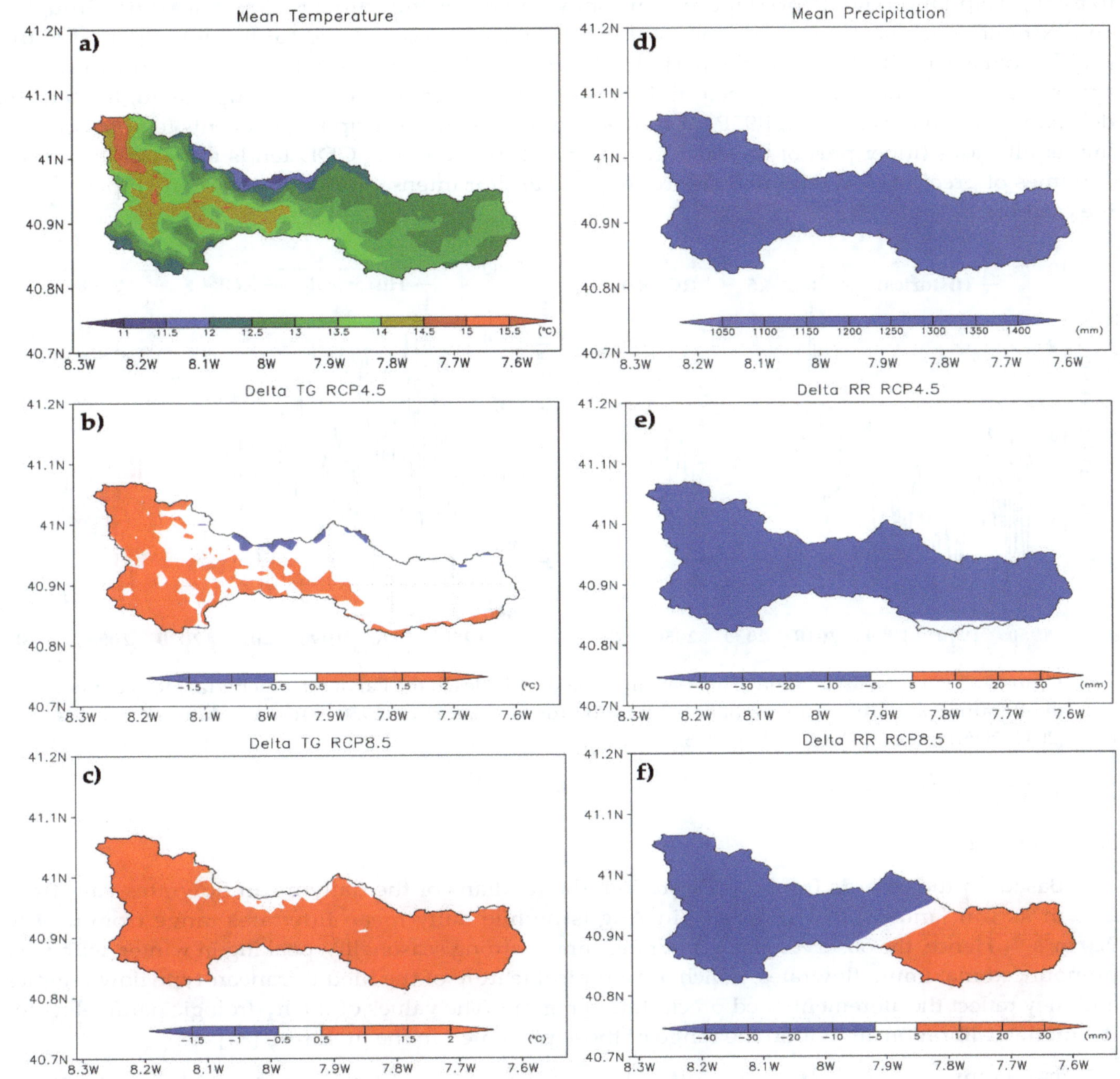

Figure 3. Spatial distribution of climate variables in the Paiva catchment: (**a**) annual mean temperature; (**b**) mean temperature change under RCP4.5; (**c**) mean temperature variation under RCP8.5; (**d**) mean annual precipitation; (**e**) mean precipitation change under RCP4.5; and, (**f**) mean precipitation change under RCP8.5.

The chronograms of the area-mean climate variables (Figure 4), averaged over the whole catchment, show an increasing trend for temperature and a decreasing for precipitation for both RCP 4.5 and RCP 8.5 scenarios in the Paiva catchment. The projected temperature shows an increase of 0.5 °C and 1.1 °C under RCP 4.5 and RCP 8.5, respectively. While projected precipitation decreases when compared to the historical period. Overall, there is an average annual decrease of 230 mm and 214 mm of precipitation for scenarios RCP 4.5 and RCP8.5, respectively.

Two indices of precipitation extremes were also computed in order to better understand the precipitation regime of the Paiva River catchment and its likely changes in the future. The number of consecutive dry days (CDD) and the relative contribution of precipitation above the 95-percentile to total precipitation (R95PTot) (Figure 5) are showed here as indicators to assess seasonal droughts and extreme precipitation events. Both of the indicators were calculated for the entire period of the gridded dataset (1950–2015) and the period 2041–2070 for both RCP scenarios. CDD shows a similar spatial pattern as the seasonal precipitation, with 105 consecutive dry days upstream, to 85 days, downstream. On the contrary, R95PTot shows that the severest precipitation events tend to occur at higher altitudes (inner part of the catchment). In future scenarios, CDD tends to decrease and the extremes of precipitation tend to be more frequent and/or intense, particularly in the upper area of the catchment.

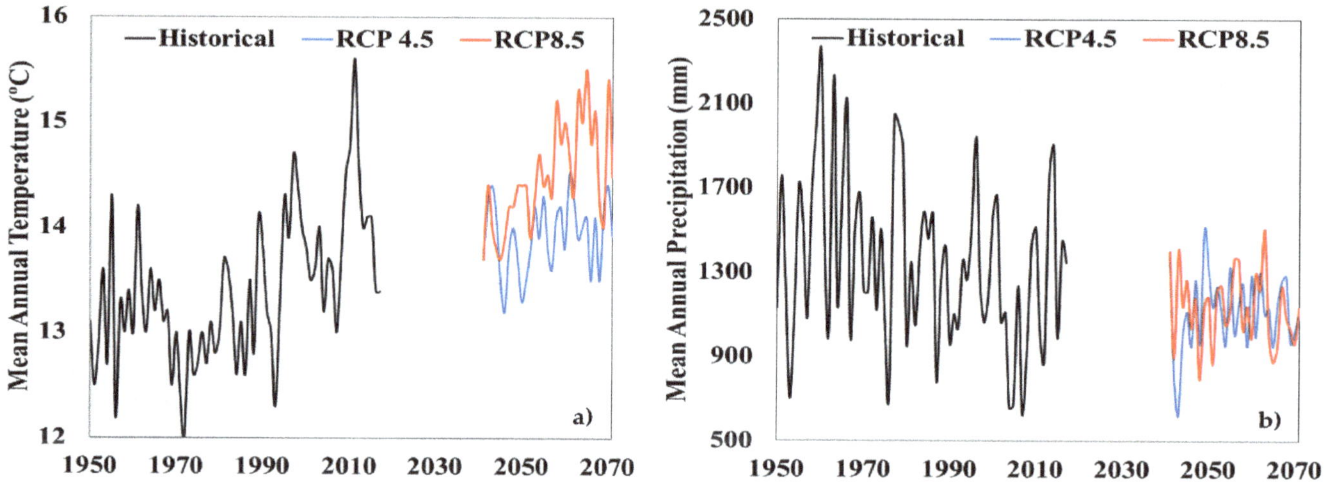

Figure 4. Temporal distribution of area-mean climate variables in the Paiva catchment: (**a**) mean annual temperature; and, (**b**) mean annual precipitation for the historical (1950–2015) and future scenarios (2041–2070) under RCP4.5 and RCP8.5.

3.2. Hydrological Characterization

3.2.1. Flowrates

Based on the records from the FT station, the medians of the daily mean flowrates vary from 25.5 m^3s^{-1}, in January, to 1.16 m^3s^{-1}, in August, while daily mean flowrates range from 0.19 to 920 m^3s^{-1}. Hence, the observed daily flowrates hint at strong seasonality, peaking in winter, with very pronounced maximum flowrates, which are a typical feature of the Mediterranean-type flow regimes and they reflect the aforementioned precipitation regime. The values of the hydrologic parameters for the model calibration are within the range of those presented in the literature [84].

The HSPF model performed well with respect to the simulation of river flow in the Paiva River (Figure 6), since the fits between observed and simulated flows present high Nash–Sutcliffe coefficients (E) values, as well as high coefficients of determination (R^2), for both calibration and validation, as shown hereafter. The deviation of volumes (Dv) also confirms the goodness-of-fit of the model, with a deviation of 3% and −11% for calibration and validation, respectively, though, for validation, the model slightly underestimates the flowrates. The Nash–Sutcliffe model efficiency coefficient shows a value of 0.95 for calibration and 0.76 validation. Based on previously published recommendations [85], model simulations with values of E above 0.75 are generally considered to be satisfactory. The coefficients of determination also confirm the goodness-of-fit of the model, with particularly good performance >0.95 for calibration and 0.82 for validation. It is important to consider many factors when modelling water quality in a catchment, of which the quality of the observed data is of the utmost importance. In the Paiva River catchment, there is considerable variability of the observed values and a general lack of homogenous and continuous data, which is an important limitation to the analysis.

Concerning the distribution of the projected precipitation on the daily timescale (Figure 7), an increase in skewness under both RCP are found (more extremes), despite the aforementioned decrease in the annual means. These changes are in general agreement with Figure 5. These changes combined lead to a decrease of flowrate variability (distributional dispersion) from the historical period to RCP4.5 (lower precipitation), and an increase of the flowrate variability from scenario RCP4.5 to RCP8.5 (more extremes).

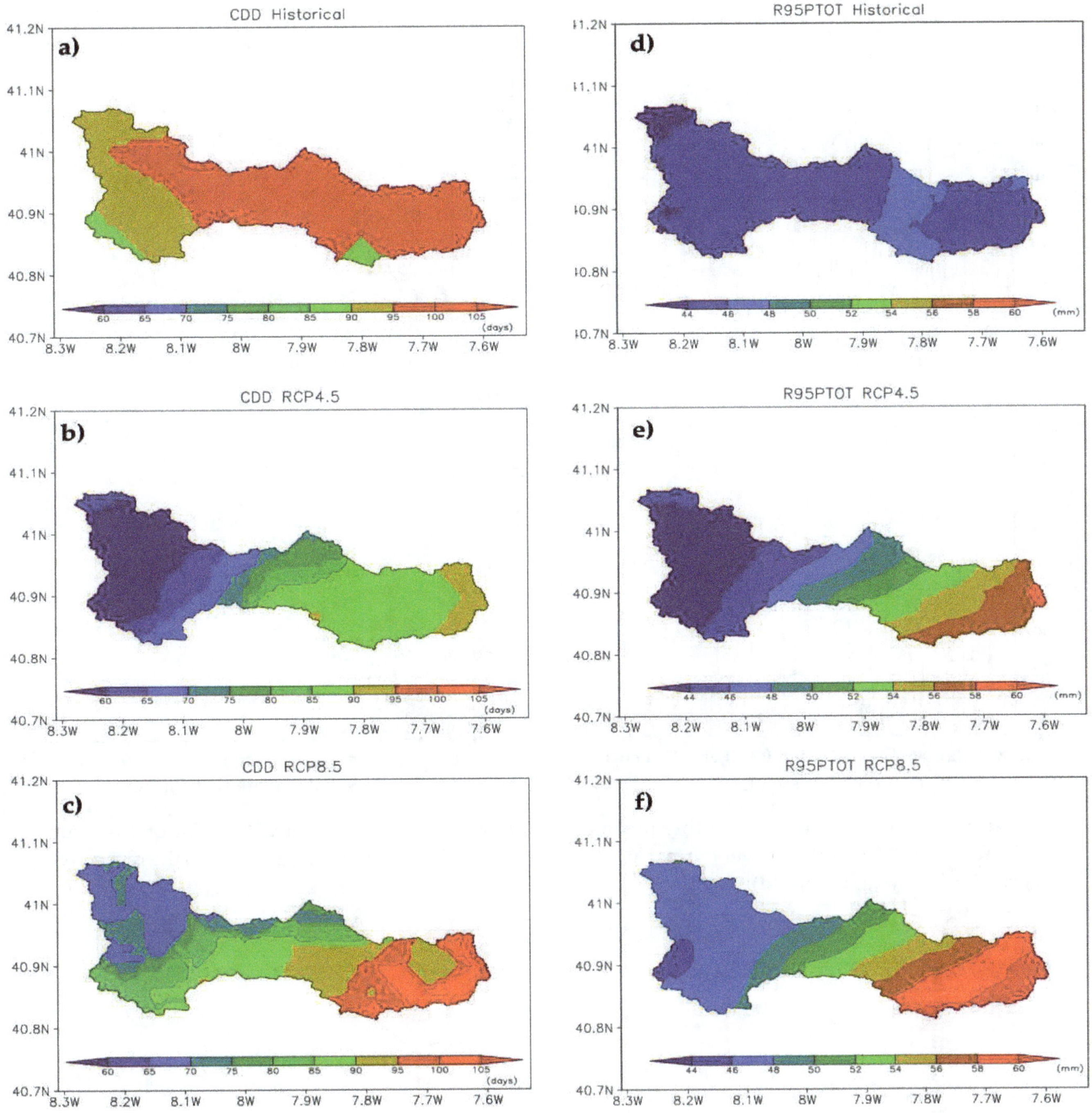

Figure 5. CDD—consecutive dry days and R95PTOT—Precipitation fraction in the Paiva catchment for (**a**,**d**) the historical period; (**b**,**e**) RCP4.5 and (**c**,**f**) RCP8.5.

3.2.2. Water Quality

In this study, the water quality constituents were simulated on a daily timescale by HSPF, while observed data were only available at the monthly timescale and with numerous gaps over the simulation period. Therefore, the calibration and validation of the model were performed in order to match the observed long-term mean against the simulated long-term mean, while displaying the monthly simulated hydrographs (Figure 8) to understand the temporal evolution and seasonality of the quality constituents.

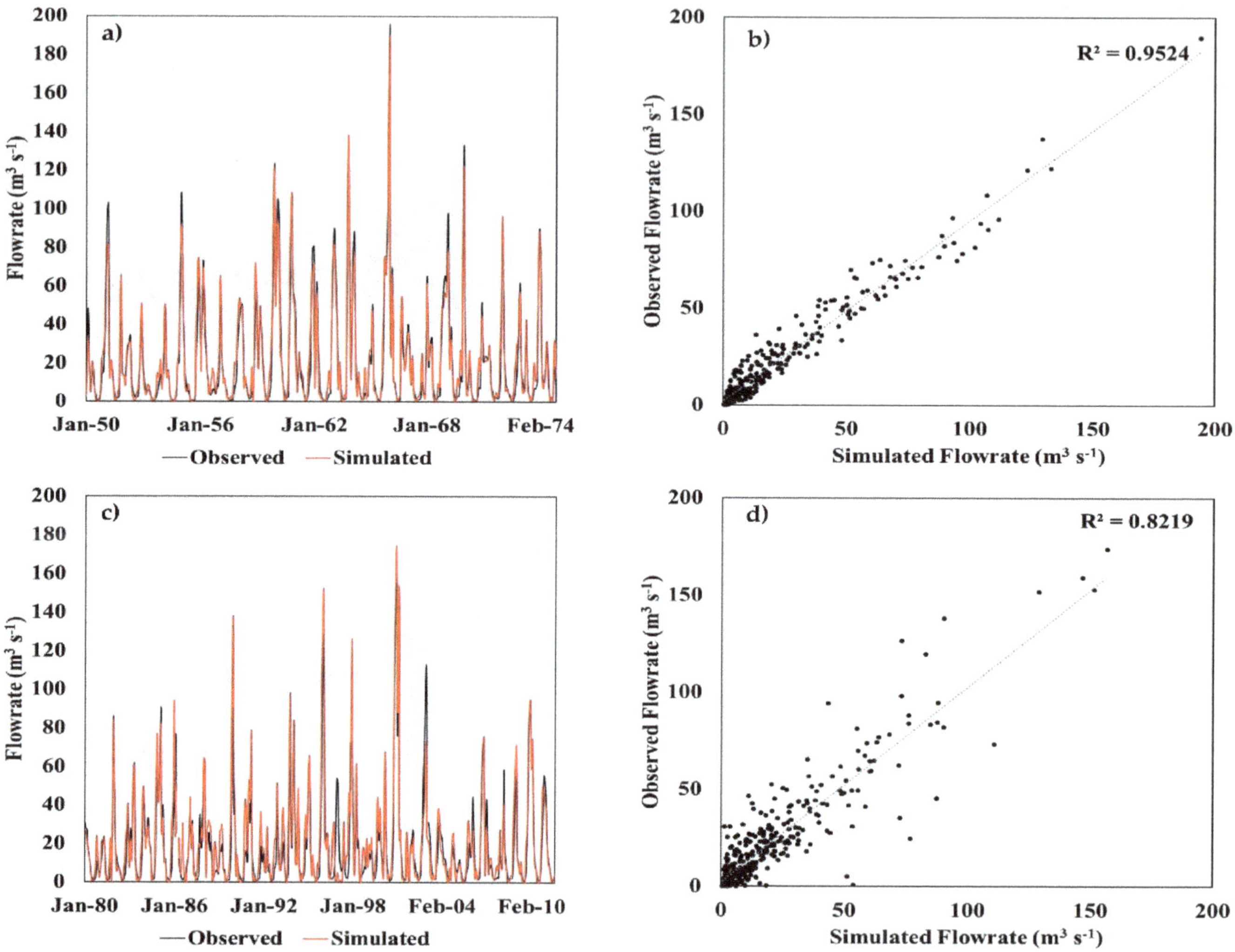

Figure 6. Chronograms of the monthly mean flows observed and Hydrological Simulation Program FORTRAN (HSPF) simulated over the (**a**) calibration and (**c**) validation periods, along with the respective regression analysis (**b**) and (**d**).

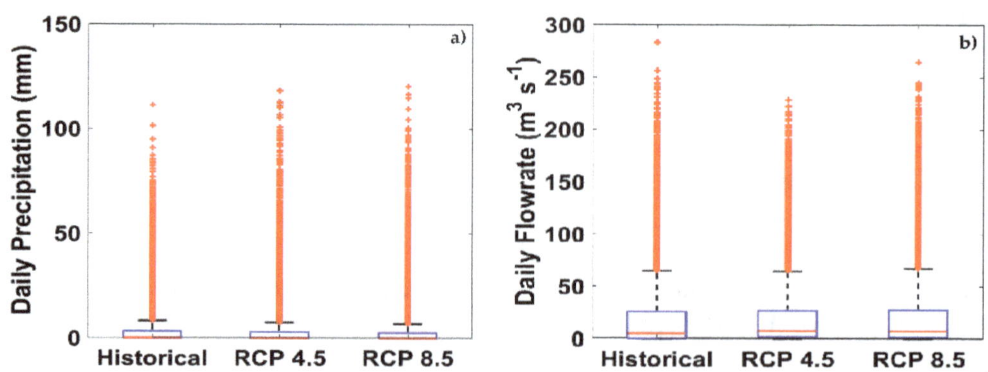

Figure 7. Boxplots of the daily (**a**) precipitation and (**b**) flowrate for the historical (1950–2015) and future scenarios (2041–2070) under RCP4.5 and RCP8.5.

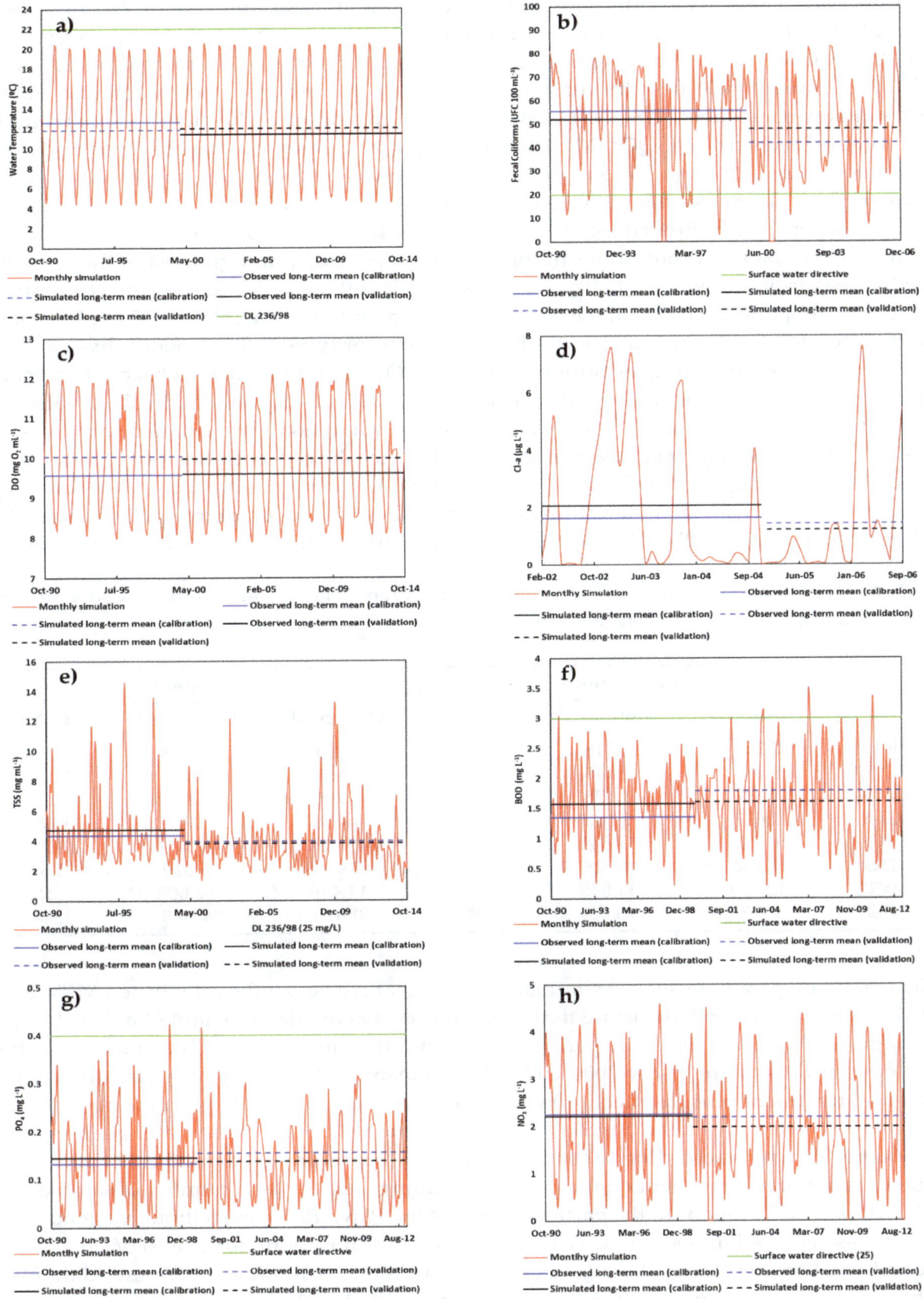

Figure 8. Monthly simulated (red line) and long-term means of observed (blue line) and simulated (black line) for the calibration and validation period (dashed lines) and maximum recommended value (MRV) of surface freshwater intended for the production of drinking water for type A1 treatment (green line) for (**a**) temperature, (**b**) faecal coliforms, (**c**) dissolved oxygen, (**d**) chlorophyll-a, (**e**) total suspended solids, (**f**) biochemical oxygen demand, (**g**) orthophosphates, and (**h**) nitrates.

Several parameters were adjusted first for calibration based on the literature [7,84,86–88]. The monthly nutrient concentration in interflow (MON-IFLW-CONC, $mg{\cdot}L^{-1}$) and groundwater (MON-GRND-CONC, $mg{\cdot}L^{-1}$) had significant impacts on phosphorus and nitrogen yields, followed by the ratio of constituent yield to sediment outflow (POTFS), the rate of surface runoff that will remove 90% of the quality constituent (WSQOP), and the maximum storage of the quality constituent that can be stored on the land surface (SQOLIM). For nitrates (NO_3) calibration, the observed and simulated concentrations were 2.257 and 2.218 $mg{\cdot}L^{-1}$, respectively. For validation, those values were 2.200 and 1.987 $mg{\cdot}L^{-1}$. These results indicate a good agreement between the observed and simulated NO_3 concentrations, with differences of 1.7% for calibration and −9.7% for validation. There is an underestimation of NO_3 concentrations in the validation period, although most of the records of the observed values from the water quality station stated less than 2. Moreover, the concentration of NO_3 for the entire period is below the maximum recommended value (MRV) (25 $mg{\cdot}L^{-1}$) of surface freshwater intended for the production of drinking water for type A1 treatment (physical treatment and disinfection) [89]. Regarding the orthophosphates (PO_4^{3-}) calibration, the observed and simulated concentrations were 0.1333 and 0.145 $mg{\cdot}L^{-1}$, respectively, whereas, for validation, these values were of 0.1555 and 0.138 $mg{\cdot}L^{-1}$, respectively. The calibration and validation were attained with differences of 8.8% and −11.2%. For the entire simulation period, only twice the concentration of PO_4^{3-} shows values above the MRV of 0.4 $mg{\cdot}L^{-1}$, 0.418 $mg{\cdot}L^{-1}$, and 0.414 $mg{\cdot}L^{-1}$ in the calibration and validation period, respectively. There is a general underestimation in the validation period of the quality parameters that are addressed in this study (Table 2), except for water temperature (WT) and dissolved oxygen (DO).

Table 2. Mean concentration values of water quality parameters for the calibration and validation period and respective relative bias (Percent Bias (PBIAS) in %).

	Calibration			Validation		
	Observed	**Simulated**	**PBIAS**	**Observed**	**Simulated**	**PBIAS**
NO_3^-	2.257	2.218	−1.7%	2.201	1.988	−9.7%
PO_4^{3-}	0.133	0.145	8.9%	0.155	0.138	−11.2%
BOD_5	1.351	1.578	16.8%	1.795	1.607	−10.5%
Cl-a	1.637	2.060	25.9%	1.438	1.211	−15.8%
FC	55.708	52.041	−6.6%	42.125	47.881	13.7%
TSS	4.352	4.728	8.6%	3.960	3.819	−3.6%
WT	12.639	11.848	−6.3%	11.419	12.008	5.2%
DO	9.589	10.049	4.8%	9.610	9.988	3.9%

The remaining parameters show simulated values that are below the established MRV, except for BOD in the calibration period, depicting three occurrences that overlap the limit (3 mg O_2 L^{-1}), and for faecal coliforms (FC) across the entire period, with values that are above 20 UFC 100 mL^{-1}. The faecal coliform concentrations are below 2000 UFC mL^{-1}, meaning that the type A2 treatment (physical and chemical treatment plus disinfection) would be necessary to comply with the above legislation. The percentage difference values (PBIAS) for calibration and validation of the water quality parameters show good performance according to the literature [85]. Furthermore, a good calibration of quality constituents is not only dependent on the frequency, timing, and the total number of samples, but also of good hydrology calibration. In the particular case of the Paiva River, even though a good hydrology calibration was achieved (based on the criteria results) at FT, it is more difficult to reach a meaningful water quality calibration, since there is only one hydrometric station with enough data for model calibration. This may explain why the HSPF results for validation showed a consistent bias towards an underestimation of the concentrations (PBIAS < 0) of most of the quality constituents, as the validation of the hydrological model also showed a deviation between the simulated and observed flowrate of −11%. Further, the observed quality constituent records are infrequent and/or below the detection limit of the legislated methodology [89].

3.3. Current Ecological Status versus Water Quality

The macroinvertebrate sampling revealed a total of 46262 specimens within 87 families. The most representative families were: Simuliidae (9860—21.7%, present in all sampling sites), Chironomidae (5688—12.5%), Philopotamidae (5449—12.0%), Elmidae (4378—9.6%), Baetidae (3550—7.8%), Hydropsychidae (3039—6.7%), Heptageniidae (2035—4.5%), and Caenidae (1034—2.3%) (Table S3). These eight families represent 77.1% of the all captured individuals and 43.2% of them pertain to the orders Ephemoptera (mayflies) and Trichoptera (caddisflies) that are pollution sensitive [90–92]. Further, Simuliidae is considered to be a reliable indicator of good water quality [93,94]. Notwithstanding, Baetidae, Caenidae (Ephemeroptera), and Hydropsychidae (Trichoptera) families exhibited tolerance to organic pollution and habitat disturbance [95,96]. According to Raburu et al. [97] and Masese et al. [98], this tolerance enables them to be broadly distributed, with higher abundance in areas with this kind of pollution. Despite this fact, it can be highlighted that the majority of sampling sites (65%) have more than 40% of EPT-Taxa [%] (min–max: 4.5%–79.0%) and all of them have more than three very intolerant EPT families (Table S3). Additionally, in agreement with Chirhart [99], the presence of moderate numbers of intolerant taxa is an indicator of good aquatic health. This corroborates the general good biological status that was verified throughout the Paiva catchment (Figure 9) using the $IPtI_N$ index. According to Marshall et al. [100], this same author, biotic indices around the family level of macroinvertebrates preclude the need for species-level identification, which can be time consuming and requires taxonomically trained specialists.

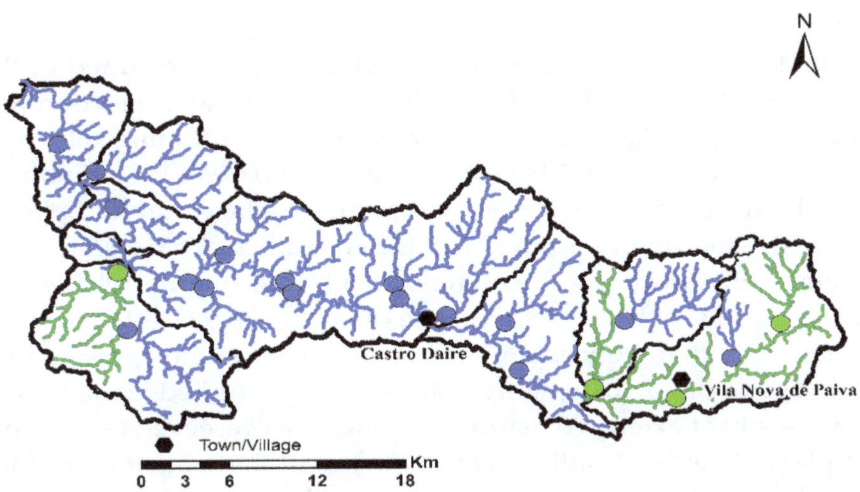

Figure 9. Ecological status of the Paiva River catchment according to $IPtI_N$ (Blu—excellent quality, Green—Good quality).

Overall, the quality is excellent, apart from the sampling sites number 4, 16, 18, and 20 (Figure 9), which reveal good ecological status. This relatively good quality had also been reported by Sousa et al. (2013) [76] when evaluating the ecological status of a *Margaritifera margaritifera* population at the Southern Edge of its Distribution, i.e., in the Paiva River.

We performed CA analyses to test the likely influence of environmental variables at different scales (LU/LC and water quality+flow) on the distribution of macroinvertebrate species and traits/metrics (Figure 10).

The invertebrate taxonomic data had a gradient of 1.4 SD units along the main axis of the CA (Figure 10). Seven variables were selected from the initial number of 25 environmental variables at reach scale, following a forward selection approach ($p < 0.05$). The environmental variables, which included water quality+flow, explained 45.92% of the variance (with a total variation of 0.67967). The first two axes of the CA explained 22.43% and 16.05% of the total variance, respectively.

CAs that were accomplished in the species matrix and water quality+flow variables revealed in Axis-2 a very clear impact gradient related to chemical and organic contamination (Figure 10a).

On the top left of the ordination space are sites (P2, P7, P14) impacted by organic contamination related with the increase of nitrates. This situation is a consequence of intensive eucalyptus afforestation observed mainly in the western part of river catchment (P2 and P7). Conversely, on the bottom, sites P1, P4, P5, P8 and P19 are associated with chemical contamination namely high concentrations of Al, Ni, and Zn, although the values obtained respect the minimum environmental quality objectives according Portuguese legislation. According to Rosseland et al. [101], aluminium acts as a toxic agent on gill-breathing animals such as fish and invertebrates, by causing loss of plasma- and haemolymph ions leading to osmoregulatory failure. Yet, Sparling et al. [102] refer that in general, aquatic invertebrates are less sensitive to Al toxicity and acidity than fish. Regarding zinc, Carvalho et al. [103] report that, even in relatively low concentrations, zinc metals can alter the reproductive pattern of freshwater invertebrates, compromising the aquatic ecosystem and its recovery capacity. Relatively to Ni, Eisler [104] state that its concentrations in unimpaired aquatic ecosystems are typically below 10 μg/L. CA analyses show that 50% of the studied sites (P1, P3, P4, P5, P6, P15, P16, P17, P18, P19) had a relationship with this heavy metal, though only detected in P4, P16, and P18 where good biological quality was achieved. While Ni is considered to be essential for a wide variety of animals' species [105], several studies revealed a Ni-related depression of immune system in both vertebrates and invertebrates [104,106]. According the Portuguese Mineral Resources and Occurrence Information System (SIORMINP), provided by National Energy and Geology Laboratory, I.P, these three metallic minerals occur in the studied area. Although the ecological status of the catchment is very good to excellent, according to the $IPtI_N$ index, CA analyses revealed minor contamination and established relationships between nutrients and macroinvertebrates.

CAs that were made with species matrix and LU/LC variables resulted in the selection of five LU/LC variables from the initial number of 11, following a forward selection approach ($p < 0.05$). This analysis showed a gradient related to agriculture in the Axis-1 (Figure 10b). The LU/LC variables only explained 24.59% of the variance (with a total variation of 0.67967). Ordination of sites on the CA plot can be interpreted by the presence of species in each assemblage. Larval Ephemerellidae (mayfly), Gomphidae (dragonflies), Chloroperlidae and Perlidae (stoneflies), Dytiscidae and Scirtidae (beetles), and Corbiculidae (clam) were the most influential taxa on Axis-1 (Figure 10b). Although this gradient that opposes agriculture and shrubs to the presence of eucalyptus (P1–P5 and P7) and urban areas (P13 and P18) has been verified, to notice, however, that both situations present intolerant families to pollution, reflecting the good biological quality obtained in the studied area. This may be justified by agricultural activities, mostly of extensive nature, and with the presence of small urban areas. In what concerns eucalyptus plantations [107], still the consequences of these plantations for the functioning of water courses are largely unknown. Fundazioa [107] refers that, despite the decomposition of its dead leaves having fewer problems in the initial time, due to the phenolic components, this material tends to be used in rivers as much as that from deciduous trees like autochthonous species. Additionally, Abelho and Graça [108], when comparing the dynamic of the organic matter (eucalyptus versus chestnut leaf litter), available for the aquatic macroinvertebrates, found that the monoculture of eucalyptus affects low order stream communities. Notwithstanding, it was verified that the impact may be attenuated if natural riparian vegetation is kept in those areas of plantation.

Finally, Figure 10d shows summary results of taxa diversity within samples. The CA highlight that the higher biodiversity was recorded in River Paiva tributaries (P2, P7, P11, and P14), which present excellent quality, linked with well-preserved areas.

The traits/metrics data had a gradient of 0.8 SD units along the main axis of the CA (Figure 11). From the initial number of 25 environmental variables at reach scale, following a forward selection approach ($p < 0.05$), six variables were selected. Water quality+flow, explained 35.54 % of the variance (with a total variation of 0.15795). The first two axes of the CA explained 30.45% and 23.15% of the total variance, respectively. Contrary to what was verified with the analyses of invertebrate taxonomic data, the traits/metrics gradient organic *versus* chemical disturbance was reflected in Axis-1 (Figure 11a). On the bottom left of the ordination space are sites (P2, P4, P7, P11, P12, P14), impacted by organic contamination also related with the increase of nitrates. On the right site of the ordination, sites P6, P13, P16, P18, P19, and P20 are associated with chemical contamination, namely high concentrations of Mn and Zn. Manganese also respects the minimum environmental quality objectives according Portuguese legislation. Manganese, which is an essential element actively assimilated and utilized by both plants and animals, can be significantly bio concentrated by aquatic biota at lower trophic levels [109]. According to this same author, freshwater mollusks and crustaceans are the most manganese-sensitive freshwater invertebrates, followed by arthropods and oligochaetes. Oligochaetes and gastropods are in the left part of the Axis-1 corroborating this idea, as it can be seen in the Figure 11a. Heal [110] refers that afforestation of upland areas increase manganese concentrations in surface waters. Additionally, has been widely documented that catchments planted with conifers that acidify soil and water are associated with enhanced manganese concentrations in surface waters [110–112]. Thus, by observing Figure 2 the coniferous woodland in mainly located at the upper part of Paiva catchment which is in agreement with the previous statement. As mentioned before, even though the ecological status is good to excellent, the CA analyses for the traits/metrics also demonstrate a correlation between species and organic contamination.

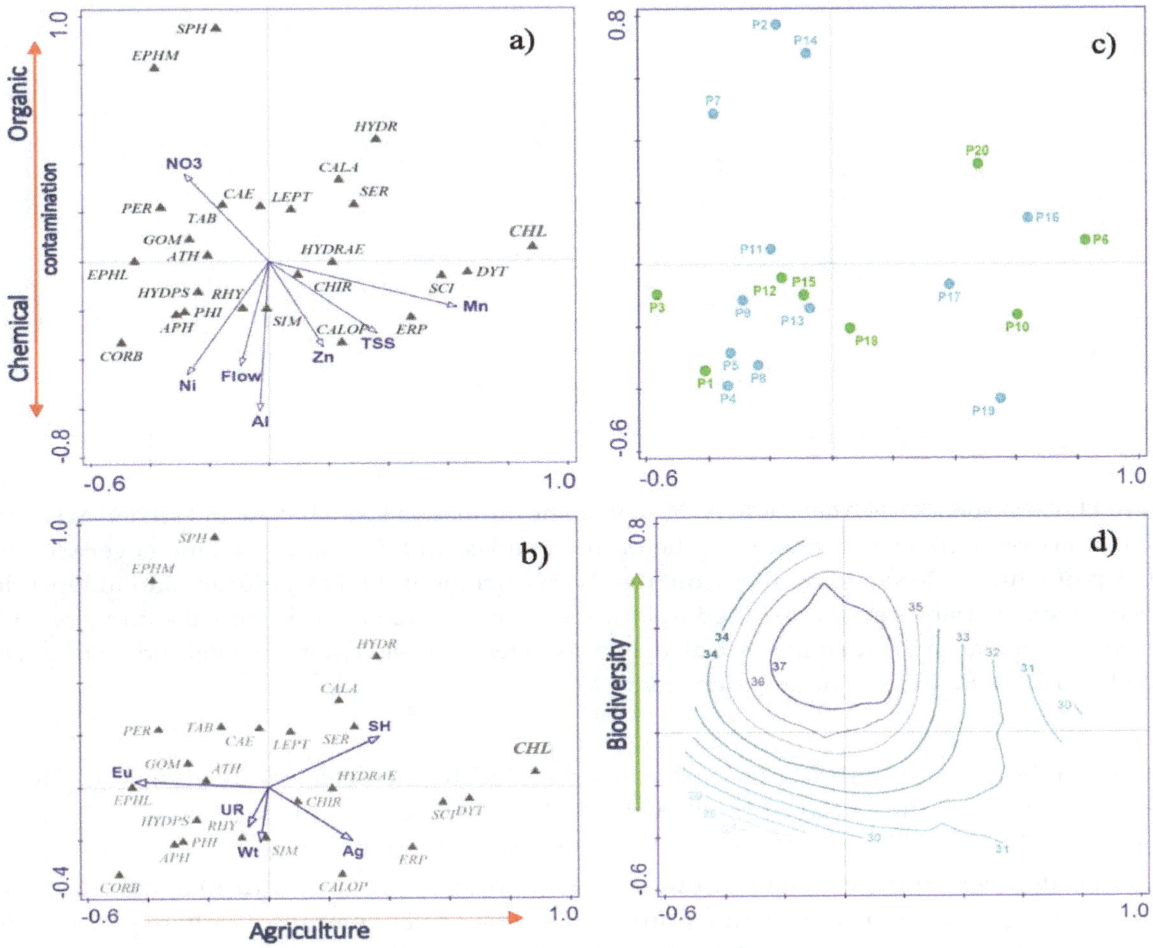

Figure 10. Correspondence Analysis (CA) bi-plot, in the ordination space of first and second axis, of the macroinvertebrate taxonomic assemblages and (**a**) physicochemical variables; and (**b**) Land-use land cover variables. (**c**) CA plot with the 20 sampling sites from the Paiva catchment; and, (**d**) contour plot of the taxa diversity within samples summarizing a fitted regression model (loess model). Only significant and independent environmental variables and more fitted taxa are shown. The red arrow shows the direction of the disturbance gradient and the green arrow the biodiversity gradient, common to all diagrams. Sites, taxa, and environmental codes are given in Table 1 and Table S2—Electronic Supplementary Material.

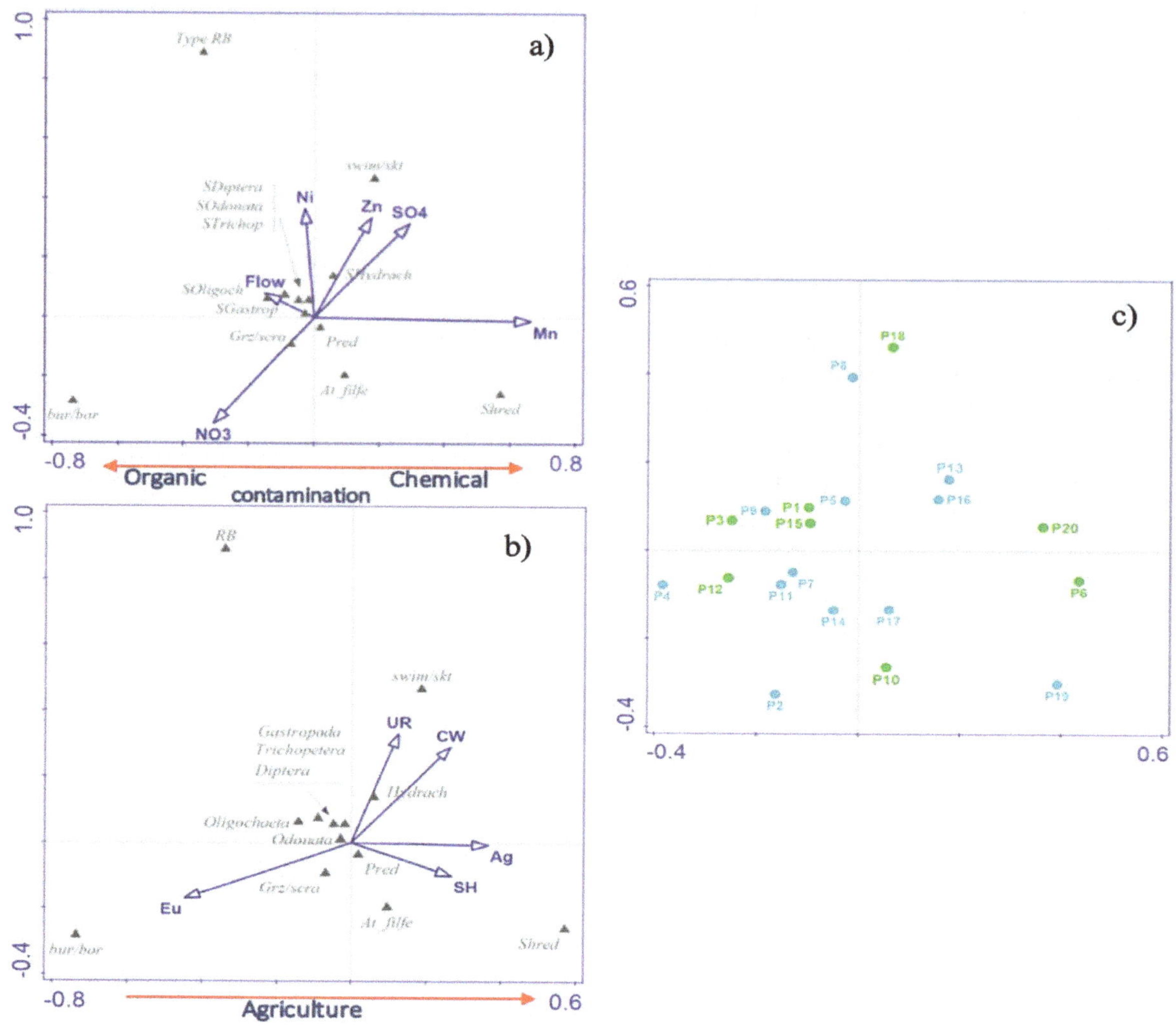

Figure 11. Correspondence Analysis (CA) bi-plot, in the ordination space of first and second axis, of the macroinvertebrate traits and (**a**) physicochemical variables; and, (**b**) Land-use land cover variables. (**c**) CA plot with the 20 sampling sites from the Paiva catchment. Only significant and independent environmental variables and more fitted traits are shown. The red arrows show the direction of the disturbance gradient and are common to all diagrams. Sites, taxa and environmental codes are given in Table 1 and Table S2—Electronic Supplementary Material.

The ordination made with LU/LC variables resulted in the selection of five (Eu, UR, CW, Ag, and SH) variables from the initial number of 11, following a forward selection approach ($p < 0.05$). A gradient related to agriculture in the Axis-1 (Figure 11b). This analysis showed a separation between the sites associated with the areas of large eucalyptus plantations (monoculture) against the remaining LU/LC in which agriculture has the greatest influence. The LU/LC variables only explained 25.03% of the variance (with a total variation of 0.15795). Shredders life-history characteristics are linked with agriculture and shrubs areas that mainly occur in the upper part (P19 and P20) and lower western part (P6) of River Paiva catchment. Masese et al. [98] stated that shredders are clearly very important in the transformation process of organic matter in forested and deforested streams. In our work, despite the shredders being associated with the presence of agriculture, contrary to the huge amount of studies that prove the decrease of these organisms with this type of uses [113–117], it should be noted that these areas are also associated with the strong presence of scrub and broadleaf woodlands. These areas contribute with great amount of leaves, their main source of food, for breakdown processes.

As verified by Larrañaga et al. [118], eucalypt plantations in a northern Spain stream has a negative effect on macroinvertebrate communities mostly on the density of shredders. This is in accordance with what was observed in Paiva catchment, where shredders were put in the opposite end in the ordination space (Axis-1) to the monoculture of eucalyptus. The burrowing/boring traits (P2, P3, and P4) were associated with this LU/LC. The presence of invertebrates that burrows in the substrate could be favored by the increase of the sedimentation in river channels [119]. Thus, it is precisely in this part, further downstream of the Paiva basin, where a greater amount of finer sediment is found in the riverbed as a consequence of the afforestation and/or harvesting of the eucalyptus monoculture that requires intervention in the soil with relatively short periods due to the exploitation of timber.

Swimming/skating were dominant in urban and coniferous areas (P13, P16, and P18). The potential for land use effects on the composition of aquatic invertebrate communities to alter the composition of behavioral related traits was also seen in research by Wang et al. [116], which found that forested streams exhibited higher relative abundances of Swimmer taxa when compared with the agricultural stream.

4. Conclusions

The HSPF model was applied to a 796 km^2 catchment in northern Portugal, the Paiva River catchment, a predominantly forest and agricultural catchment, aiming to validate its performance in simulating hydrologic and water quality responses. The flowrate and water quality components were calibrated against the observed data that were collected at the hydrological station of "Fragas da Torre" and extrapolated for the entire watershed. Overall, the runoff volumes predicted by HSPF agreed well with the observed data at the monitoring site within the catchment. Both model calibration and validation show satisfactory criteria values. Furthermore, because simulated water quality processes are based on hydrologic processes, the quality of the hydrology calibration has a major impact on the quality of the calibration of the constituent concentrations. Nonetheless, a satisfactory calibration was achieved.

The status of water quality is considered to be good when compared with the maximum recommended value for surface freshwater intended for the production of drinking water. According to the HSPF model output results, only BOD, PO_4, and faecal coliforms would require a secondary treatment to comply with the legislation in force. Regarding the ecological status, an overall excellent status is observed in the Paiva River catchment, with only four sites disclosing good ecological status.

Climate change is likely to have extensive effects on water quality, in the particular case of Paiva River catchment, due to increase of extreme events in both precipitation and temperature, both of which will affect the concentration of water pollutants and erosion, especially for scenario RCP8.5. The simulated flowrates for future scenarios show an increasing trend that may contribute to a better water quality status (providing the current anthropogenic pressures), thus reflecting a continuous likelihood of good ecological status.

Correspondence analyses highlight some potential predictors of macroinvertebrate populations, which can be used to implement population distribution models in the Paiva River catchment. These models can be of foremost relevance for developing projections for macroinvertebrates under future regional climate change scenarios.

While the present study outcomes refer to a specific watershed, they may be generalized to other watersheds in Portugal and in the Mediterranean basin, as the climate change projections for these regions indicate future drier and warmer climates.

Nature-based solutions can offer sustainable approaches in order to cope with climate change mitigation and adaptation challenges. Namely, nature-based climate adaptation aims to preserve ecosystem services and reduce the impact of the projected negative effects of climate change (e.g., floods, droughts, and heat waves).

Author Contributions: In the overall context of this paper, the contributions are as follow: climate assessment done by A.R.F. and J.A.S., Hydrological model by A.R.F., macroinvertebrates sampling and water analyses done by A.R.F., E.C., S.G.P.V. and S.M.M., data analyses by E.C., J.L.M., R.M.V.C. and S.G.P.V. All authors provided critical feedback and helped shape the research, analysis and manuscript. All authors have read and agreed to the published version of the manuscript.

Acknowledgments: This work is supported by National Funds by FCT—Portuguese Foundation for Science and Technology, under the project UIDB/04033/2020.

References

1. Baptista, C.; Santos, L. Water quality monitoring in the Paul do Boquilobo Biosphere Reserve. *Phys. Chem. Earth Parts A/B/C* **2016**, *94*, 180–187. [CrossRef]
2. Nichols, S.J.; Robinson, W.A.; Norris, R.H. Sample variability influences on the precision of predictive bioassessment. *Hydrobiologia* **2006**, *572*, 215–233. [CrossRef]
3. Verdonschot, P.; Spears, B.; Feld, C.; Brucet, S.; Keizer–Vlek, H.; Borja, A.; Elliott, M.; Kernan, M.; Johnson, R. A comparative review of recovery processes in rivers, lakes, estuarine and coastal waters. *Hydrobiologia* **2013**, *704*, 453–474. [CrossRef]
4. Singh, J.; Knapp, H.V.; Arnold, J.; Demissie, M. Hydrological modeling of the Iroquois river watershed using HSPF and SWAT 1. *JAWRA* **2005**, *41*, 343–360. [CrossRef]
5. Al–Abed, N.; Al–Sharif, M. Hydrological modeling of Zarqa River basin–Jordan using the hydrological simulation program–FORTRAN (HSPF) model. *Water Resour. Manag.* **2008**, *22*, 1203–1220. [CrossRef]
6. Albek, M.; Öğütveren, Ü.B.; Albek, E. Hydrological modeling of Seydi Suyu watershed (Turkey) with HSPF. *J. Hydrol.* **2004**, *285*, 260–271. [CrossRef]
7. Fonseca, A.; Ames, D.P.; Yang, P.; Botelho, C.; Boaventura, R.; Vilar, V. Watershed model parameter estimation and uncertainty in data–limited environments. *Environ. Model. Softw.* **2014**, *51*, 84–93. [CrossRef]
8. Gallagher, M.; Doherty, J. Parameter estimation and uncertainty analysis for a watershed model. *Environ. Model. Softw.* **2007**, *22*, 1000–1020. [CrossRef]
9. Buytaert, W.; Beven, K. Models as multiple working hypotheses: Hydrological simulation of tropical alpine wetlands. *Hydrol. Process.* **2011**, *25*, 1784–1799. [CrossRef]
10. Gupta, H.V.; Sorooshian, S.; Yapo, P.O. Toward improved calibration of hydrologic models: Multiple and noncommensurable measures of information. *Water Resour. Res.* **1998**, *34*, 751–763. [CrossRef]
11. Beven, K.; Freer, J. Equifinality, data assimilation, and uncertainty estimation in mechanistic modelling of complex environmental systems using the GLUE methodology. *J. Hydrol.* **2001**, *249*, 11–29. [CrossRef]
12. Liu, Y.; Gupta, H.V. Uncertainty in hydrologic modeling: Toward an integrated data assimilation framework. *Water Resour. Res.* **2007**, *43*. [CrossRef]
13. Götzinger, J.; Bárdossy, A. Generic error model for calibration and uncertainty estimation of hydrological models. *Water Resour. Res.* **2008**, *44*. [CrossRef]
14. Stedinger, J.R.; Vogel, R.M.; Lee, S.U.; Batchelder, R. Appraisal of the generalized likelihood uncertainty estimation (GLUE) method. *Water Resour. Res.* **2008**, *44*. [CrossRef]
15. Beven, K. Environmental Modelling: An Uncertain Future? CRC Press: Boca Raton, FL, USA, 2018.
16. Fonseca, A.; Botelho, C.; Boaventura, R.A.R.; Vilar, V.J.P. Integrated hydrological and water quality model for river management: A case study on Lena River. *Sci. Total Environ.* **2014**, *485–486*, 474–489. [CrossRef] [PubMed]
17. Fonseca, A.R.; Sanches Fernandes, L.F.; Fontainhas–Fernandes, A.; Monteiro, S.M.; Pacheco, F.A.L. From catchment to fish: Impact of anthropogenic pressures on gill histopathology. *Sci. Total Environ.* **2016**, *550*, 972–986. [CrossRef] [PubMed]
18. Fonseca, A.R.; Santos, M.; Santos, J.A. Hydrological and flood hazard assessment using a coupled modelling approach for a mountainous catchment in Portugal. *Stoch. Environ. Res. Risk Assess.* **2018**, *32*, 2165–2177. [CrossRef]

19. Fonseca, A.; Botelho, C.; Boaventura, R.A.R.; Vilar, V.J.P. Global Warming Effects on Faecal Coliform Bacterium Watershed Impairments in Portugal. *River Res. Appl.* **2015**, *31*, 1344–1353. [CrossRef]
20. Fonseca, A.R.; Santos, J.A. A new very high–resolution climatological dataset in Portugal: Application to hydrological modeling in a mountainous watershed. *Phys. Chem. Earth* **2019**, *109*, 2–8. [CrossRef]
21. Fonseca, A.R.; Santos, J.A. Predicting hydrologic flows under climate change: The Tâmega Basin as an analog for the Mediterranean region. *Sci. Total Environ.* **2019**, *668*, 1013–1024. [CrossRef]
22. Barnett, T.P.; Adam, J.C.; Lettenmaier, D.P. Potential impacts of a warming climate on water availability in snow–dominated regions. *Nature* **2005**, *438*, 303–309. [CrossRef] [PubMed]
23. Stocker, T.F.; Qin, D.; Plattner, G.; Tignor, M.; Allen, S.; Boschung, J.; Nauels, A.; Xia, Y.; Bex, V.; Midgley, P. Climate change 2013: The physical science Basis. In *Intergovernmental Panel on Climate Change, Working Group I Contribution to the IPCC Fifth Assessment Report (AR5)*; Cambridge University Press: New York, NY, USA, 2013.
24. Cornes, R.C.; van der Schrier, G.; van den Besselaar, E.J.; Jones, P.D. An ensemble version of the E-OBS temperature and precipitation data sets. *J. Geophys. Res. Atmos.* **2018**, *123*, 9391–9409. [CrossRef]
25. Kite, G. Modelling the Mekong: Hydrological simulation for environmental impact studies. *J. Hydrol.* **2001**, *253*, 1–13. [CrossRef]
26. Azorin–Molina, C.; Vicente–Serrano, S.M.; McVicar, T.R.; Jerez, S.; Sanchez–Lorenzo, A.; López–Moreno, J.I.; Revuelto, J.; Trigo, R.M.; Lopez–Bustins, J.A.; Espírito–Santo, F. Homogenization and assessment of observed near–surface wind speed trends over Spain and Portugal, 1961–2011. *J. Clim.* **2014**, *27*, 3692–3712. [CrossRef]
27. Vieira, J.; Fonseca, A.; Vilar, V.J.; Boaventura, R.A.; Botelho, C.M. Water quality in Lis river, Portugal. *Environ. Monit. Assess.* **2012**, *184*, 7125–7140. [CrossRef] [PubMed]
28. Vieira, J.; Fonseca, A.; Vilar, V.J.; Boaventura, R.A.; Botelho, C.M. Water quality modelling of Lis river, Portugal. *Environ. Sci. Pollut. Res.* **2013**, *20*, 508–524. [CrossRef] [PubMed]
29. Wahren, F.T.; Julich, S.; Nunes, J.P.; Gonzalez–Pelayo, O.; Hawtree, D.; Feger, K.H.; Keizer, J.J. Combining digital soil mapping and hydrological modeling in a data scarce watershed in north–central Portugal. *Geoderma* **2016**, *264*, 350–362. [CrossRef]
30. Andersen, J.; Refsgaard, J.C.; Jensen, K.H. Distributed hydrological modelling of the Senegal River Basin—model construction and validation. *J. Hydrol.* **2001**, *247*, 200–214. [CrossRef]
31. World Health Organization. *The World Health Report 2006: Working Together for Health*; World Health Organization: Geneva, Switzerland, 2006.
32. Stanford, J.A.; Lorang, M.; Hauer, F. The shifting habitat mosaic of river ecosystems. *Int. Ver. Theor. Angew. Limnol. Verh.* **2005**, *29*, 123–136. [CrossRef]
33. Stocker, T.F.; Qin, D.; Plattner, G.-K.; Tignor, M.M.; Allen, S.K.; Boschung, J.; Nauels, A.; Xia, Y.; Bex, V.; Midgley, P.M. *Climate Change 2013: The Physical Science Basis. Contribution of Working Group I to the Fifth Assessment Report of IPCC the Intergovernmental Panel on Climate Change*; Cambridge University Press: New York, NY, USA, 2014.
34. Robin, L.; Vannote, G.; Minshall, W.; Cummins, K.; Sedell, J.; Cushing, C. The river continuum concept. *Can. J. Fish Aquat. Sci.* **1980**, *37*, 130–137.
35. Mattson, R.A.; Merritt, R.W.; Cummins, K.W. An Introduction to the Aquatic Insects of North America. *J. N. Am. Benthol. Soc.* **1996**, *15*, 401–403. [CrossRef]
36. Lenat, D.R.; Rosenberg, D.; Resh, V.H. Freshwater Biomonitoring and Benthic Macroinvertebrates. *J. N. Am. Benthol. Soc.* **1993**, *12*, 220–222. [CrossRef]
37. Barbour, M.T.; Gerritsen, J.; Snyder, B.D.; Stribling, J.B. *Rapid Bioassessment Protocols for Use in Streams and Wadeable Rivers: Periphyton, Benthic Macroinvertebrates and Fish*; US Environmental Protection Agency, Office of Water: Washington, DC, USA, 1999; Volume 339.
38. Cabecinha, E.; Hughes, S.; Cortes, R. Consistent, congruent or redundant? Lotic community and organisational response to disturbance. *Ecol. Indic.* **2018**, *89*, 175–187. [CrossRef]
39. Extence, C.A.; Chadd, R.P.; England, J.; Dunbar, M.J.; Wood, P.J.; Taylor, E.D. The assessment of fine sediment accumulation in rivers using macro–invertebrate community response. *River Res. Appl.* **2011**, *29*, 17–55. [CrossRef]
40. Colas, F.; Archaimbault, V.; Pain-Devin, S. Scale-dependency of macroinvertebrate communities: Responses to contaminated sediments within run-of-river dams. *Sci Total Environ.* **2011** *409*, 1336–1343. [CrossRef]

41. Friberg, N.; Bonada, N.; Bradley, D.C.; Dunbar, M.J.; Edwards, F.K.; Grey, J.; Hayes, R.B.; Hildrew, A.G.; Lamouroux, N.; Trimmer, M. Biomonitoring of human impacts in freshwater ecosystems: The good, the bad and the ugly. In *Advances in Ecological Research*; Elsevier: Amsterdam, The Netherlands, 2011; Volume 44, pp. 1–68.
42. Murphy, J.F.; Davy–Bowker, J.; McFarland, B.; Ormerod, S.J. A diagnostic biotic index for assessing acidity in sensitive streams in Britain. *Ecol. Indic.* **2013**, *24*, 562–572. [CrossRef]
43. Lillie, R.A.; Szczytko, S.W.; Miller, M.A. *Macroinvertebrate Data Interpretation Guidance Manual*; Wisconsin Department of Natural Resources: Madison, WI, USA, 2003.
44. Fitzpatrick, F.A.; Scudder, B.C.; Lenz, B.N.; Sullivan, D.J. Effects of Multi–Scale Environmental characteristics on agricultural stream biota in Easterns Wisconsin. *JAWRA* **2001**, *37*, 1489–1507. [CrossRef]
45. Peterson, R.; Van Eeckhaute, L. Distributions of Ephemeroptera, Plecoptera, and Trichoptera of three maritime catchments differing in pH. *Freshw. Biol.* **1992**, *27*, 65–78. [CrossRef]
46. Zarger, T.; Scanlon, D.; Nicholson, C.; Brown, S.; Starnes, L.; Harned, W. *Ecological Recovery after Reclamation of Toxic Spoils Left By Coal Surface Mining: Phase 2, An Assessment of Environmental Changes Following Intensive Remedial Treatments*; Tennessee Valley Authority, Norris (USA): Knoxville, TN, USA, 1987.
47. Boyer, C.; Chaumont, D.; Chartier, I.; Roy, A.G. Impact of climate change on the hydrology of St. Lawrence tributaries. *J. Hydrol.* **2010**, *384*, 65–83. [CrossRef]
48. Brown, T.C.; Kroll, S. Avoiding an uncertain catastrophe: Climate change mitigation under risk and wealth heterogeneity. *Clim. Chang.* **2017**, *141*, 155–166. [CrossRef]
49. Besacier Monbertrand, A.-L.; Timoner, P.; Rahman, K.; Burlando, P.; Fatichi, S.; Gonseth, Y.; Moser, F.; Castella, E.; Lehmann, A. Assessing the vulnerability of aquatic macroinvertebrates to climate warming in a mountainous watershed: Supplementing presence–only data with species traits. *Water* **2019**, *11*, 636. [CrossRef]
50. Poff, N.L. Landscape filters and species traits: Towards mechanistic understanding and prediction in stream ecology. *J. N. Am. Benthol. Soc.* **1997**, *16*, 391–409. [CrossRef]
51. Bonada, N.; Doledec, S.; Statzner, B. Taxonomic and biological trait differences of stream macroinvertebrate communities between mediterranean and temperate regions: Implications for future climatic scenarios. *Glob. Chang. Biol.* **2007**, *13*, 1658–1671. [CrossRef]
52. Charvet, S.; Statzner, B.; Usseglio-Polatera, P.; Dumont, B. Traits of benthic macroinvertebrates in semi-natural French streams: An initial application to biomonitoring in Europe. *Freshw. Biol.* **2000**, *43*, 277–296. [CrossRef]
53. Phillips, N. Stream biomonitoring using species traits. *Water Atmos.* **2004**, *12*, 8–9.
54. Vieira, N.K.; Poff, N.; Carlisle, D.M.; Moulton, S.R.; Koski, M.L.; Kondratieff, B.C. A database of lotic invertebrate traits for North America. *US Geol. Surv. Data Ser.* **2006**, *187*, 1–15.
55. Schuwirth, N.; Kattwinkel, M.; Stamm, C. How stressor specific are trait–based ecological indices for ecosystem management? *Sci. Total Environ.* **2015**, *505*, 565–572. [CrossRef]
56. Kuzmanovic, M.; Dolédec, S.; de Castro–Catala, N.; Ginebreda, A.; Sabater, S.; Muñoz, I.; Barceló, D. Environmental stressors as a driver of the trait composition of benthic macroinvertebrate assemblages in polluted Iberian rivers. *Environ. Res.* **2017**, *156*, 485–493. [CrossRef]
57. APA. Plano de Gestão da Região Hidrográfica do Douro RH3. In *Relatório de Base*; APA: Porto, Portugal, 2016; Volume Parte 2.
58. ICNF. *Programa Regional de Ordenamento Florestal*; Instituto da Conservação da Natureza e das Florestas, Ed.; ICNF: Lisboa, Portugal, 2016.
59. Fonseca, A.R.; Santos, J.A. High–resolution temperature datasets in Portugal from a geostatistical approach: Variability and extremes. *J. Appl. Meteorol. Climatol.* **2018**, *57*, 627–644. [CrossRef]
60. Belo-Pereira, M.; Dutra, E.; Viterbo, P. Evaluation of global precipitation data sets over the Iberian Peninsula. *J. Geophys. Res. Atmos.* **2011**, *116*. [CrossRef]
61. Bicknell, B.; Imhoff, J.; Kittle, J.; Jobes, T.; Donigian, A.; Johanson, R. Hydrological Simulation Program FORTRAN: HSPF version 12 user's Manual. In *Cooperation with the US Geological Survey and US Environmental Protection Agency*; AQUA TERRA Consultants: Mountain View, CA, USA, 2001.
62. Donigian, A.; Davis, H. *Agricultural Runoff Management (Arm): Users Manual*; Report EPA–600/3–78–080; US EPA: Athens, GA, USA, 1978.

63. Donigian, A.S.; Crawford, N.H. *Modeling Nonpoint Pollution From the Land Surface*; US Environmental Protection Agency, Office of Research and Development: Washington, DC, USA, 1976.
64. Crawford, N.H.; Linsley, R.K. *Digital Simulation in Hydrology'Stanford Watershed Model 4*; TRID: Stanford, CA, USA, 1966.
65. Stern, M.A.; Flint, L.E.; Minear, J.T.; Flint, A.L.; Wright, S.A. Characterizing Changes in Streamflow and Sediment Supply in the Sacramento River Basin, California, Using Hydrological Simulation Program—FORTRAN (HSPF). *Water* **2016**, *8*, 432. [CrossRef]
66. Donigian, A.S.; Huber, W.C. *Modeling of Nonpoint–Source Water Quality in Urban and Non–Urban Areas*; AQUA TERRA Consultants: Mountain View, CA, USA, 1991.
67. Bergman, M.; Green, W.; Donnangelo, L. Calibration of storm loads in the South Prong watershed, Florida, using basins/HSPF 1. *JAWRA* **2002**, *38*, 1423–1436.
68. Zhang, J.; Ross, M.; Trout, K.; Zhou, D. Calibration of the HSPF model with a new coupled FTABLE generation method. *Prog. Nat. Sci.* **2009**, *19*, 1747–1755. [CrossRef]
69. Nash, J.E.; Sutcliffe, J.V. River flow forecasting through conceptual models part I—A discussion of principles. *J. Hydrol.* **1970**, *10*, 282–290. [CrossRef]
70. Cortes, R.; Varandas, S.; Teixeira, A.; Hughes, S.; Magalhaes, M.; Barquín, J.; Alvarez–Cabria, M.; Fernández, D. Effects of landscape metrics and land use variables on macroinvertebrate communities and habitat characteristics. *Limnetica* **2011**, *30*, 347–362.
71. Alves, M.H.; Ferreira, J.; Pádua, J.; Martins, J.P.; Rafael, T.; Tomaz, G.; Figueiredo, H.; Bernardo, J.M.; Pinto, P.; Morais, M. Tipologia de rios em portugal continental no âmbito da implementação da directiva quadro da água. In *I–Caracterização Abiótica*; Ministério do Ambiente, do Ordenamento do Território e do Desenvolvimento Regional: Lisboa, Portugal, 2008.
72. Oliveira, S. *Avaliação da Integridade Ecológica em Ecossistemas Aquáticos. Tese de Doutoramento*; UTAD: Vila Real, Portugal, 2006.
73. Fernandes, A.C.P.; Fernandes, L.F.S.; Terêncio, D.P.S.; Cortes, R.M.V.; Pacheco, F.A.L. Seasonal and scale effects of anthropogenic pressures on water quality and ecological integrity: A study in the Sabor River basin (NE Portugal) using partial least squares–path modeling. *Water* **2019**, *11*, 1941. [CrossRef]
74. Fernandes, A.; Ferreira, A.; Fernandes, L.F.S.; Cortes, R.; Pacheco, F.A.L. Path modelling analysis of pollution sources and environmental consequences in river basins. *Water Pollut. XIV* **2018**, *228*, 79–87. [CrossRef]
75. Junior, R.F.V.; Varandas, S.G.; Pacheco, F.A.; Pereira, V.R.; Santos, C.F.; Cortes, R.M.; Fernandes, L.F.S. Impacts of land use conflicts on riverine ecosystems. *Land Use Policy* **2015**, *43*, 48–62. [CrossRef]
76. Sousa, R.; Amorim, A.; Sobral, C.; Froufe, E.; Varandas, S.; Teixeira, A.; Lopes-Lima, M. Ecological Status of a Margaritifera margaritifera (Linnaeus, 1758) Population at the Southern Edge of its Distribution (River Paiva, Portugal). *Environ. Manag.* **2013**, *52*, 1230–1238. [CrossRef]
77. Fernandes, A.; Fernandes, L.F.S.; Cortes, R.V.; Pacheco, F. The role of landscape configuration, season, and distance from contaminant sources on the degradation of stream water quality in urban catchments. *Water* **2019**, *11*, 2025. [CrossRef]
78. Instituto da Água, IP. Critérios para a Classificação do Estado das Massas de Água Superficiais–Águas de Transição e Costeira. In *Ministério do Ambiente, do Ordenamento do Território e do Desenvolvimento Regional*; Instituto da Água, IP: Lisboa, Portugal, 2010.
79. Šmilauer, P.; Lepš, J. *Multivariate Analysis of Ecological Data Using CANOCO 5*; Cambridge University Press: Cambridge, UK, 2014.
80. Jalut, G.; Amat, A.E.; Bonnet, L.; Gauquelin, T.; Fontugne, M. Holocene climatic changes in the Western Mediterranean, from south-east France to south-east Spain. *Palaeogeogr. Palaeoclim. Palaeoecol.* **2000**, *160*, 255–290. [CrossRef]
81. Gavilán, R.G.; Fernández–González, F.; Blasi, C. Climatic classification and ordination of the Spanish Sistema Central: Relationships with potential vegetation. *Plant Ecol.* **1998**, *139*, 1–11. [CrossRef]
82. Santos, J.A.; Costa, R.; Fraga, H. New insights into thermal growing conditions of Portuguese grapevine varieties under changing climates. *Theor. Appl. Clim.* **2018**, *135*, 1215–1226. [CrossRef]
83. Santos, J.A.; Andrade, C.; Corte-Real, J.; Leite, S.M. The role of large-scale eddies in the occurrence of winter precipitation deficits in Portugal. *Int. J. Clim.* **2008**, *29*, 1493–1507. [CrossRef]
84. USEPA. *BASINS Technical Note 6: Estimating Hydrology and Hydraulic Parameters for HSPF*; Office of Water, US Environmental Protection Agency: Washington, DC, USA, 2000.

85. Moriasi, D.N.; Arnold, J.G.; Van Liew, M.W.; Bingner, R.L.; Harmel, R.D.; Veith, T.L. Model Evaluation Guidelines for Systematic Quantification of Accuracy in Watershed Simulations. *Trans. ASABE* **2007**, *50*, 885–900. [CrossRef]
86. El-Kaddah, D.N.; Carey, A.E. Water quality modeling of the Cahaba River, Alabama. *Environ. Earth Sci.* **2004**, *45*, 323–338. [CrossRef]
87. Tong, S.T.; Liu, A.J.; Goodrich, J.A. Assessing the water quality impacts of future land-use changes in an urbanising watershed. *Civ. Eng. Environ. Syst.* **2009**, *26*, 3–18. [CrossRef]
88. Liu, Z. Using HSPF to Model the Hydrologic and Water Quality Impacts of Riparian Land-Use Change in a Small Watershed. *J. Environ. Informatics* **2011**, *17*, 1–14. [CrossRef]
89. Ambiente, M..d. Decreto–Lei n° 236/98. *Diário República I Série A* **1998**, 3676–3722.
90. Bonada, N.; Prat, N.; Resh, V.H.; Statzner, B. Developments in aquatic insect biomonitoring: A comparative analysis of recent approaches. *Annu. Rev. Entomol.* **2006**, *51*, 495–523. [CrossRef]
91. Chapman, D.V. *Water Quality Assessments: A Guide to the Use of Biota, Sediments and Water in Environmental Monitoring*; CRC Press: Boca Raton, FL, USA, 1996.
92. Sibley, P.; Raby, M.; Wirtz, J.; McCoole, M.; Lagadic, L.; Soucek, D.; Norberg–King, T.; Roessink, I. *Evaluating the Relative Sensitivity of the "P&T" in EPT: Implications for Standardized Toxicity Testing*; SETAC: Sacramento, CA, USA, 2018.
93. Rabha, B.; Dhiman, S.; Yadav, K.; Hazarika, S.; Bhola, R.K.; Veer, V. Influence of water physicochemical characteristics on Simuliidae (Diptera) prevalence in some streams of Meghalaya, India. *J. Vector Borne Dis.* **2013**, *50*, 18. [PubMed]
94. Hamid, N.A.A.; Lardizabal, M.L.T.; Takaoka, H.; Stephen, E.; Mahali, M. The relationship between water quality and black fly (Diptera: Simulidae) abundance in Tambunan district, Sabah. *Serangga* **2016**, *21*, 121–131.
95. THORNE, R.; WILLIAMS, P. The response of benthic macroinvertebrates to pollution in developing countries: A multimetric system of bioassessment. *Freshw. Biol.* **1997**, *37*, 671–686. [CrossRef]
96. Minaya, V.; McClain, M.E.; Moog, O.; Omengo, F.; Singer, G.A. Scale–dependent effects of rural activities on benthic macroinvertebrates and physico–chemical characteristics in headwater streams of the Mara River, Kenya. *Ecol. Indic.* **2013**, *32*, 116–122. [CrossRef]
97. Raburu, P.O.; Masese, F.O.; Mulanda, C.A. Macroinvertebrate Index of Biotic Integrity (M–IBI) for monitoring rivers in the upper catchment of Lake Victoria Basin, Kenya. *Aquat. Ecosyst. Health Manag.* **2009**, *12*, 197–205. [CrossRef]
98. Masese, F.O.; Kitaka, N.; Kipkemboi, J.; Gettel, G.M.; Irvine, K.; McClain, M.E. Litter processing and shredder distribution as indicators of riparian and catchment influences on ecological health of tropical streams. *Ecol. Indic.* **2014**, *46*, 23–37. [CrossRef]
99. Chirhart, J. *Development of a Macroinvertebrate Index of Biological Integrity (MIBI) for Rivers and Streams of the St. Croix River Basin in Minnesota*; Minnesota Pollution Control Agency: Saint Paul, MN, USA, 2003.
100. Marshall, J.C.; Steward, A.L.; Harch, B.D. Taxonomic resolution and quantification of freshwater macroinvertebrate samples from an Australian dryland river: The benefits and costs of using species abundance data. *Hydrobiologia* **2006**, *572*, 171–194. [CrossRef]
101. Rosseland, B.O.; Eldhuset, T.D.; Staurnes, M. Environmental effects of aluminium. *Environ. Geochem. Health* **1990**, *12*, 17–27. [CrossRef]
102. Sparling, D.; Lowe, T.; Campbell, P. *Ecotoxicology of Aluminum to Fish and Wildlife*; USGS: Washington, DC, USA, 1997.
103. Carvalho, M.M.; Lira, V.S.; Watanabe, C.H.; Fracacio, R. Study of metals toxicity (zinc and cadmium) to Ceriodaphnia dubia, for multi–exposition and biological recovery of offspring. *Eng. Sanit. E Ambient.* **2017**, *22*, 961–968. [CrossRef]
104. Eisler, R. *Nickel Hazards to Fish, Wildlife, and Invertebrates: A Synoptic Review*; Geological Survey: Washington, DC, USA, 1998.
105. Muyssen, B.T.; Brix, K.; DeForest, D.; Janssen, C. Nickel essentiality and homeostasis in aquatic organisms. *Environ. Rev.* **2004**, *12*, 113–131. [CrossRef]
106. Sun, H.-X.; Dang, Z.; Xia, Q.; Tang, W.-C.; Zhang, G. The effect of dietary nickel on the immune responses of Spodoptera litura Fabricius larvae. *J. Insect Physiol.* **2011**, *57*, 954–961. [CrossRef] [PubMed]

107. Fundazioa, E. Impact of Eucalyptus Plantations on the Ecology of Rivers. Available online: https://cordis.europa.eu/article/id/114549-impact-of-eucalyptus-plantations-on-the-ecology-of-rivers (accessed on 15 September 2020).
108. Abelho, M.; Graça, M. Effects of eucalyptus afforestation on leaf litter dynamics and macroinvertebrate community structure of streams in Central Portugal. *Hydrobiologia* **1996**, *324*, 195–204. [CrossRef]
109. Howe, P.; Malcolm, H.; Dobson, S. *Manganese and its Compounds: Environmental Aspects. Concise International Chemical Assessment Document 63*; World Health Organization: Geneva, Switzerland, 2004.
110. Heal, K.V. Manganese and land-use in upland catchments in Scotland. *Sci. Total Environ.* **2001**, *265*, 169–179. [CrossRef]
111. Neal, C.; Reynolds, B.; Neal, M.; Wickham, H.; Hill, L.; Pugh, B. The Impact of Conifer Harvesting on Stream Water Quality: A Case Study in Mid-Wales. *Water Air Soil Pollut. Focus* **2003**, *3*, 119–138. [CrossRef]
112. Watmough, S.A.; Eimers, M.C.; Dillon, P.J. Manganese cycling in central Ontario forests: Response to soil acidification. *Appl. Geochem.* **2007**, *22*, 1241–1247. [CrossRef]
113. Nedeau, E.J.; Merritt, R.W.; Kaufman, M.G. The effect of an industrial effluent on an urban stream benthic community: Water quality vs. habitat quality. *Environ. Pollut.* **2003**, *123*, 1–13. [CrossRef]
114. Smith, R.F.; Lamp, W.O. Comparison of insect communities between adjacent headwater and main-stem streams in urban and rural watersheds. *J. N. Am. Benthol. Soc.* **2008**, *27*, 161–175. [CrossRef]
115. Sterling, J.L.; Rosemond, A.D.; Wenger, S.J. Watershed urbanization affects macroinvertebrate community structure and reduces biomass through similar pathways in Piedmont streams, Georgia, USA. *Freshw. Sci.* **2016**, *35*, 676–688. [CrossRef]
116. Wang, L.; Gao, Y.; Han, B.-P.; Fan, H.; Yang, H. The impacts of agriculture on macroinvertebrate communities: From structural changes to functional changes in Asia's cold region streams. *Sci. Total Environ.* **2019**, *676*, 155–164. [CrossRef]
117. Oliveira, P.C.D.R.; Kraak, M.H.; Pena-Ortiz, M.; Van Der Geest, H.G.; Verdonschot, P.F. Responses of macroinvertebrate communities to land use specific sediment food and habitat characteristics in lowland streams. *Sci. Total Environ.* **2020**, *703*, 135060. [CrossRef]
118. Larrañaga, A.; Basaguren, A.; Pozo, J. Impacts ofEucalyptus globulusPlantations on Physiology and Population Densities of Invertebrates Inhabiting Iberian Atlantic Streams. *Int. Rev. Hydrobiol.* **2009**, *94*, 497–511. [CrossRef]
119. Dolédec, S.; Phillips, N.R.; Townsend, C. Invertebrate community responses to land use at a broad spatial scale: Trait and taxonomic measures compared in New Zealand rivers. *Freshw. Biol.* **2011**, *56*, 1670–1688. [CrossRef]

2

Comparison of River Basin Water Level Forecasting Methods: Sequential Neural Networks and Multiple-Input Functional Neural Networks

Chih-Chiang Wei

Department of Marine Environmental Informatics & Center of Excellence for Ocean Engineering, National Taiwan Ocean University, Keelung 20224, Taiwan; ccwei@ntou.edu.tw

Abstract: To precisely forecast downstream water levels in catchment areas during typhoons, the deep learning artificial neural networks were employed to establish two water level forecasting models using sequential neural networks (SNNs) and multiple-input functional neural networks (MIFNNs). SNNs, which have a typical neural network structure, are network models constructed using sequential methods. To develop a network model capable of flexibly consolidating data, MIFNNs are employed for processing data from multiple sources or with multiple dimensions. Specifically, when images (e.g., radar reflectivity images) are used as input attributes, feature extraction is required to provide effective feature maps for model training. Therefore, convolutional layers and pooling layers were adopted to extract features. Long short-term memory (LSTM) layers adopted during model training enabled memory cell units to automatically determine the memory length, providing more useful information. The Hsintien River basin in northern Taiwan was selected as the research area and collected relevant data from 2011 to 2019. The input attributes comprised one-dimensional data (e.g., water levels at river stations, rain rates at rain gauges, and reservoir release) and two-dimensional data (i.e., radar reflectivity mosaics). Typhoons Saola, Soudelor, Dujuan, and Megi were selected, and the water levels 1 to 6 h after the typhoons struck were forecasted. The results indicated that compared with linear regressions (REG), SNN using dense layers (SNN-Dense), and SNN using LSTM layers (SNN-LSTM) models, superior forecasting results were achieved for the MIFNN model. Thus, the MIFNN model, as the optimal model for water level forecasting, was identified.

Keywords: water level; streamflow; remote sensing; deep learning; neural networks; modeling

---

1. Introduction

Taiwan is located in the Northwest Pacific, where typhoons frequently strike, and most of its rivers are short and steep. During typhoons, short-term heavy rains and hyperconcentrated streamflows cause river surges in downstream areas, which increases the possibility of flood loss in areas along rivers. The Hsintien River basin (Figure 1) in northern Taiwan is the main source of service water to the Taipei Metropolitan Area (with a population of approximately four million); hence, water source management and conservation within the basin are essential. Typhoon-related disasters in the Hsintien River basin, such as Typhoon Soudelor in 2015, prompted us to develop highly accurate water level forecasting models to inform flood prevention.

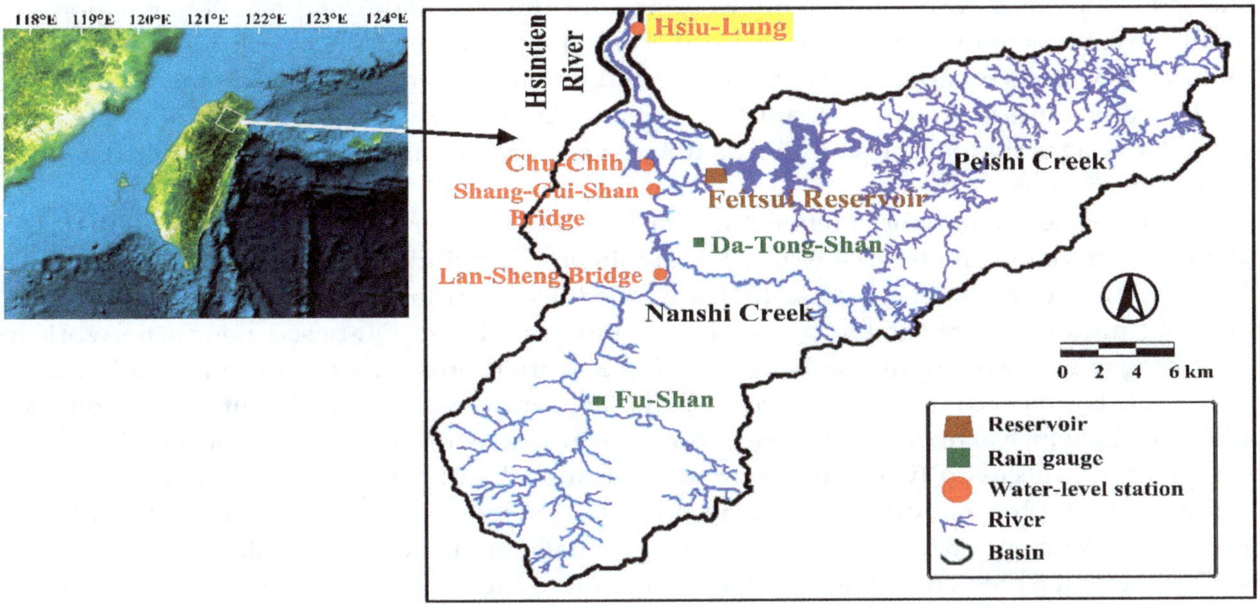

Figure 1. Map of the Hsintien River Basin.

In research on flood-stage estimation and prediction, physically-based models often yield accurate results, but developing such a forecasting system requires extensive computational time and various hydrogeomorphological data [1]. Examples of such models are MIKE Flood [2], MODFLOW [3], HEC-RAS [4,5], and CCCMMOC [6,7]. Another common method is data-driven forecasting models; Rainfall-runoff and river-stage forecasting models are established using machine learning (ML). The advantages of these methods are that they are less subject to the constraints of physical mechanisms and that they effectively identify and learn correlated patterns between input data sets and the corresponding target values [8,9]. Such ML-based methods can be used to construct various algorithm models, such as support vector regression [10], multilayer feedforward perceptron [11], self-organizing map [12], k-nearest neighbor [13], and random forest [14], on the basis of rainfall data at precipitation stations and water level or water flow data at hydrographic stations.

In general, artificial neural network (ANN)-based models can process all types of sequential data. Time series data, such as streamflow and river stage, are a special type of sequence data ordered by timestamps [15,16]. Among ANN-based models, sophisticated ML-based time series models, including recurrent neural networks (RNNs), have been developed to solve the problem of time series data extending the memory length on the time axis. Accordingly, on the basis of RNN, models capable of automatically determining the memory length (e.g., LSTM neural networks) have been established. Researchers have employed RNN-based models to solve flood-stage prediction problems [17–19].

In most studies on flood-stage estimation and prediction, ML-based models are created using ground observation data, namely one-dimensional (1-D) data such as rainfall, water flow, and water level. According to Gires et al. [20] and Ochoa-Rodriguez et al. [21], rain gauges provide relatively accurate point rainfall estimates near the ground surface; nonetheless, they cannot effectively capture the spatial variability of rainfall, which significantly affects hydrological systems and thus runoff modeling. With the advancement and popularization of remote-sensing equipment, remote-sensing data have become cheaper and more accessible. Therefore, radar-sensing data can add value to hydrological data. Remote-sensing data (e.g., ground radar reflectivity data, which are commonly applied) can be employed to estimate ground rainfall conditions. Radar reflectivity is the signal reflected by the electromagnetic wave actively emitted by radar through water vapor particles in the atmosphere [22–24]. Radar reflectivity images exhibit various colors that display the intensity of the echo signals received by radars, and these images provide information on water vapor in space. Researchers over the past few decades have adopted radar reflectivity data for rainfall estimation [25–31].

Radar reflectivity data are essential remote-sensing data that facilitate the prediction of regional ground rainfall as well as the rainfall-runoff process. Borga [32] described the promising potential of radar rainfall observations because their high spatial and temporal resolution and extensive areal coverage provide detailed information on precipitation events. Accordingly, the application of two-dimensional (2-D) radar images compensates for the insufficient 1-D spatial rainfall data collected from land-based observation stations.

ML-based neural network architecture has been utilized for image recognition, object detection, and computer vision. Networks with such architecture, including convolutional neural networks (CNNs) [33] and deep belief networks [34], are called deep learning (DL) models because they require additional hidden layers for the processing of images. These DL-based neural network models have emerged as powerful methods for learning feature representations automatically from image data [35,36]. The most commonly employed CNNs are composed mainly of multiple convolutional and pooling layers, which extract and strengthen image features to further identify objects. CNN algorithms have also been successfully applied to hydrological problems. For example, Wang et al. [37] and Kimura et al. [38] have adopted CNNs and input 1-D water level data for water level forecasting. Van et al. [39] developed a CNN with 1-D data for rainfall-runoff modeling. Kabir et al. [40] developed a CNN model for the prediction of fluvial flood inundation, and the CNN model was trained with outputs from a hydraulic model to predict water depths. In the aforementioned findings, 1-D time-series data retrieved from land-based observation stations were initially collated into image data sets during preprocessing and were subsequently analyzed using CNN models. In addition, Wang et al. [41] developed a CNN for flood susceptibility mapping by using various 2-D images (such as altitude, curvature, distance to rivers, and rainfall).

The above-mentioned studies have used either 2-D images or 1-D attribute data alone to construct DL-based models. To the best of my knowledge, in the literature on water level prediction for typhoons, using both 2-D (e.g., radar reflectivity images) and 1-D (e.g., attributes obtained from land-based observation stations) data simultaneously have seldom been used as input variables for DL-based neural networks.

The development of a DL-based model using data from multiple sources and in multiple dimensions is thus valuable. Accordingly, this study used DL-based neural network models for accurately forecasting the downstream water level in the catchment area during typhoons and selected the Hsintien River basin as the research area. In general, multiple-input functional neural networks (MIFNNs) developed in this study offer the following contributions:

(1) The increasing accessibility of ground observation and remote-sensing data enabled us to flexibly consolidate data from multiple sources in two dimensions (1-D and 2-D) for training DL models. The ground rainfall, water level, and reservoir release for 1-D data and radar reflectivity images for 2-D data were adopted.

(2) Feature extractions are required during the construction of MIFNN models when image files (e.g., radar reflectivity images) serve as input attributes; this provides appropriate feature maps for model training. Hence, the convolutional and pooling layers for extracting image features were employed.

(3) Because of time delays in water levels, this study applied LSTM layers, which enabled memory block units to automatically determine the memory length during the training of ANN-based models to guarantee the most effective memory length for time series data on the time axis.

2. Region and Material

The Hsintien River is located in northern Taiwan (Figure 1), with a drainage area of approximately 916 km^2 and a total length of 84.6 km [42]. The Hsintien River catchment area contains two main tributaries, Nanshi Creek and Peishi Creek. The source of the Peishi Creek is at an altitude of approximately 600 m, and the Feitsui Reservoir is located within its basin. The Feitsui Reservoir, the second-largest reservoir in Taiwan with a total storage capacity of approximately 406 million m^3, was established in 1981 to ensure the stability of the tap water supply to the Taipei Metropolitan Area [43]. The source of Nanshi Creek is at an altitude of approximately 1300 m. It passes along the Lan-Sheng Bridge and the Shang-Gui-Shan Bridge water level stations and then the Chu-Chih and Hsiu-Lung stations after merging with the Peishi Creek. Table 1 lists the elevation information regarding these water stations. Located in the densely populated Taipei Metropolitan Area, the Hsiu-Lung station monitors the river flood stage to prevent overflows and floods. Therefore, the water level measured at the Hsiu-Lung station as the research objective was selected.

Table 1. Water level stations in Hsintien River.

Water-Level Station	River Bed Elevation (m)	Design Defense Flood Level (m)	Embankment Top Elevation (m)
Hsiu-Lung	0.00	13.03	15.04
Chu-Chih	40.50	57.80	54.30
Shang-Gui-Shan Bridge	55.24	66.69	68.88
Lan-Sheng Bridge	107.50	119.95	121.45

2.1. *Typhoons and Raw Data*

The water level data after 2011 collected at the water level stations located in the Hsintien River basin are comprehensive; hence, this study retrieved the hydrological data related to historical typhoon events from 2011 to 2019. As demonstrated in Table 2, 16 typhoons caused major rainfall in the research area, which substantially increased the downstream water level in the catchment area.

Table 2. Rainfall and water levels associated with typhoons in the study area.

Date	Typhoon	Max Water Level at Hsiu-Lung Station (m)	Da-Tong-Shan Station		Fu-Shan Station		Total Release from Feitsui Reservoir ($10^6 m^3$)
			Max Rain Rate (mm/h)	Total Precipitation (mm)	Max Rain Rate (mm/h)	Total Precipitation (mm)	
26–31 August 2011	Namadol	4.5	14	187	51	269	9.9
29 July–8 August 2012	Saola	11.4	50	666	84	878	135.3
11–16 July 2013	Soulik	8.1	36	210	10	41	57.3
19–23 August 2013	Trami	6.9	59	543	35	471	19.0
31 August–6 September 2013	Kong-Rey	4.8	55	319	11	83	43.2
21–27 July 2014	Matmo	5.9	13	127	29	272	44.4
21–29 September 2014	Fung-Wong	4.2	59	336	70	310	13.3
7–11 August 2015	Soudelor	12.0	69	538	82	731	59.1
25 September–6 October 2015	Dujuan	9.8	48	443	76	725	138.0
13–15 September 2016	Meranti	3.9	12	70	17	130	11.4
16–21 September 2016	Malakas	5.0	18	243	28	279	40.9
26–30 September 2016	Megi	9.2	41	456	68	513	104.5
28 July–1 August 2017	Nesat	3.4	18	110	26	155	11.3
6–12 August 2019	Lekima	4.8	21	210	29	331	21.8
22–26 August 2019	Bailu	3.3	9	38	14	225	19.9
26 September–4 October 2019	Mitag	7.6	33	408	25	467	91.2

This study collected the water level records of typhoons from the Hsiu-Lung, Chu-Chih, Shang-Gui-Shan Bridge, and Lan-Sheng Bridge stations using 1 h as the time interval, and the statistical properties of the collected data are displayed in Table 3. However, the rainfall data in the catchment area of Nanshi Creek were separately collected because it is an unregulated river. The data from two rain gauges in the catchment area, namely the Da-Tong-Shan gauge (at an elevation of 500 m) and the Fu-Shan gauge (at an elevation of 916 m), were obtained. This study also examined reservoir-regulated flows discharged from the Feisui Reservoir, which affected Peishi Creek. The rainfall data and the reservoir release values are listed in Table 3.

The water level stations and rain gauges within the research area are currently within the jurisdiction of the 10th River Management Office of the Water Resources Agency, and the Feitsui Reservoir is governed by the Taipei Feitsui Reservoir Administration of Taipei City Government.

Table 3. Statistical properties of rainfall data and reservoir release values.

Attribute	Name	Unit	Mean	Range	Standard Deviation
Water level	Hsiu-Lung	m	3.07	1.86–12.04	1.25
	Chu-Chih	m	49.23	47.23–52.57	0.59
	Shang-Gui-Shan Bridge	m	57.52	55.32–68.00	1.37
	Lan-Sheng Bridge	m	110.23	107.83–120.30	1.42
Rain	Da-Tong-Shan	mm/h	1.90	0–69	6.01
	Fu-Shan	mm/h	2.28	0–84	7.45
Reservoir release	Feitsui	m^3/s	89.05	12.15–733.59	106.18

2.2. Radar Imagery

The 2952 radar mosaics with resolutions of 1024 × 1024 pixels were collected. The hourly radar reflectivity images were obtained from the Central Weather Bureau of Taiwan. As depicted in Table 2, the following four typhoons had the highest maximum water levels at the Hsiu-Lung station, in descending order: Typhoon Soudelor in 2015 (maximum water level = 12.0 m), Typhoon Saola in 2012 (11.4 m), Typhoon Dujuan in 2015 (9.8 m), and Typhoon Megi in 2016 (9.2 m). The radar mosaics for these typhoons are illustrated in Figure 2.

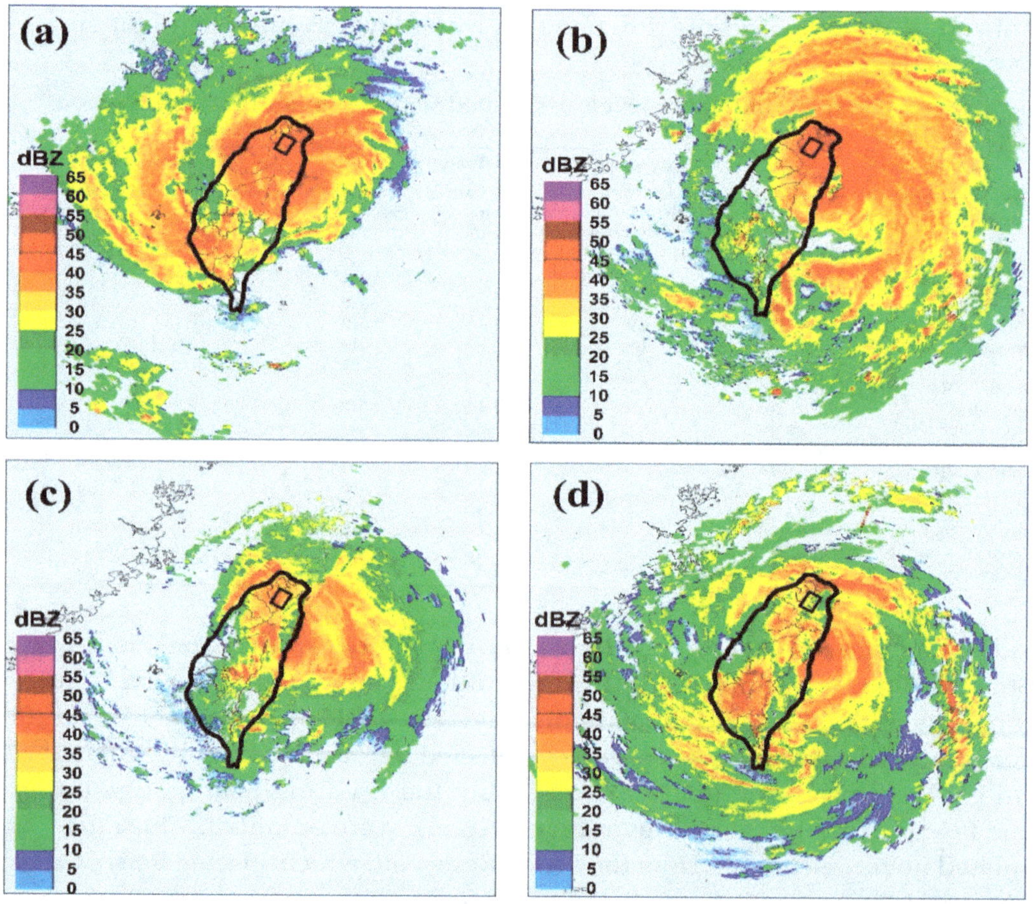

Figure 2. Radar reflectivity images from typhoons' approach to Taiwan: (**a**) Typhoon Saola (2 August 2012, 0300 local standard time (LST)), (**b**) Typhoon Soudelor (8 August 2015, 0400 LST), (**c**) Typhoon Dujuan (28 September 2015, 1800 LST), and (**d**) Typhoon Megi (27 September 2016, 1400 LST).

2.3. *Typhoon Paths*

The influence of typhoon paths on the rainfall and water level in the research area were examined. As depicted in Figure 3a, the path of the eye of Typhoon Saola, one of the aforementioned four typhoons, moves northward after touching the east coast of Taiwan and then moves along the northeast coast of Taiwan. In the radar reflectivity image in Figure 2a, the typhoon circulation covered the location of the research area. Because of the relatively long time for which it hovered over northern Taiwan, total rainfall values of 666 and 878 m were measured at the two precipitation stations (i.e., the Da-Tong-Shan and Fu-Shan stations), which were the highest rainfall values of the 16 typhoons. Moreover, the paths of typhoons Soudelor, Dujuan, and Megi were similar to those presented in Figure 3b–d. They continued to travel to the west after landing on the east coast of Taiwan. When typhoon circulations are lifted by mountain terrains, heavy rains are likely to occur in windward areas, resulting in exceptionally high rainfall in the catchment area.

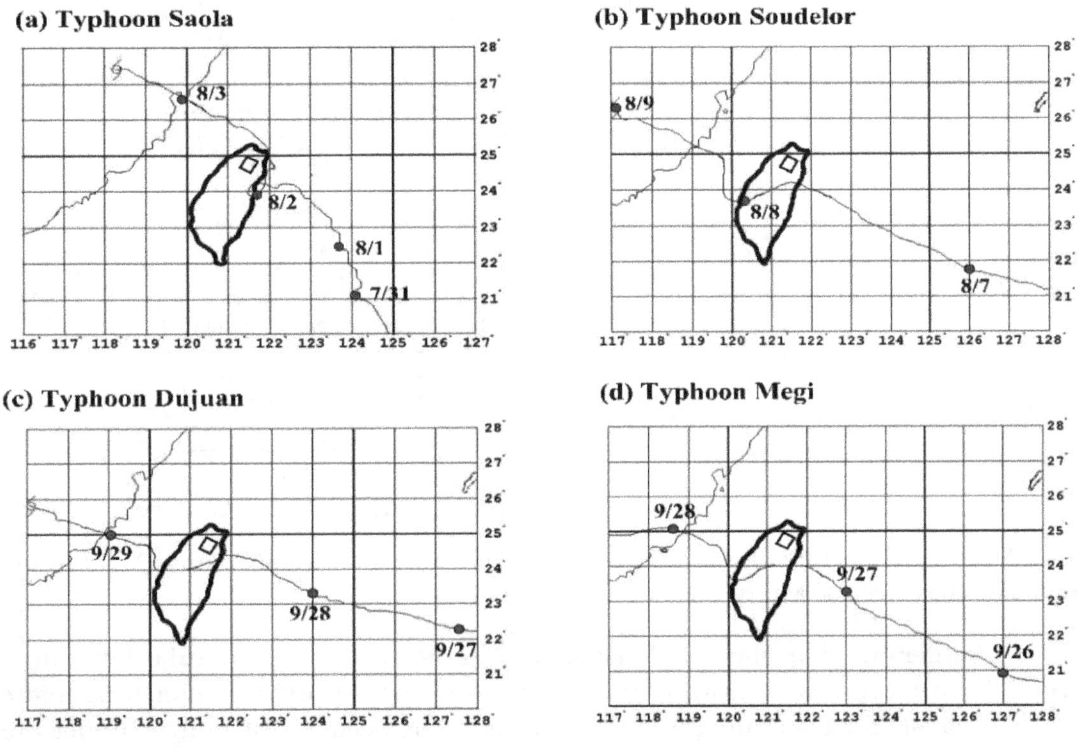

Figure 3. Typhoon center tracks: (**a**) Typhoon Saola in 2012, (**b**) Typhoon Soudelor in 2015, (**c**) Typhoon Dujuan in 2015, and (**d**) Typhoon Megi in 2016.

3. Model Development

The SNN and MIFNN forecasting models for water level estimation during typhoons were developed (the frameworks are displayed in Figure 4).

3.1. *SNNs*

SNNs, a type of network model composed of sequential multilayers, have a typical neural network structure. As depicted in Figure 4a, the input layer is available for the input of 1-D array data (i.e., the attribute data of water level, rainfall, and reservoir discharge). Multiple dense layers, or LSTM layers, can be applied as the hidden layers (or core layers) in the middle [44]. After the dropout layer, the simulation results are produced from the output layer. The dropout layer is to randomly drop the weights of a certain proportion of nodes between the hidden layers to prevent overfitting.

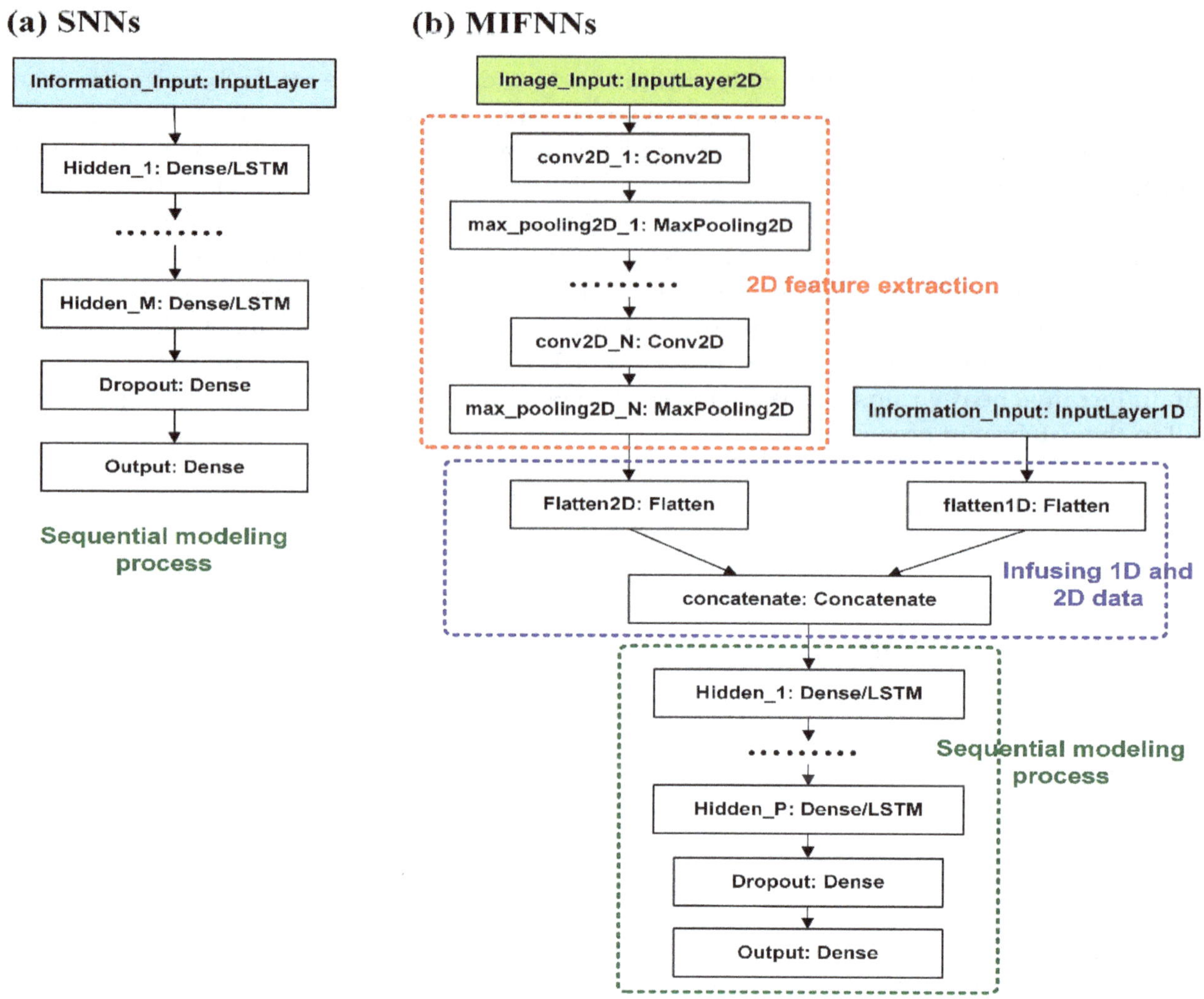

Figure 4. Model architectures for (**a**) sequential neural networks (SNNs) and (**b**) multiple-input functional neural networks (MIFNNs).

Core layers are generally applied as dense layers. A dense layer is a regular layer of neurons in a neural network. Each neuron receives input from all neurons in the previous layer, thus ensuring a dense connection. Dense layers are also called fully connected layers because the neuron nodes are connected to the neurons in the previous layer. Dense layer implements the operation: output = activation {dot product of (input, kernel) + bias}, where activation is the element-wise activation function passed as the activation argument, the kernel is a weight matrix created by the layer, and the bias is a bias vector created by the layer [45,46]. Dense layers are not appropriate for all types of problems, particularly time sequence problems.

RNNs are suitable for sequential information. In traditional neural networks such as multiple layer perceptrons, all inputs and outputs are assumed to be independent. RNNs are termed recurrent because they perform the same task for every element of a sequence, with the output depending on the previous computations [47,48]. However, RNNs are difficult to train and are detrimentally affected by vanishing or exploding gradients; therefore, they cannot solve the long-term dependency problem [49]. LSTM neural networks are a type of RNN-based model. An LSTM network has LSTM cell blocks instead of standard hidden layers [50]. The components of these cells are known as the input gate; the forget gate, and the output gate. The structural details of LSTM networks were elaborated by [51,52]. In LSTM networks, a gate mechanism is introduced to prevent backpropagated errors from vanishing or exploding, and this approach has subsequently been proved to be more effective than the use of conventional RNNs [53].

3.2. MIFNNs

As illustrated in Figure 4b, MIFNNs could be used for modeling with 1-D hydrographic data and 2-D radar imagery. Accordingly, radar reflectivity images were employed to enhance the rainfall data captured in the upstream catchment area because the application of only rainfall data retrieved at rain gauges in SNN models was insufficient to comprehensively describe the rainfall signals within the catchment area.

The input data had multiple sources and two dimensions; hence, this study developed a model capable of merging and integrating data into single inputs. As depicted in Figure 4b, the two input layers separately served as entrances for the 1-D and 2-D data. First, 2-D images were processed. The input layer received the original bitmaps and extracted features through image-processing techniques to provide more effective feature maps for model training. As revealed in Figure 4b, in the first stage (2-D feature extraction, which is marked by red dotted lines), a network structure comprising convolutional and pooling layers are generally adopted for feature extraction. For convolutional layers, input images are implemented through convolution by using various kernels [54]. Convolution is implemented in two steps: sliding and calculating dot products; that is, using filters that slide onto input images and continuing to calculate matrices to obtain dot products. The convoluted images are called feature maps [55]. However, features generated in convolutional layers are often excessively accurate; hence, the sensitivity of convolutional layer–generated features to edges decreases after passing through pooling layers, which reduces the number of subsequent calculations required. Detailed equation derivation can be found in the studies of [56,57].

The second stage (marked by blue dotted lines) in the figure is to infuse 1-D and 2-D data. First, flattened layers are used to flatten both 1-D and 2-D data; specifically, dimension transmissions are processed by flattened layers. Subsequently, concatenate layers (also called shared layers) are employed to convert data into 1-D arrays. Eventually, sequential modeling (marked by green-dotted lines) is conducted to verify model parameters. The structure in this part is identical to that of the SNN model.

4. Model Constructions

The presented models were implemented with the open-source scikit-learn and Keras libraries in Python 3.7 (Python Software Foundation, Wilmington, DE, USA) [58,59]. The MIFNN and SNN models were constructed with a functional application programming interface (API) and a Keras sequential API, respectively. Because the proposed DL-based models use two-dimensional matrices to calculate images, the advanced computer processing units of graphics cards are required. The computation environment comprised GeForce RTX 2080 Ti graphics card in the study.

4.1. Data Splitting

Before model construction, data sets were divided into training, validation, and testing data set. Four typhoons (i.e., typhoons Saola, Soudelor, Dujuan, and Megi) were selected as the testing data set that significantly influenced the water level at the Hsiu-Lung station. As presented in Table 2, the maximum flood stages of the remaining 12 typhoons not in the testing data set were relatively low, which may compromise the accuracy of the estimated maximum flood-stage values if used for model training.

Several new cross-validation procedures were introduced to overcome the defects of the old ones [60]. One of the methods, the leave-p-out method (where p denotes the cardinality of the testing data set), could resolve the problem I encountered during data splitting. The leave-p-out method involves the use of p observations as the testing data set and the use of the remaining observations as the training–validation data set. This is repeated in all ways to cut the original sample on a testing data set of p observations and a training–validation data set. To increase the data in the training–validation data set, I let p equal 1. That is, one of the four biggest typhoons as the test data and the remaining 15 typhoons were used as the training and validation data sets. The remaining 15 typhoons were then used to train the models through 10-fold cross-validation. Therefore, four models were built.

4.2. Selection of Lag Times

In this study, the water levels at the Hsiu-Lung station were influenced by the natural processes of rainfall-runoff and control flow from upstream reservoirs. Accordingly, for the proposed water level prediction models, the hourly water levels predicted were assumed to be a function of the respective related attributes and could be expressed as:

$$L_{H,t+i} = f\left(P_{F,t-\Delta t_a}, P_{F,t-\Delta t_b}, R_{D,t-\Delta t_c}, L_{H,t-\Delta t_d}, L_{C,t-\Delta t_e}, L_{S,t-\Delta t_f}, L_{L,t-\Delta t_g}\right) \quad (1)$$

where $L_{H,t+i}$ represents the forecast water levels at the Hsiu-Lung station for the *i*-h ahead forecast (*i* ranges from 1 to 6 h in this research); P_F and P_D denote the rain rates at Da-Tong-Shan and Fu-Shan rain gauges, respectively; R_F is the release from Feitsui Reservoir; L_H, L_C, L_S, and L_L are the water levels at Hsiu-Lung, Chu-Chih, Shang-Gui-Shan Bridge, and Lan-Sheng Bridge stations, respectively; and t and Δt_a–Δt_g are the indices of the time period and lag-time length, respectively.

Note that in Taiwan, the most common release strategy for reservoir flood control operations involves predefined operation rules, in which releases are expressed as a function of reservoir variables (water levels) and hydrological inputs (reservoir inflows) [61,62]. Moreover, numerous studies presented in-depth reviews of reservoir operations and modeling on this topic [63–67]. Because this study focused on developing the downstream water levels prediction models, for simplicity, the records of reservoir outflow were employed as an attribute when the modeling process.

Acceptable lag times (Δt_a–Δt_g) for the water level forecasts were first identified. Figure 5 illustrates the correlation coefficient (R) between water level with lead times (1- to 6-h ahead forecast) at Hsiu-Lung station and various attributes with lag times. The results indicated that a lag time of 1 h yielded the highest R values for the four water level stations and the Feitsui Reservoir because of the small catchment. For the two rain gauges, lag times of 4, 3, 2, 1, 0, and 0 h, respectively, 1 to 6 h ahead yielded the highest R values. Note that lag time = 0 refers to the current time.

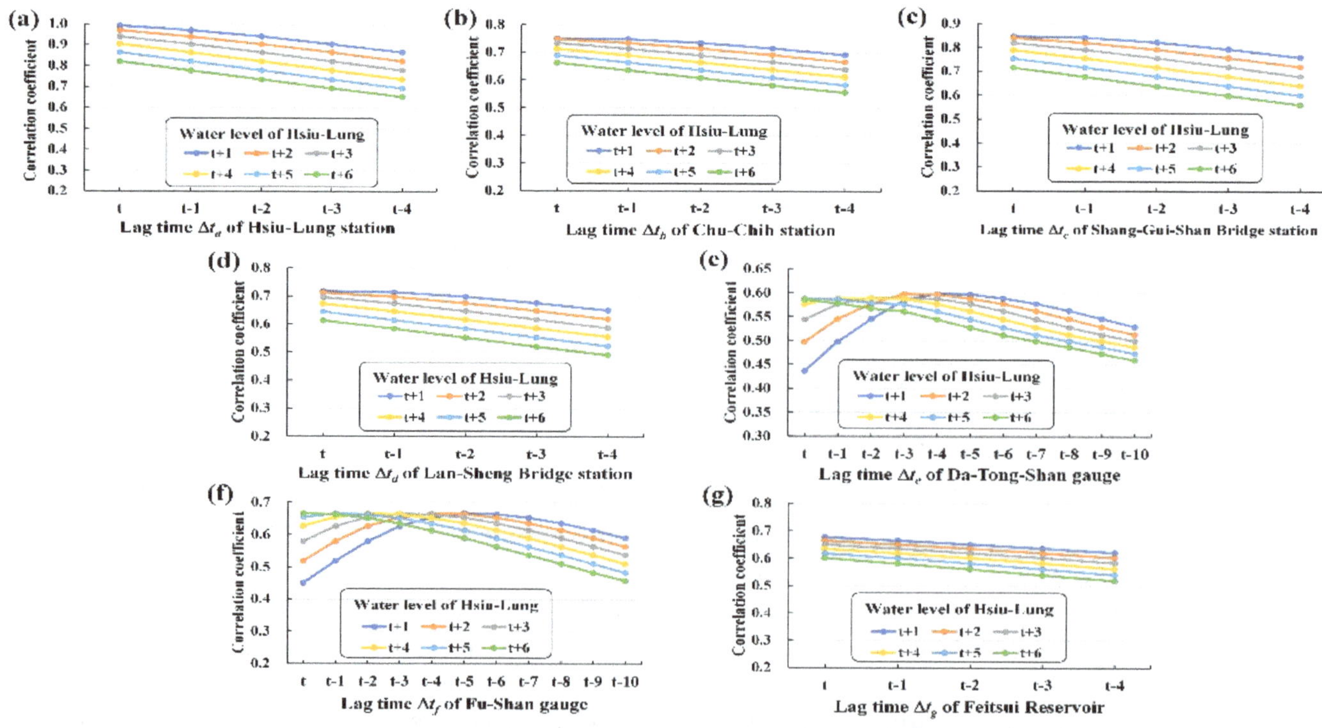

Figure 5. Correlation coefficient between water level with lead times of Hsiu-Lung station and various attributes with lag times: (**a**) Hsiu-Lung station, (**b**) Chu-Chih station, (**c**) Shang-Gui-Shan Bridge station, (**d**) Lan-Sheng Bridge station, (**e**) Da-Tong-Shan gauge, (**f**) Fu-Shan gauge, and (**g**) Feitsui Reservoir.

4.3. Training SNNs

Section 3.1 details the use of dense and LSTM layers as hidden layers of SNN models. For the modeling in this section, this study adopted two types of hidden layers to establish models and named them the SNN-Dense and SNN-LSTM models. The calculation parameters of the SNN-Dense model were set as follows: the loss function of dense layers = mean squared errors; activation function = sigmoid; the dropout rate of the dropout layer = 0.2. The parameters of the SNN-LSTM model were set as follows: the loss function of LSTM layers = mean squared error, cell activation function = rectified linear units; and gate activation function = sigmoid. The settings of the output layer of the SNN-Dense and SNN-LSTM models were loss function = mean squared error and activation function = sigmoid.

The optimizer method used in SNN-based models was the adaptive moment estimation (Adam) optimization algorithm. Adam [68] optimizes the momentum and learning rate. The merit of the Adam optimizer is its ability to compute individual adaptive learning rates for various parameters from estimates of the first and second moments of the gradients [69].

Hyperparameters were calibrated for both SNN-based models, comprising the number of hidden layers and neuron nodes in a hidden layer. In Figure 6, the testing typhoon, Saola, was used as an example. The number of hidden layers was calibrated from 1 to 4. The optimal lengths of the hidden layers were mostly 3 and 2 for SNN-Dense and SNN-LSTM models, respectively (Figure 6a,b). Subsequently, the number of nodes in a hidden layer, adjusting them to values between 10 and 100, were calibrated. In Figure 6c,d, the optimal numbers of nodes were within the ranges of 20–30 and 30–60 for the SNN-Dense and SNN-LSTM models, respectively.

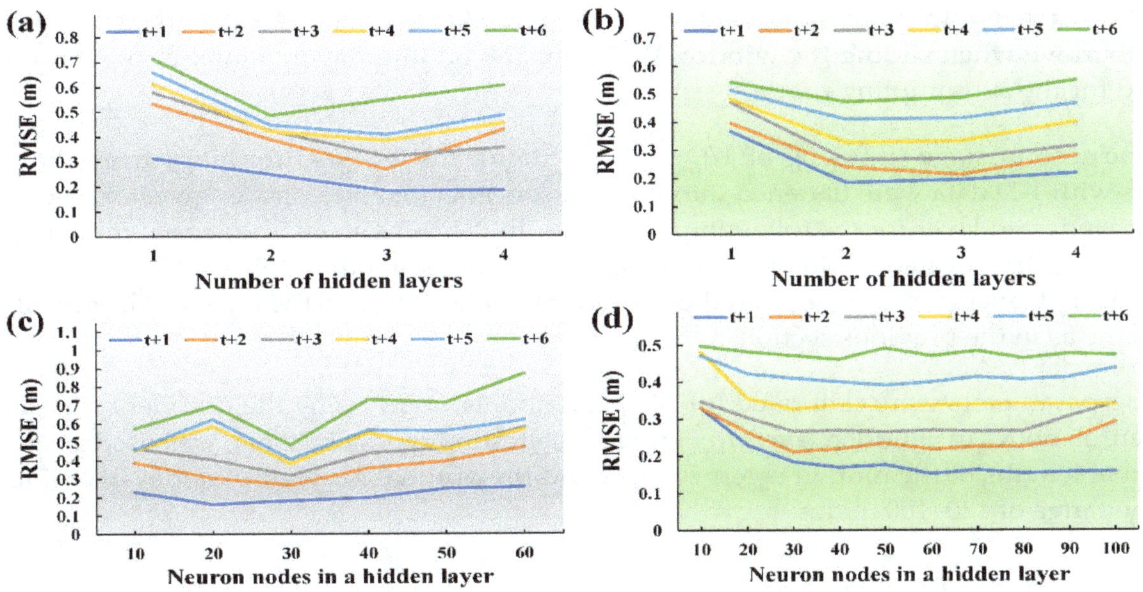

Figure 6. Hyperparameter calibration for sequential neural networks (SNN) models: (**a**) number of dense layers in SNN (SNN-Dense), (**b**) number of LSTM layers in SNN (SNN-LSTM), (**c**) neuron nodes of a dense layer in SNN-Dense and (**d**) neuron nodes of a LSTM layer in SNN-LSTM.

4.4. Training MIFNNs

In Section 3.2, 2-D radar reflectivity images were incorporated into the MIFNN model. As described in Section 2.1, because the Peishi Creek is regulated by the Feitsui Reservoir, I cropped the corresponding position of the image according to the longitudinal and latitudinal range of the Nanshi Creek catchment. The longitude of the Nanshi Creek catchment is 121.43° E–121.60° E, and its latitude is 24.68° N–24.89° N. Its corresponding area on radar images could be represented by approximately 28 × 28 pixels. Figure 7 presents the cropped radar reflectivity images of the Nanshi Creek catchment for 16 consecutive hours on 8 August 8, 2015, when Typhoon Soudelor made landfall in Taiwan on 8 August 2015, from 0000 LST to 1500 LST. The figure illustrates how the intensity of the radar reflectivity signals dBZ changes with the interaction between typhoon circulation and the land.

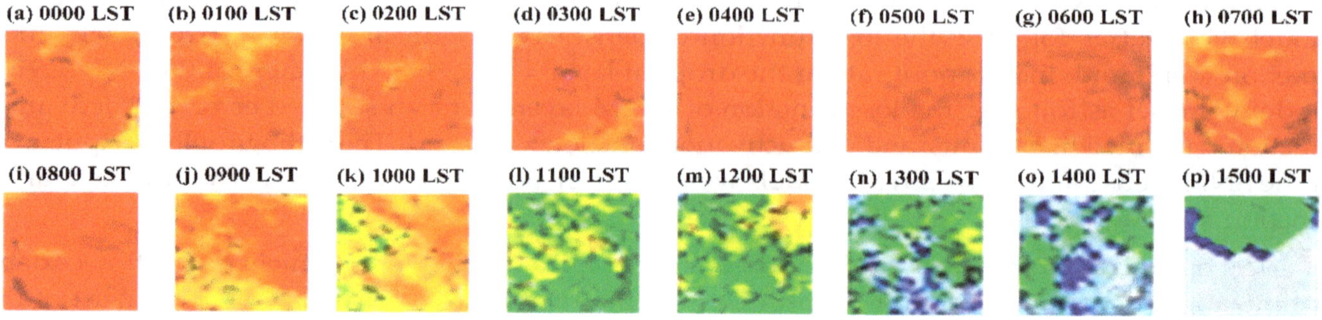

Figure 7. Typhoon Soudelor: cropped images at 8 August 2015 from 0000 LST to 1500 LST at 1 h sampling frequency: (**a**) 0000 LST, (**b**) 0100 LST, (**c**) 0200 LST, (**d**) 0300 LST, (**e**) 0400 LST, (**f**) 0500 LST, (**g**) 0600 LST, (**h**) 0700 LST, (**i**) 0800 LST, (**j**) 0900 LST, (**k**) 1000 LST, (**l**) 1100 LST, (**m**) 1200 LST, (**n**) 1300 LST, (**o**) 1400 LST, (**p**) 1500 LST.

The parameter settings of the MIFNN model are as follows: In the first stage (2-D feature extraction), the input layer receives a radar reflectivity of 28 × 28 pixels as input; this study then used a sequence of three convolutional and pooling layers as feature extractors. For the 1- to 6-h predictions, this study employed the difference input image on the basis of the closest correlation of lag times. According to Section 4, lag times of 4, 3, 2, 1, 0, and 0 h, respectively at 1 to 6 h ahead, yielded the greatest R values. For the convolutional layer setting, the kernel size = 3 × 3, and stride (the number of grids to move when sliding the window) = 1. For the pooling layer, a max-pooling method was employed for the subsampling process.

In the second stage (infusing of 1-D and 2-D data), the 2-D feature maps from these feature extractors with 1-D data were flattened into vectors, concatenated into one long vector, and passed on to a fully connected layer for the following modeling. In the third stage (sequential modeling), LSTM layers were applied because during the modeling of SNN-based models; this study discovered that the errors for LSTM layers were lower than those for dense layers. The LSTM layer environmental setups are the same as in the previous section.

Hyperparameters were calibrated for MIFNN models, comprising the number of hidden layers and of neuron nodes in a hidden layer. In Figure 8, the testing typhoon, Saola, was used as an example. The optimal lengths of the hidden layers were 3, and the optimal number of nodes in a hidden layer was in the range of 120–160 nodes.

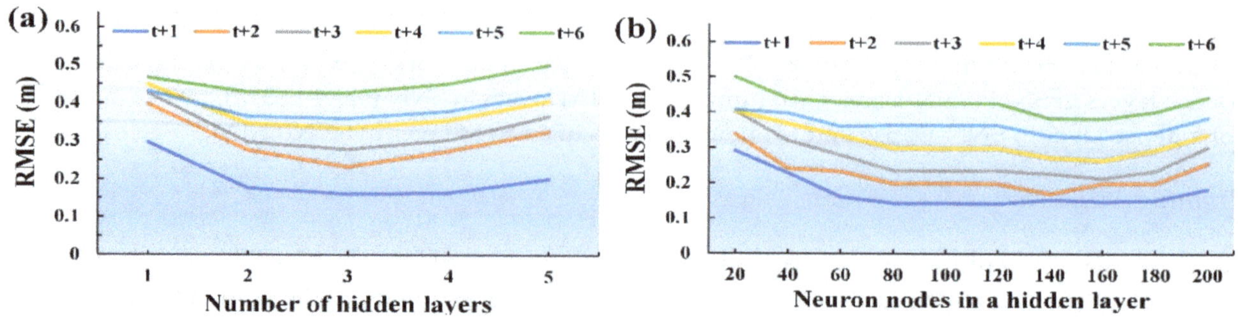

Figure 8. Hyperparameter calibration for multiple-input functional neural networks (MIFNN) models of (**a**) number of hidden layers and (**b**) neuron nodes in a hidden layer.

5. Simulations

A simulation analysis of each of the selected typhoons was implemented. Four testing typhoons were employed to identify the most suitable water level prediction models in this section. Subsequently, a conventional regression-based (REG) water level forecasting model was established to serve as a benchmark model.

5.1. Testing Typhoons and Predictions

Water level changes at the Hsiu-Lung station mainly result from the upstream rainfall and the water released by the reservoir. Hence, a rainfall and reservoir release time-series image was developed. Figure 9 (typhoons Saola and Soudelor) and Figure 10 (typhoons Dujuan and Megi) depict the hyetograph at the Da-Tong-Shan gauge and the Fu-Shan gauge, the reservoir released at the Feitsui Reservoir, and the water levels at the Hsiu-Lung station during each typhoon.

Figures 9 and 10 show that the peak water level of the Hsiu-Lung station is contributed primarily by upstream rainfall (because the peak water level at the Hsiu-Lung station occurred after maximum rainfall) and secondarily by reservoir release (because the maximum water discharge times of the four tested typhoons all occurred after the water level peaked at the Hsiu-Lung station). This study further confirmed that, for the four testing typhoons, the difference between the maximum precipitation time and the peak water level time at the Hsiu-Lung station (i.e., the lag time) was minimal (between 3 and 4 h). Furthermore, this study discovered that, for the four testing typhoons, the maximum precipitation values estimated at the Da-Tong-Shan gauge and the Fu-Shan gauge differed greatly. The results were as follows: the maximum rainfall values at the Da-Tong-Shan gauge and the Fu-Shan gauge were (50, 84 mm/h) for Typhoon Saola, (69, 82 mm/h) for Typhoon Soudelor, (48, 76 mm/h) for Typhoon Dujuan, and (41, 68 mm/h) for Typhoon Megi. Accordingly, the spatial distribution of rainfall within the catchment area varied considerably.

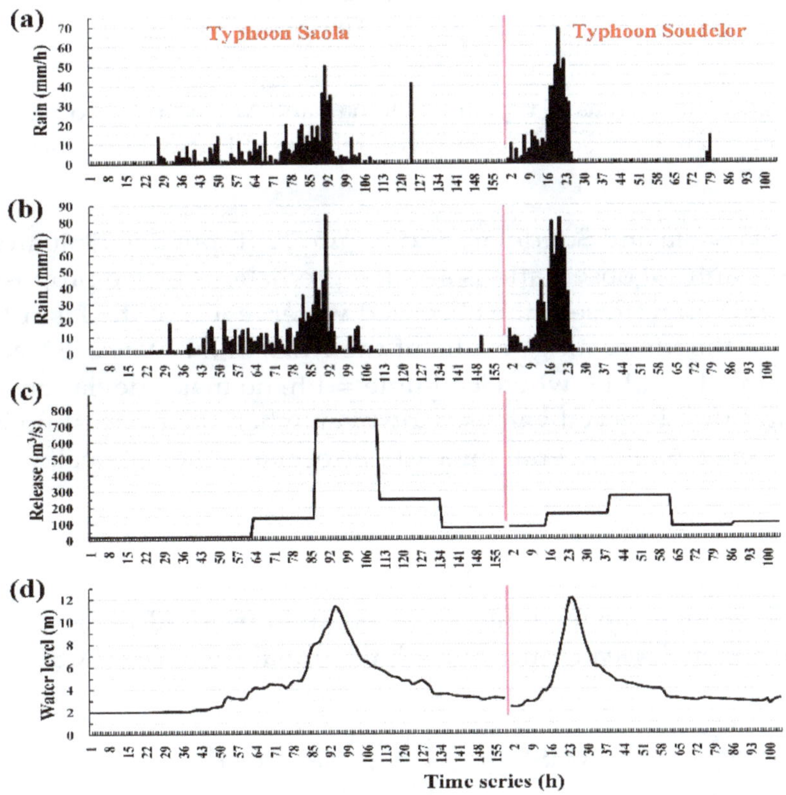

Figure 9. Hydrologic observations of typhoons Saola and Soudelor: (**a**) hyetograph at Da-Tong-Shan gauge, (**b**) hyetograph at Fu-Shan gauge, (**c**) reservoir releases at Feitsui Reservoir, and (**d**) water levels at Hsiu-Lung station.

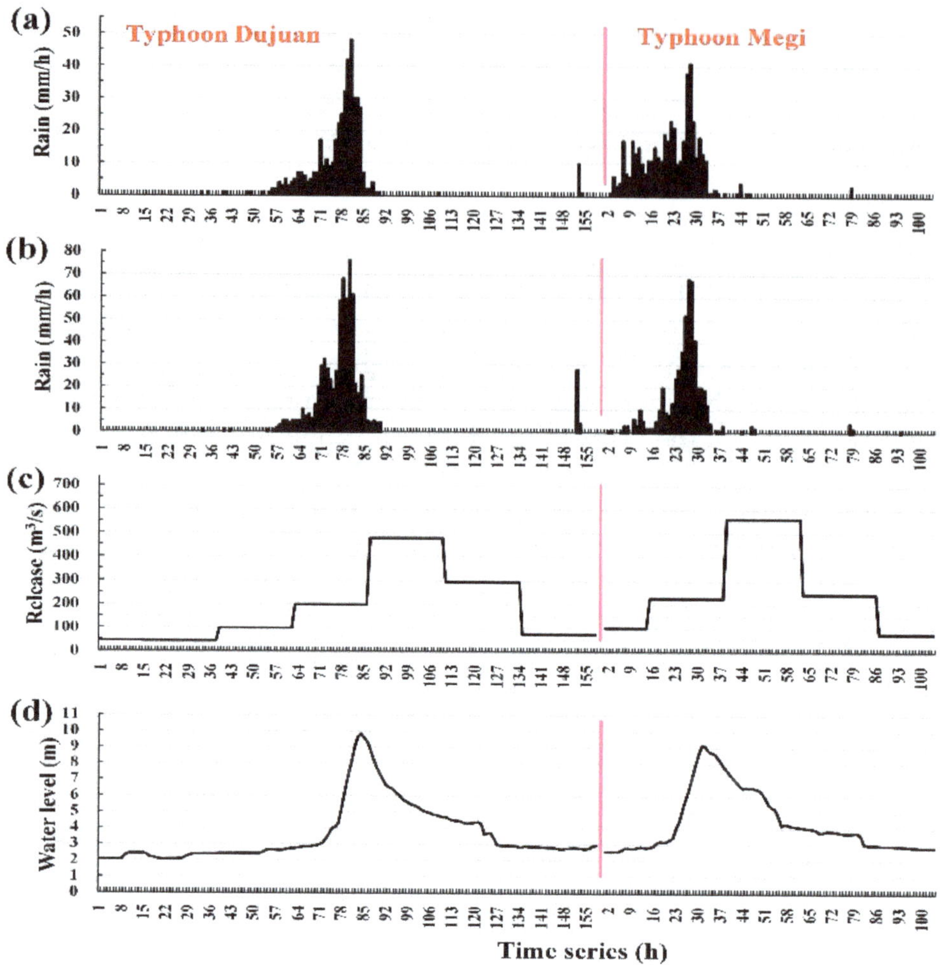

Figure 10. Hydrologic observations of typhoons Dujuan and Megi: (**a**) hyetograph at Da-Tong-Shan gauge, (**b**) hyetograph at Fu-Shan gauge, (**c**) reservoir releases at Feitsui Reservoir, and (**d**) water levels at Hsiu-Lung Station.

Figure 11 (typhoons Saola and Soudelor) and Figure 12 (typhoons Dujuan and Megi) depict the water level time series graphs of observations and the predictions of the proposed models. The thick gray solid lines (observation) represent the observed water levels of the Hsin-Lung station, and the remaining four lines are the forecasting results of the REG, SNN-Dense, SNN-LSTM, and MIFNN models. As seen in Figures 11 and 12, when lead time = 1 h, no major deviations were visible between the predicted results and the observed values. However, when the forecasting horizons were longer, the differences between the models' overall forecast performance, actual peak water levels, and arrival times increased.

5.2. *Evaluation*

To evaluate the deviations between the predicted and observed values, the root-mean-squared error (RMSE) deviation results for lead times of 1–6 h were analyzed. The RMSE is calculated as follows:

$$\text{RMSE} = \sqrt{\frac{1}{n}\sum_{i=1}^{n}(P_t - O_t)^2} \quad (2)$$

where n is the total number of observations, P_t is the predicted water levels at time t, and O_t is the observed water levels at time t.

For forecasting horizons of 1–6 h, the RMSE evaluation values demonstrated that the ordering of the sizes of the models' deviations (Figure 13) was as follows, from largest to smallest: the REG model, the SNN-Dense and SNN-LSTM models, and then the MIFNN model. Figure 14 illustrates the scatter diagrams of the observed and predicted values for all tested typhoons. The figure reveals that, when the lead time was larger, the MIFNN model's predicted values were closer to its observed values compared with those for the REG, SNN-Dense, and SNN-LSTM models. In terms of correlation, the ordering of the values for the models' coefficients of determination (R^2) in the figure was as follows, from largest to smallest: the MIFNN model, the SNN-LSTM and SNN-Dense models, and then the REG model.

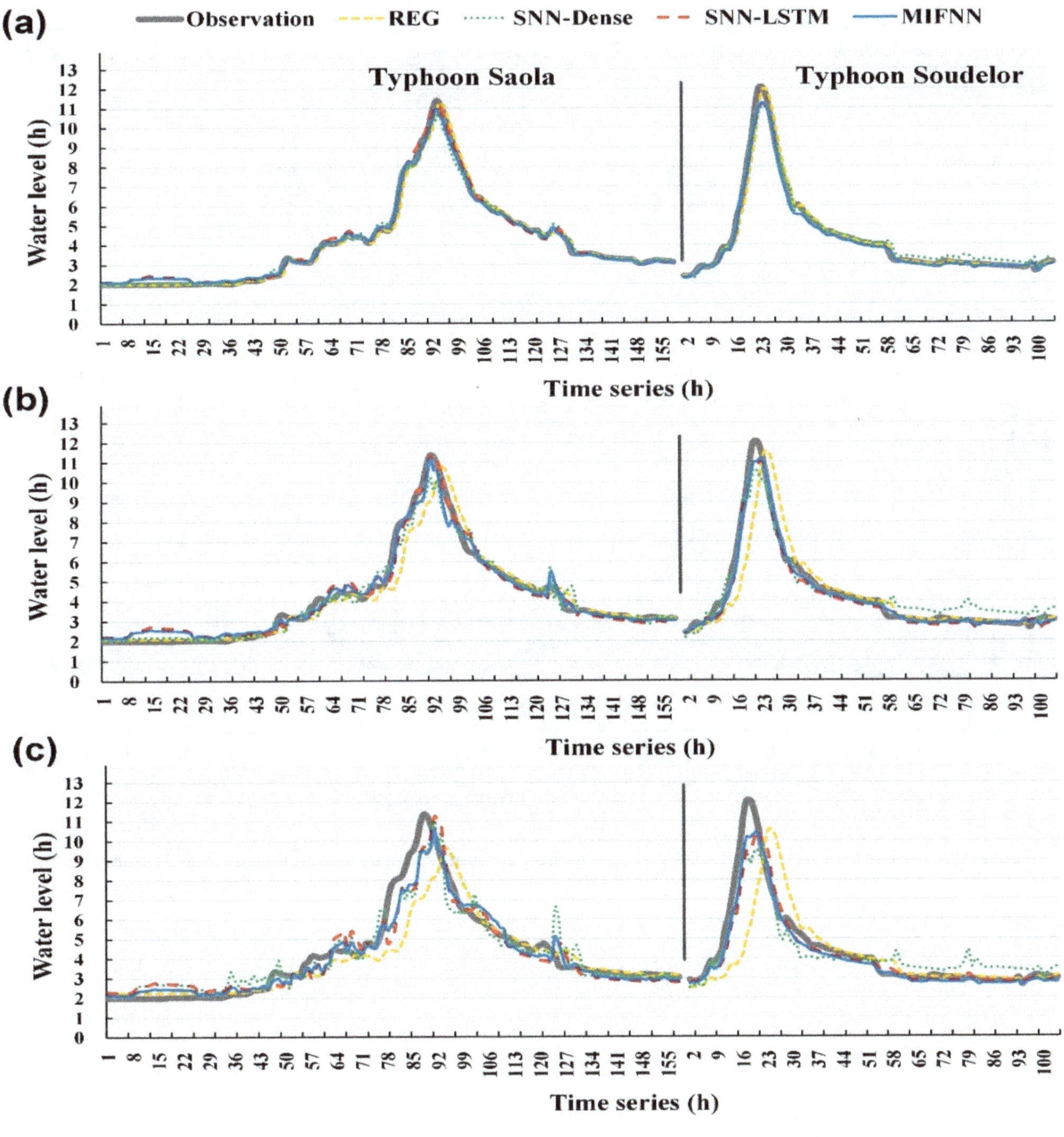

Figure 11. Simulation of water levels for typhoons Saola and Soudelor at lead times of (**a**) 1 h, (**b**) 3 h, and (**c**) 6 h.

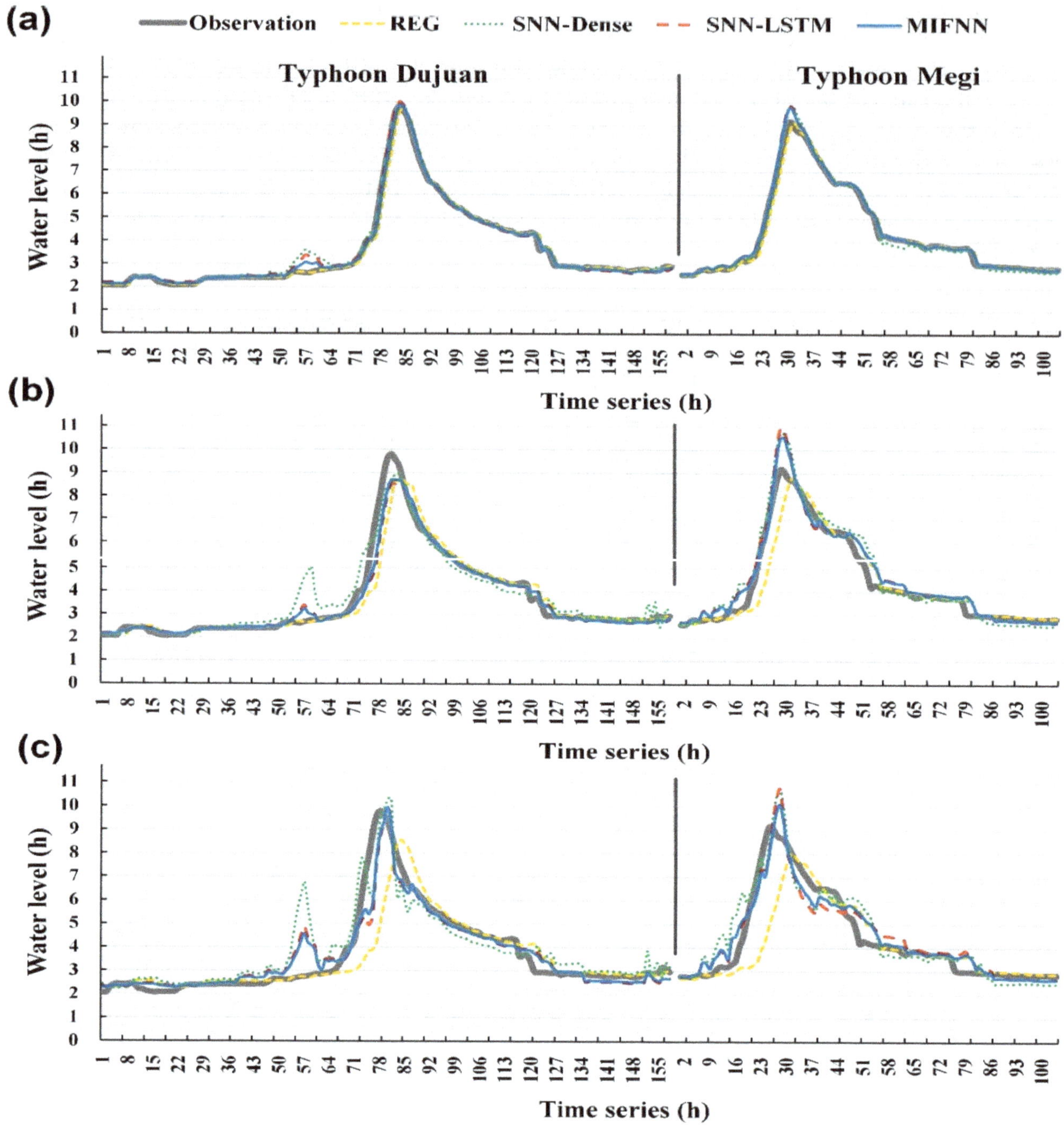

Figure 12. Simulation of water levels for typhoons Dujuan and Megi at lead times of (**a**) 1 h, (**b**) 3 h, and (**c**) 6 h.

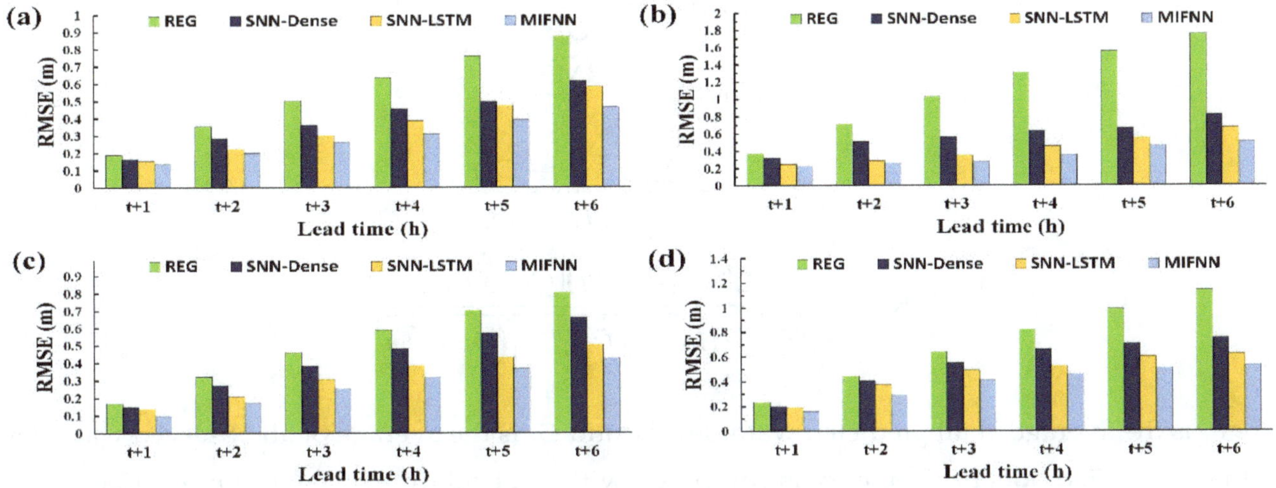

Figure 13. Performance for testing typhoon: (**a**) Saola, (**b**) Soudelor, (**c**) Dujuan, and (**d**) Megi.

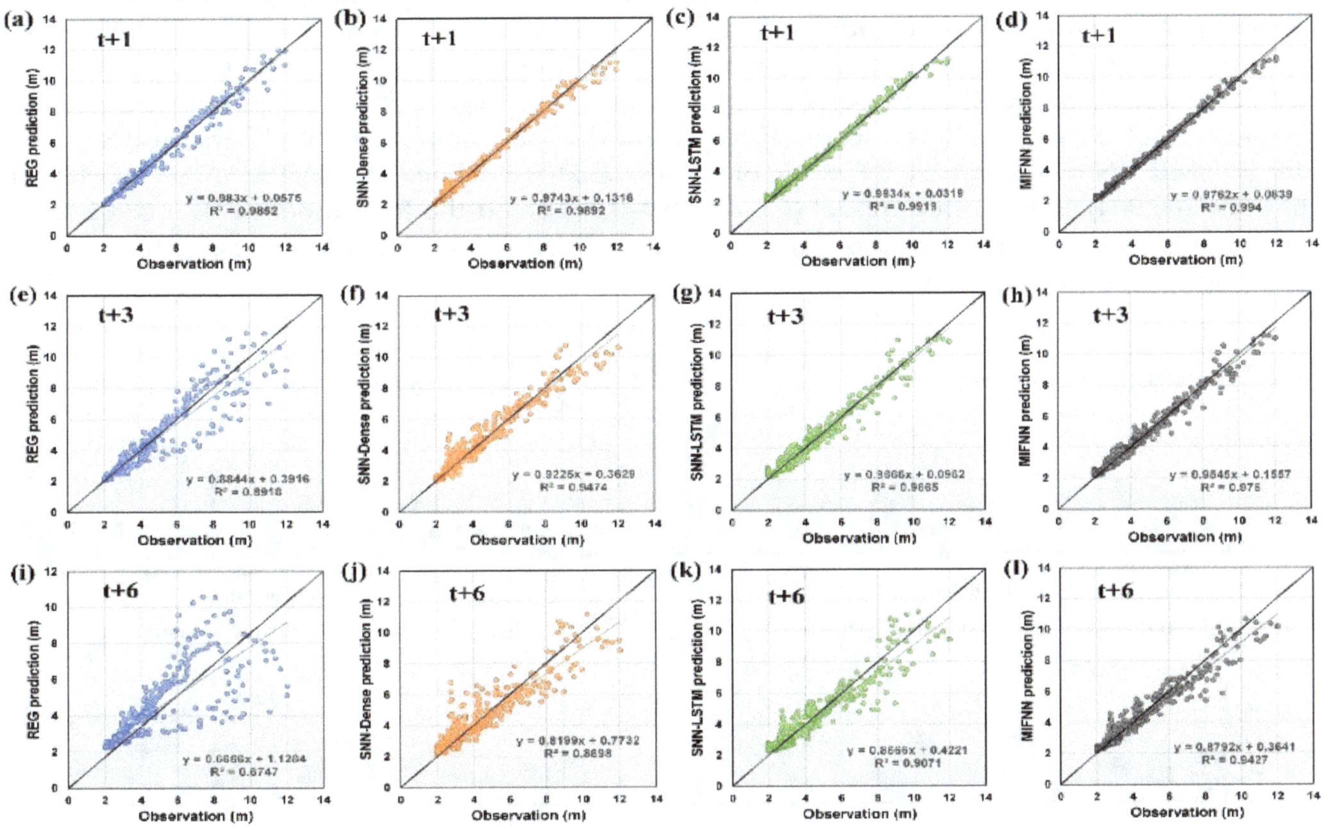

Figure 14. Observations versus predictions using REG, SNN-Dense, SNN-LSTM, and multiple-input functional neural networks (MIFNN) models for lead times of 1 h (**a–d**), 3 h (**e–h**), and 6 h (**i–l**).

5.3. Overall, Performance

To estimate the model deviation for all four testing typhoons, the mean absolute error (MAE), relative MAE (rMAE), RMSE, and relative RMSE (rRMSE) were adopted as indicators. Moreover, the efficiency coefficient (CE) indicating the goodness-of-fit measures and R were used. The relevant equations are as follows:

$$\text{MAE} = \frac{1}{n}\sum_{i=1}^{n}|P_t - O_t| \tag{3}$$

$$\text{rMAE} = \frac{\text{MAE}}{\overline{O_t}} \tag{4}$$

$$\text{rRMSE} = \frac{\text{RMSE}}{\overline{O_t}} \tag{5}$$

$$\text{CE} = 1 - \frac{\sum_{i=1}^{n} (P_t - O_t)^2}{\sum_{i=1}^{n} (O_t - \overline{O_t})^2} \tag{6}$$

$$\text{R} = \frac{\sum_{i=1}^{n} (P_t - \overline{P_t})(O_t - \overline{O_t})}{\sqrt{\sum_{i=1}^{n} (P_t - \overline{P_t})^2 \sum_{i=1}^{n} (O_t - \overline{O_t})^2}} \tag{7}$$

where $\overline{P_t}$ is the average of all predicted water levels, and $\overline{O_t}$ is the average of all observed water levels.

In Figure 15, model performance is compared within a lead time of 1–6 h. Figure 15a–d reveals that in terms of the MAE, rMAE, RMSE, and rRMSE indicators, the MIFNN model outperformed the REG, SNN-Dense, and SNN-LSTM models. In the deviation comparison, the SNN-LSTM model was inferior to the MIFNN model but superior to the REG and SNN-Dense models; the REG model outperformed the SNN-Dense model only when the lead time was 1–4 h, yet its remaining deviation indicators were all higher than those of other models. The ordering of the performance in terms of CE (Figure 15e), according to the scores obtained by the four models, is as follows, in descending order: the MIFNN model (with a CE range of 0.994–0.936 for 1–6 h lead time), the SNN-LSTM model (CE = 0.992–0.904), SNN-Dense model (CE = 0.989–0.864), and the REG model (CE = 0.985–0.671). In addition, according to the R results (Figure 15f), the MIFNN and REG models, respectively, yielded the best and worst performance.

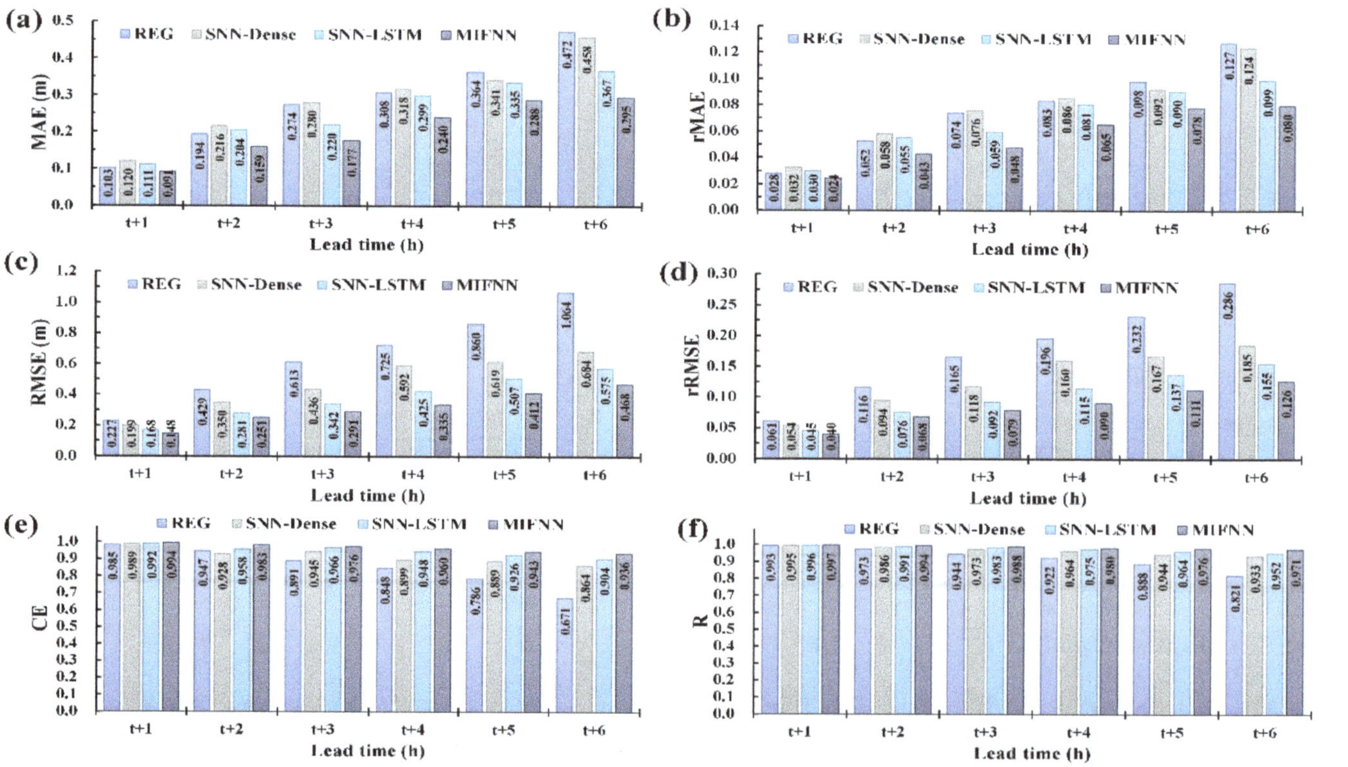

Figure 15. Comparisons of model predictions: (**a**) mean absolute error (MAE), (**b**) relative MAE (rMAE), (**c**) root-mean-squared error (RMSE), (**d**) relative RMSE (rRMSE), (**e**) CE, and (**f**) correlation coefficient (R).

Table 4 presents the mean measured values of the deviation indicators within a lead time of 1–6 h. Table 4 indicates that among the four models, the MIFNN model most effectively reduced forecasting errors. The improvement rate of the REG model by using its CE results was defined:

$$\mathrm{IR} = (\mathrm{CE} - \mathrm{CE}_{\mathrm{REG}})/\mathrm{CE}_{\mathrm{REG}} \tag{8}$$

After calculation, the SNN-Dense, SNN-LSTM, and MIFNN models exhibited respective improvements of 7.49%, 10.99%, and 12.87% compared with the REG model.

Table 4. Average performance levels for 1–6 h predictions.

Measure	REG	SNN-Dense	SNN-LSTM	MIFNN
MAE (m)	0.286	0.289	0.256	0.208
rMAE	0.077	0.078	0.069	0.056
RMSE (m)	0.653	0.480	0.383	0.317
rRMSE	0.176	0.130	0.103	0.086
CE	0.855	0.919	0.949	0.965
R	0.924	0.966	0.977	0.984
$\mathrm{EW_P}$ (%)	7.822	10.979	7.686	6.150
$\mathrm{TW_P}$ (h)	3.333	1.792	1.417	1.167

Moreover, to examine peak water levels and the deviations of their arrival times, the error percentage of peak water levels ($\mathrm{EW_P}$) and time error of peak water levels ($\mathrm{TW_P}$) were defined:

$$\mathrm{EW_P} = \frac{1}{m}\sum_{k=1}^{m}\left|\frac{P_k - O_k}{O_k}\right| \times 100\% \tag{9}$$

$$\mathrm{TW_P} = \frac{1}{m}\sum_{k=1}^{m}\left|T_{P_k} - T_{O_k}\right| \tag{10}$$

where P_k is the predicted peak water level for typhoon event k; O_k is the observed peak water level for typhoon event k; m is the total number of typhoon events; T_{P_k} is the arrival time of the predicted peak water level for typhoon event k; and T_{O_k} is the arrival time of the observed peak water level for typhoon event k.

Figure 16 depicts the results of $\mathrm{EW_P}$ and $\mathrm{TW_P}$. (1) According to the $\mathrm{EW_P}$ values of the models, the peak water level deviation rate of the REG model was within the range of 3.87%–13.19%, that of the SNN-Dense model was within 5.46%–11.21%, that of the SNN-LSTM model was within 4.74%–8.47%, and that of the MIFNN model was within 3.66–7.97%. (2) According to the models' $\mathrm{TW_P}$ values, the arrival times of peak water levels of the REG model were within 1–6 h for a lead time of 1–6 h, those of the SNN-Dense were within 0.75–3.5 h, those of the SNN-LSTM model were within 0.5–2.75 h, and those of the MIFNN model were within 0.25–2.5 h. Table 4 further indicates that, on average, the MIFNN model more accurately forecasts peak water levels and arrival times than REG, SNN-Dense, and SNN-LSTM models.

5.4. Discussion

As described, the results obtained from the MIFNN model indicated considerable changes in the spatial distribution of rainfall in the catchment area examined. However, the limited ground observation data failed to comprehensively address the spatial changes in rainfall. This study thus employed radar reflectivity images to effectively describe the spatial changes in water vapor and to provide additional hydrophilic data for the catchment area. For REG, SNN-Dense, SNN-LSTM, and MIFNN models, the RMSE values for our water level forecasts were 0.653, 0.480, 0.383, and 0.317 m, respectively. The results indicate that the MIFNN had superior forecasting accuracy to the REG, SNN-Dense, and SNN-LSTM models without the incorporation of radar reflectivity images.

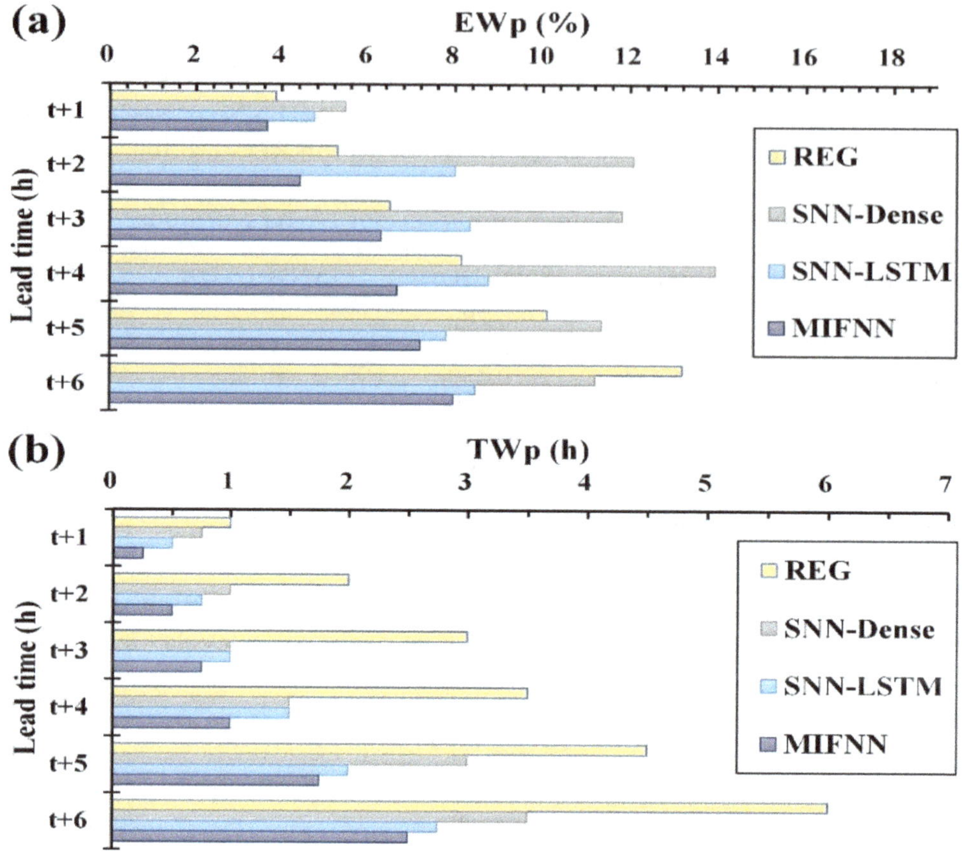

Figure 16. Comparisons of peak water level performance for (**a**) percentage of peak water levels (EW_p) and (**b**) time error of peak water levels (TW_p).

In addition, the application of LSTM layers as the network structure of the MIFNN model enhanced the forecasting ability of the algorithm. The LSTM layer is a neural network containing memory cell blocks. Each memory cell block comprises memory cell units that retain state across time-steps as well as three types of specialized gate units that learn to protect, utilize, or destroy this state as appropriate [44,70]. Memory blocks have a gate for determining whether the input data are valuable enough to be remembered and output. Such blocks can memorize information from various time lengths to automatically determine the amount of information that is useful [71]. Therefore, in typhoon water level forecasting through time series data analysis, determining for how long preceding memories should be retained is crucial. By contrast, because the REG and SNN-Dense models lacked memory functions, limited data were available for use. The experimental results also verified that the two aforementioned models yielded less accurate forecasting performance compared with the SNN-LSTM and MIFNN models, which had memory functions.

Limitations and suggestions are as follows: in the present study, this study used the trial-and-error method to calibrate the model hyperparameters. According to Maier et al. [72], as the number of model parameters increases, it is the difficulty in finding the parameter value or combination of parameter values that results in the smallest model error. Therefore, future studies are likely to focus on conducting a suitable optimization algorithm, such as the genetic algorithm, to increase the ability to find global optima, although this is generally at the expense of computational efficiency.

In the modeling process, this study used data that were collected from 2-D images of weather radars. Radar instruments can easily monitor and provide information on the spatial distribution of rainfall over larger areas [73]. However, the radar rainfall estimates face many acquisition limitations, such as the radar coverage is limited due to terrain complexity and the shading problem [74]. In addition, there are huge areas where only satellite observations available, so any global solution for precipitation prediction needs to incorporate satellite data [75]. Hence, the use of data from satellite measurements or data fusion of radar and satellite measurements is suggested for constructing DL-based models in the future.

Furthermore, in Taiwan, other existing forecast methods were using physical-based models such as hydrodynamics-hydrology models (e.g., SOBEK model [76]) and conceptual models (e.g., PWRI tank model [77]). Wei [78] demonstrated the multilayer perceptron neural networks achieved superior prediction accuracy compared with the conceptual PWRI rainfall-runoff model for flood forecasting in the Tsengwen Reservoir watershed in Southern Taiwan. The inputs of multilayer perceptron neural networks comprised the observed rainfalls, reservoir inflows, and ground weather data. This was similar to the model case of SNN-Dense in the paper. This study may suggest that applying the proposed models using deep learning combined with radar reflectivity images in other regions or compared the physical-based models with proposed models in the Hsintien River catchment in future research.

6. Conclusions

To precisely predict the downstream water level in catchment areas during typhoons, the water level–forecasting models were developed by using DL-based ANN models. In addition, due to the increasing accessibility of ground observation and remote-sensing data, methods are required for integrating data from multiple sources and in two dimensions (1-D and 2-D data) for DL-model training.

Two types of models: SNN and MIFNN models, were designed. The SNN, a typical neural network structure, was a network model constructed using sequential methods. The MIFNN model, however, could process multisource or multidimensional data, meaning that a network model was required to flexibly consolidate data. Specifically, when image files (e.g., radar reflectivity images) were employed as input attributes, feature extraction was necessary to provide effective feature maps for model training. Therefore, this study adopted convolutional and pooling layers to extract features and applied LSTM layers during model training to enable memory cell units to automatically determine the length of memory and to provide more useful preliminary information during training.

The Hsintien River in northern Taiwan was selected as the research area, and data concerning the changes in rainfall and water levels caused by 16 typhoons in the area from 2011 to 2019 were collected. Furthermore, the following input attributes were applied: 1-D data (i.e., water levels at river station, rain rate at rain gauges, and reservoir release of Feitsui Reservoir) and 2-D data (i.e., radar reflectivity mosaics obtained from the central weather bureau). I established models to forecast the water level at the Hsiu-Lung station and examined the results for Typhoon Saola in 2012, Typhoon Soudelor in 2015, Typhoon Dujuan in 2015, and Typhoon Megi in 2015.

This study forecasted the results for lead times of 1–6 h and reached the following conclusions. First, in terms of the predictions of water level for a typhoon, on average, the RMSE values of the REG, SNN-Dense, SNN-LSTM, and the MIFNN models were, respectively 0.653, 0.480, 0.383, and 0.317 m. This indicated that the MIFNN model had an apparently superior forecasting performance to the REG, SNN-Dense, and SNN-LSTM models. Compared with the REG model, the SNN-Dense, SNN-LSTM, and MIFNN models achieved improvements in the efficiency of 7.49%, 10.99%, and 12.87%, respectively. In addition, the peak water levels and their arrival times estimated by the REG, SNN-Dense, SNN-LSTM, and MIFNN models were 7.832 m and 3.333 h, 10.979 m and 1.792 h, 7.686 m and 1.417 h, and 6.150 m and 1.167 h, respectively. In brief, this study successfully developed a MIFNN model able to simultaneously process data from multiple sources and in multiple dimensions, provided a new technique for establishing water level forecasting models, and further enhanced these models' accuracy.

Acknowledgments: The authors acknowledge data provided by Taiwan's Central Weather Bureau, which are available at https://rdc28.cwb.gov.tw/and by Water Resources Agency, which are available at https://gweb.wra.gov.tw/Hydroinfo.

References

1. Phan, T.T.H.; Nguyen, X.H. Combining statistical machine learning models with ARIMA for water level forecasting: The case of the Red river. *Adv. Water Resour.* **2020**, *142*, 103656. [CrossRef]
2. Yan, J.; Jin, J.; Chen, F.; Yu, G.; Yin, H.; Wang, W. Urban flash flood forecast using support vector machine and numerical simulation. *J. Hydroinformatics* **2018**, *20*, 221–231. [CrossRef]
3. Ou, G.; Chen, X.; Kilic, A.; Bartelt-Hunt, S.; Li, Y.; Samal, A. Development of a cross-section based streamflow routing package for MODFLOW. *Environ. Model. Softw.* **2013**, *50*, 132–143. [CrossRef]
4. Drake, J.; Bradford, A.; Joy, D. Application of HEC-RAS 4.0 temperature model to estimate groundwater contributions to Swan Creek, Ontario, Canada. *J. Hydrol.* **2010**, *389*, 390–398. [CrossRef]
5. Leon, A.S.; Goodell, C. Controlling HEC-RAS using MATLAB. *Environ. Model. Softw.* **2016**, *84*, 339–348. [CrossRef]
6. Lai, C. Numerical modeling of unsteady open-channel flow. *Adv. Hydrosci.* **1986**, *14*, 161–333.
7. Wei, C.C.; Hsu, N.S. Multireservoir flood-control optimization with neural-based linear channel level routing under tidal effects. *Water Resour. Manag.* **2008**, *22*, 1625–1647. [CrossRef]
8. Bazartseren, B.; Hildebrandt, G.; Holz, K.P. Short-term water level prediction using neural networks and neuro-fuzzy approach. *Neurocomputing* **2003**, *55*, 439–450. [CrossRef]
9. Wei, C.C. Wavelet kernel support vector machines forecasting techniques: Case study on water-level predictions during typhoons. *Expert Syst. Appl.* **2012**, *39*, 5189–5199. [CrossRef]
10. Kisi, O.; Cimen, M. A wavelet-support vector machine conjunction model for monthly streamflow forecasting. *J. Hydrol.* **2011**, *399*, 132–140. [CrossRef]
11. Wagena, M.B.; Goering, D.; Collick, A.S.; Bock, E.; Fuka, D.R.; Buda, A.; Easton, Z.M. Comparison of short-term streamflow forecasting using stochastic time series, neural networks, process-based, and Bayesian models. *Environ. Model. Softw.* **2020**, *126*, 104669. [CrossRef]
12. Chang, L.C.; Shen, H.Y.; Chang, F.J. Regional flood inundation nowcast using hybrid SOM and dynamic neural networks. *J. Hydrol.* **2014**, *519*, 476–489. [CrossRef]
13. Wei, C.C. Comparing lazy and eager learning models for water level forecasting in river-reservoir basins of inundation regions. *Environ. Model. Softw.* **2015**, *63*, 137–155. [CrossRef]
14. Xiang, Z.; Demir, I. Distributed long-term hourly streamflow predictions using deep learning: A case study for State of Iowa. *Environ. Model. Softw.* **2020**, *131*, 104761. [CrossRef]
15. Dietterich, T.G. Machine learning for sequential data: A review. In *Structural, Syntactic, and Statistical Pattern Recognition*; Springer: Berlin/Heidelberg, Germany, 2002; pp. 15–30.
16. Längkvist, M.; Amy Loutfi, L.K. A review of unsupervised feature learning and deep learning for time-series modeling. *Pattern Recognit. Lett.* **2014**, *42*, 11–24. [CrossRef]
17. Bai, Y.; Bezak, N.; Sapač, K.; Klun, M.; Zhang, J. Short-term streamflow forecasting using the feature-enhanced regression model. *Water Resour. Manag.* **2019**, *33*, 4783–4797. [CrossRef]
18. Bi, X.; Li, B.; Lu, W.; Zhou, X. Daily runoff forecasting based on data-augmented neural network model. *J. Hydroinform.* **2020**, *22*, 900–915. [CrossRef]
19. Hrnjica, B.; Bonacci, O. Lake level prediction using feed forward and recurrent neural networks. *Water Resour. Manag.* **2019**, *33*, 2471–2484. [CrossRef]
20. Gires, A.; Onof, C.; Maksimović, Č.; Schertzer, D.; Tchiguirinskaia, I.; Simoes, N. Quantifying the impact of small scale unmeasured rainfall variability on urban runoff through multifractal downscaling: A case study. *J. Hydrol.* **2012**, *442*, 117–128. [CrossRef]
21. Ochoa-Rodriguez, S.; Wang, L.P.; Willems, P.; Onof, C. A review of radar-rain gauge data merging methods and their potential for urban hydrological applications. *Water Resour. Res.* **2019**, *55*, 6356–6391. [CrossRef]
22. Chang, P.L.; Lin, P.F.; Jou, B.J.D.; Zhang, J. An application of reflectivity climatology in constructing radar hybrid scans over complex terrain. *J. Atmos. Ocean. Technol.* **2009**, *26*, 1315–1327. [CrossRef]

23. Seliga, T.A.; Bringi, V.N. Potential use of radar differential reflectivity measurements at orthogonal polarizations for measuring precipitation. *J. Appl. Meteorol. Climatol.* **1976**, *15*, 69–76. [CrossRef]
24. Stout, G.E.; Neill, J.C. Utility of radar in measuring areal rainfall. *Bull. Am. Meteorol. Soc.* **1953**, *34*, 21–27. [CrossRef]
25. Ciach, G.J.; Krajewski, W.F.; Anagnostou, E.N.; Baeck, M.L.; Smith, J.A.; McCollum, J.R.; Kruger, A. Radar rainfall estimation for ground validation studies of the Tropical Rainfall Measuring Mission. *J. Appl. Meteorol. Climatol.* **1997**, *36*, 735–747. [CrossRef]
26. Georgakakos, K.P. Covariance propagation and updating in the context of real-time radar data assimilation by quantitative precipitation forecast models. *J. Hydrol.* **2000**, *239*, 115–129. [CrossRef]
27. Kidd, C.; Kniveton, D.R.; Todd, M.C.; Bellerby, T.J. Satellite rainfall estimation using combined passive microwave and infrared algorithms. *J. Hydrometeorol.* **2003**, *4*, 1088–1104. [CrossRef]
28. Lee, J.; Paz, I.; Schertzer, D.; Lee, D.I.; Tchiguirinskaia, I. Multifractal analysis of rainfall-rate datasets obtained by radar and numerical model: The case study of Typhoon Bolaven (2012). *J. Appl. Meteorol. Climatol.* **2020**, *59*, 819–840. [CrossRef]
29. Sauvageot, H. Rainfall measurement by radar: A review. *Atmos. Res.* **1994**, *35*, 27–54. [CrossRef]
30. Sun, J.; Zhang, Y.; Ban, J.; Hong, J.S.; Lin, C.Y. Impact of combined assimilation of radar and rainfall data on short-term heavy rainfall prediction: A case study. *Mon. Weather Rev.* **2020**, *148*, 2211–2232. [CrossRef]
31. Wei, C.C. Simulation of operational typhoon rainfall nowcasting using radar reflectivity combined with meteorological data. *J. Geophys. Res. Atmos.* **2014**, *119*, 6578–6595. [CrossRef]
32. Borga, M. Accuracy of radar rainfall estimates for streamflow simulation. *J. Hydrol.* **2002**, *267*, 26–39. [CrossRef]
33. LeCun, Y.; Boser, B.; Denker, J.S.; Henderson, D.; Howard, R.E.; Hubbard, W.; Jackel, L.D. Backpropagation applied to handwritten zip code recognition. *Neural Comput.* **1989**, *1*, 541–551. [CrossRef]
34. Hinton, G.E.; Osindero, S.; Teh, Y.W. A fast learning algorithm for deep belief nets. *Neural Comput.* **2006**, *18*, 1527–1554. [CrossRef] [PubMed]
35. Hinton, G.E.; Salakhutdinov, R. Reducing the dimensionality of data with neural networks. *Science* **2006**, *313*, 504–507. [CrossRef] [PubMed]
36. Liu, L.; Ouyang, W.; Wang, X.; Fieguth, P.; Chen, J.; Liu, X.; Pietikäinen, M. Deep learning for generic object detection: A survey. *Int. J. Comput. Vis.* **2020**, *128*, 261–318. [CrossRef]
37. Wang, J.H.; Lin, G.F.; Chang, M.J.; Huang, I.H.; Chen, Y.R. Real-time water-level forecasting using dilated causal convolutional neural networks. *Water Resour. Manag.* **2019**, *33*, 3759–3780. [CrossRef]
38. Kimura, N.; Yoshinaga, I.; Sekijima, K.; Azechi, I.; Baba, D. Convolutional neural network coupled with a transfer-learning approach for time-series flood predictions. *Water* **2020**, *12*, 96. [CrossRef]
39. Van, S.P.; Le, H.M.; Thanh, D.V.; Dang, T.D.; Loc, H.H.; Anh, D.T. Deep learning convolutional neural network in rainfall-runoff modelling. *J. Hydroinform.* **2020**, *22*, 541–561. [CrossRef]
40. Kabir, S.; Patidar, S.; Xia, X.; Liang, Q.; Neal, J.; Pender, G. A deep convolutional neural network model for rapid prediction of fluvial flood inundation. *J. Hydrol.* **2020**, *590*, 125481. [CrossRef]
41. Wang, Y.; Fang, Z.; Hong, H.; Peng, L. Flood susceptibility mapping using convolutional neural network frameworks. *J. Hydrol.* **2020**, *582*, 124482. [CrossRef]
42. Water Resources Agency. Available online: https://eng.wra.gov.tw/ (accessed on 1 April 2020).
43. Taipei Water Management Office. Available online: https://www.wratb.gov.tw/ (accessed on 1 April 2020).
44. Hochreiter, S.; Schmidhuber, J. Long short-term memory. *Neural Comput.* **1997**, *9*, 1735–1780. [CrossRef] [PubMed]
45. TensorFlow. TensorFlow Core v2.3.0. 2020. Available online: https://www.tensorflow.org/api_docs/python/tf/keras/layers/Dense (accessed on 1 July 2020).
46. Park, J.; Edington, D.W. A sequential neural network model for diabetes prediction. *Artif. Intell. Med.* **2001**, *23*, 277–293. [CrossRef]
47. Britz, D. Recurrent Neural Networks Tutorial, Part 1–Introduction to RNNs. 2015. Available online: http://www.wildml.com/2015/09/recurrent-neural-networks-tutorial-part-1-introduction-to-rnns/ (accessed on 1 April 2020).
48. Merrill, W. Sequential Neural Networks as Automata. In Proceedings of the Workshop on Deep Learning and Formal Languages, Building Bridges, Florence, Italy, 2 August 2019; pp. 1–13.
49. LeCun, Y.; Bengio, Y.; Hinton, G. Deep learning. *Nature* **2015**, *521*, 436–444. [CrossRef] [PubMed]

50. Graves, A. *Supervised Sequence Labelling with Recurrent Neural Networks*; Springer: Berlin/Heidelberg, Germany, 2012.
51. Graves, A.; Schmidhuber, J. Framewise phoneme classification with bidirectional LSTM and other neural network architectures. *Neural Netw.* **2005**, *18*, 602–610. [CrossRef] [PubMed]
52. Graves, A. Generating sequences with recurrent neural networks. *arXiv* **2013**, arXiv:1308.0850.
53. Zhang, Q.; Wang, H.; Dong, J.; Zhong, G.; Sun, X. Prediction of sea surface temperature using long short-term memory. *IEEE Geosci. Remote Sens. Lett.* **2017**, *14*, 1745–1749. [CrossRef]
54. Wei, C.C.; Hsieh, P.Y. Estimation of hourly rainfall during typhoons using radar mosaic-based convolutional neural networks. *Remote Sens.* **2020**, *12*, 896. [CrossRef]
55. Sainath, T.N.; Kingsbury, B.; Saon, G.; Soltau, H.; Mohamed, A.; Dahl, G.; Ramabhadran, B. Deep convolutional neural networks for large-scale speech tasks. *Neural Netw.* **2015**, *64*, 39–48. [CrossRef]
56. LeCun, Y.; Bengio, Y. Convolutional Networks for Images, Speech, and Time-series. In *The Handbook of Brain Theory and Neural Networks*; MIT Press: Cambridge, MA, USA, 1995.
57. LeCun, Y.; Bottou, L.; Bengio, Y.; Haffner, P. Gradient-based Learning Applied to Document Recognition. *Proc. IEEE* **1998**, *86*, 2278–2324. [CrossRef]
58. Chollet, F. Keras: Deep Learning Library for Theano and Tensorflow. 2015. Available online: https://keras.io/ (accessed on 1 February 2020).
59. Pedregosa, F.; Varoquaux, G.; Gramfort, A.; Michel, V.; Thirion, B.; Grisel, O.; Blondel, M.; Prettenhofer, P.; Weiss, R.; Dubourg, V.; et al. Scikit-learn: Machine learning in Python. *J. Mach. Learn. Res.* **2011**, *12*, 2825–2830.
60. Celisse, A. Optimal cross-validation in density estimation with the L^2-loss[1]. *Ann. Stat.* **2014**, *42*, 1879–1910. [CrossRef]
61. Wei, C.C.; Hsu, N.S. Optimal tree-based release rules for real-time flood control operations on a multipurpose multireservoir system. *J. Hydrol.* **2009**, *365*, 213–224. [CrossRef]
62. Taipei City Government. *Guidelines of Feitsui Reservoir Operations*; Taipei City Government: Taipei, Taiwan, 2004. (In Chinese)
63. Yeh, W.W.G. Reservoir management and operation models: A state-of-the-art review. *Water Resour. Res.* **1985**, *21*, 1797–1818. [CrossRef]
64. Labadie, J.W. Optimal operation of multireservoir systems: State-of-the-art review. *J. Water Resour. Plan. Manag.* **2004**, *130*, 93–111. [CrossRef]
65. Rani, D.; Moreira, M.M. Simulation-optimization modeling: A survey and potential application in reservoir systems operation. *Water Resour. Manag.* **2010**, *24*, 1107–1138. [CrossRef]
66. Hsu, N.S.; Wei, C.C. A multipurpose reservoir real-time operation model for flood control during typhoon invasion. *J. Hydrol.* **2007**, *336*, 282–293. [CrossRef]
67. Wei, C.C. Discretized and continuous target fields for the reservoir release rules during floods. *Water Resour. Manag.* **2012**, *26*, 3457–3477. [CrossRef]
68. Kingma, D.P.; Ba, J.L. Adam: A Method for Stochastic Optimization. In Proceedings of the International Conference on Learning Representations, San Diego, CA, USA, 5–8 May 2015; pp. 1–13.
69. Ruder, S. An overview of gradient descent optimization algorithms. *arXiv* **2017**, arXiv:1609.04747.
70. Monner, D.; Reggia, J.A. A generalized LSTM-like training algorithm for second-order recurrent. *Neural Netw.* **2012**, *25*, 70–83. [CrossRef]
71. Wei, C.C. Development of stacked long short-term memory neural networks with numerical solutions for wind velocity predictions. *Adv. Meteorol.* **2020**, *2020*, 5462040. [CrossRef]
72. Maier, H.R.; Jain, A.; Dandy, G.C.; Sudheer, K.P. Methods used for the development of neural networks for the prediction of water resource variables in river systems: Current status and future directions. *Environ. Model. Softw.* **2010**, *25*, 891–909. [CrossRef]
73. Gilewski, P.; Nawalany, M. Inter-comparison of rain-gauge, radar, and satellite (IMERG GPM) precipitation estimates performance for rainfall-runoff modeling in a mountainous catchment in Poland. *Water* **2018**, *10*, 1665. [CrossRef]
74. McKee, J.L.; Binns, A.D. A review of gauge–radar merging methods for quantitative precipitation estimation in hydrology. *Can. Water Resour. J.* **2016**, *41*, 186–203. [CrossRef]
75. Ivashkin, V.; Lebedev, V. Spatiotemporal data fusion for precipitation nowcasting. *arXiv* **2018**, arXiv:1812.10915.

76. Chang, C.H.; Chung, M.K.; Yang, S.Y.; Hsu, C.T.; Wu, S.J. A case study for the application of an operational two-dimensional real-time flooding forecasting system and smart water level gauges on roads in Tainan City, Taiwan. *Water* **2018**, *10*, 574. [CrossRef]
77. Kimura, N.; Chiang, S.; Wei, H.-P.; Su, Y.-F.; Chu, J.; Cheng, C.-T.; Liou, J.-J.; Chen, Y.-M.; Lin, L.-Y. Tsengwen Reservoir Watershed hydrological flood simulation under global climate change using the 20 km mesh Meteorological Research Institute Atmospheric General Circulation Model (MRI-AGCM). *Terr. Atmos. Ocean. Sci.* **2014**, *25*, 449–461. [CrossRef]
78. Wei, C.C. Comparing single- and two-segment statistical models with a conceptual rainfall-runoff model for river streamflow prediction during typhoons. *Environ. Model. Softw.* **2016**, *85*, 112–128. [CrossRef]

3

Long-Term Downstream Effects of a Dam on a Lowland River Flow Regime: Case Study of the Upper Narew

Paweł Marcinkowski * and Mateusz Grygoruk

Department of Hydraulic Engineering, Faculty of Civil and Environmental Engineering, Warsaw University of Life Sciences, Nowoursynowska 166 Street, 02-787 Warsaw, Poland; m.grygoruk@levis.sggw.pl
* Correspondence: p.marcinkowski@levis.sggw.pl

Abstract: Most European riverine ecosystems suffer from the negative influence of impoundments on flow regime. Downstream effects of dams lead to a number of environmental and socioeconomic risks and, therefore, should be thoroughly examined in specific contexts. Our study aims to quantify the downstream effects of the Siemianówka Reservoir (Upper Narew, Poland), using statistical analysis of key elements of the river's flow regime, such as the flow duration and recurrence of floods and droughts. In a comparative study on control catchments not influenced by impoundments (the Supraśl and Narewka Rivers), we revealed the following downstream effects of the analyzed dam: significant shortening of spring floods, reduction of the duration and depth of summer droughts, decrease of the maximum discharge, and homogenization of the discharge hydrographs. Although we determined a significant decrease in the duration of summer floods in the "before" and "after" dam function periods, we showed that this issue is regional, climate-related, and replicated in control catchments, rather than an evident downstream effect of the dam. We conclude that significant hydrological downstream effects of the Siemianówka dam–reservoir system could have been the main driver inducing the deterioration of the anastomosing stretch of the Narew River downstream of the dam.

Keywords: Siemianówka; hydrology; Narew River; dam; reservoir; discharge; flow regime

1. Introduction

The vast majority of rivers in the northern hemisphere have been segmented by dams, which poses the risk of significant deterioration of downstream riverine and riparian ecosystems [1]. The influence of dams on river systems is reportedly several times greater than, for example, the influence of climate change [2]; therefore, the analysis of downstream effects of dams (reservoirs) remains an important issue in international scientific literature [3–5]. Based on the modification of river discharge on an hourly and daily basis, seasonal changes of flow regimes [6,7], and water temperature modification [8] in impounded rivers, dams can be considered a key element affecting the hydrology of downstream reaches of rivers.

Dams negatively influence all elements of riverine ecosystems. First and foremost, downstream effects of dams induce sediment transport and river channel sedimentation/erosion balance [9]. Changes of the river channel caused by sediment trapping by dams and reservoirs affect the distribution of flow velocities and induce incisions in river channels [10]. Changed water exchange and in-stream biogeochemical cycles affect the water quality of impounded rivers [11]. The changes in the longitudinal connectivity of river reaches and unstable flow regime and water temperature affect fish communities [12,13]. The alteration of the river baseflow due to the operation of the dam negatively affects macroinvertebrate communities by changing their trophic structure [14]. The changed flow regime, including modified inundation periods and seasons, also affects the riparian vegetation [15]. Complex interactions of riverine ecohydrological processes affected by dams need to be addressed in river management strategies [16] because observed and documented downstream effects of dams influence the broad spectrum of the anthroposphere, challenging livelihoods and economies [17].

Large dams have significant downstream effects [3,4,13,16,18–20] and are thus more frequently studied than small ones [8,21]. However, small- and medium-sized dams and associated reservoirs influence the vast majority of the world's rivers [1]. The role of these dams in shaping the global environment remains widespread and locally critical.

In Europe, north-east Poland is known for its high environmental awareness; the area is home to four national parks, five landscape parks, and multiple environmental reserves. Nearly 56% of the area of the Podlaskie Voivodeship province—the highest rate among all regions of the country—is covered under the EU environmental conservation program Natura 2000. The cultural landscape, consisting of a mixture of agricultural land, forests, wetlands, and settlements, is called the "Green Lungs" of Poland. The core of this unique region is the Narew River, which remains the backbone of local agricultural and environmental issues. Part of the middle reach of the Narew River formed the continent's unique anastomosing system of river branches [22,23], which was the main reason for establishing a national park in this region. The uppermost reach of the Narew River was dammed in 1990 when the Siemianówka Reservoir (SR) was constructed. Since then, specific studies have revealed changes in discharge and water-level dynamics [24–28] and biodiversity [29–32] of the riverine ecosystem. These changes have been attributed to the SR. Some observed biogeochemical changes in the downstream anastomosing river system have also been attributed to the dam [33]. The most recent negative hydrological and geomorphological changes of the anastomosing stretch of the Narew River are thought to be accelerated by the downstream effects of the SR [23]. However, no comprehensive long term data-supported hydrological research on the downstream effects of the SR has been conducted so far.

A dataset of more than 20 years of hydrological discharge data for the Narew River after the SR was established is available. Therefore, we aim to conduct a data-based and statistically supported assessment of the river's flow regime based on comparative before–after control–impact comparisons. We investigated the impact of the reservoir construction on the flow regime of downstream reaches of the Narew River. Statistical analyses of the precipitation pattern, daily discharge records (minimum, maximum, 1st, 2nd, and 3rd quartile), and occurrence and recurrence of floods and droughts were conducted for two different time periods: 1951–1989 (pre-dam) and 1990–2012 (post-dam). Our study provides hydrological evidence of the influence of the SR downstream of the Narew River; thus, we open new avenues for the interpretation of biological, ecological, hydrological, and geomorphological research.

2. Materials and Methods

2.1. Study Area

The study area, the Upper Narew Catchment (NE Poland), is the sub-catchment of the largest Polish river basin, namely, the Vistula (Figure 1). The analyzed catchment covers an area of 6656 km^2 (of which 17% belongs to Belarus). The region was formed by glacial erosion and accumulation in the Pleistocene. It is characterized by a flat relief with an average elevation of 152 m above sea level.

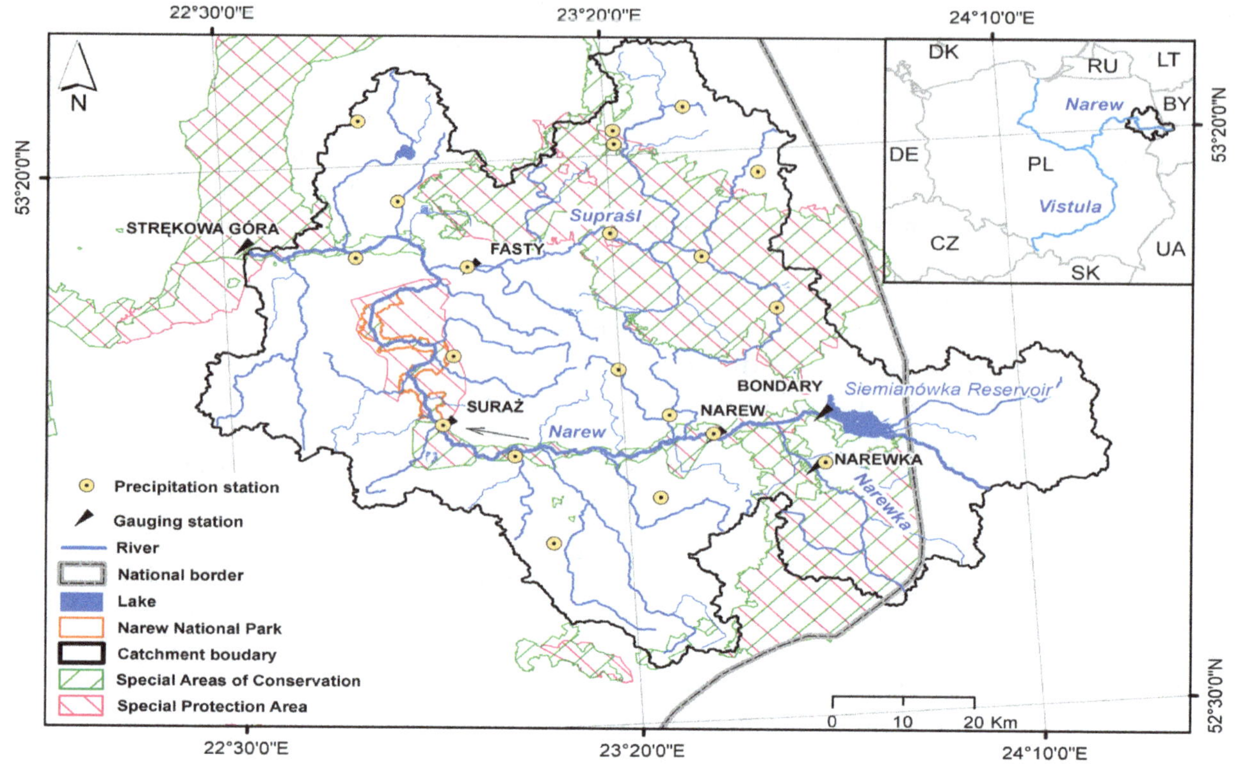

Figure 1. Case study location: Siemianówka Reservoir (SR) and Upper Narew River Catchment, and location of water gauges used to record the data used for the analysis.

The climate of the Upper Narew Catchment is continental; it is influenced by cold polar air masses originating in Russia and Scandinavia. This is reflected in the mean annual air temperature, which equals 7.1 °C, and annual magnitude of air temperature changes that can reach up to 55 °C according to the data of the Institute of Meteorology and Water Management-National Research Institute (IMGW-PIB). Mean annual sums of precipitation oscillate at about 600 mm. Land cover in the Upper Narew Catchment is dominated by agriculture (53%) with arable lands (39%) and pastures (14%) [34]. Urban and settled areas cover less than 3% of the catchment. Dominant types of soil are pure and loamy sands; very heavy impermeable soils (clay, clay loam, and silt loam) are rare [34]. The topographical setting of the catchment determines the typical lowland character of the river valley, which in primeval conditions was supposed to be exposed to regular and long-lasting inundation, most frequently during the early spring snowmelt season.

The rivers of this region of Europe are subject to thaw floods, which are considered to be the most important regular extreme phenomenon shaping riparian wetlands, including the anastomosing channel of the Narew downstream of the SR [22]. Substantially lower rain-event driven floods occur occasionally in the summer season.

The Upper Narew Catchment, being a part of the Narew Basin, is surrounded by multiple environmental protection sites, including national parks and Natura 2000 sites (Birds and Habitat Directive; Figure 1). Among the protected areas in the catchment, the Narew National Park (NNP), Upper Narew Valley Refuge Special Area of Conservation (SAC), and Upper Narew Valley Special Protection Area (SPA) are the most important due to their location in an area downstream of the SR. All three protected areas embrace unique wetland habitats shaped by and dependent on regular inundations.

2.2. *Siemianówka Reservoir*

The SR has a total capacity of 79.5 million m^3 and was created through dam construction in the course of the River Narew, (432 km, measured from the outlet). The first construction started in 1977, whereas the complete filling dates back to 1992 when the SR became entirely operational [35]. The dimensions of the reservoir are as follows: 11 km length, 0.8 to 4.5 km width, 2.5 m mean depth, 7 m maximum depth, and 1050 km^2 catchment area. The reservoir and dam were created for multiple purposes including the enhancement of local tourism and recreation, energy production, flood protection, fishing, and irrigation of agricultural lands [36]. According to reservoir management instructions [35], one of the main objectives is to increase the low flow during the summer season in the Narew to maintain biological life, and to mitigate high peak flow to reduce flood risk in the valley during the spring season. Reference catchments used as control sites in our research, that is, from the Narew River down to the Narewka gauge station, and from the Supraśl River down to the Fasty gauge station (Figure 1; Table 1), are located in the same region. Their physiographic features are similar to the analyzed catchment of the Narew River.

Table 1. Gauge stations in the Upper Narew Catchment. The asterisk '*'denotes catchments of major tributaries of the Narew River, which are used as references in this study.

Gauge Station	River	Flow Data Availability	Catchment Area (km^2)
Narewka *	Narewka	1951–2013	590.4
Bondary	Narew	1963–2013	1094.6
Narew	Narew	1951–2013	1978.0
Suraż	Narew	1951–2013	3376.5
Fasty *	Supraśl	1975–2013	1818.0
Strękowa Góra	Narew	1951–2012	6656.0

2.3. *Hydrometeorological Analysis*

This study used daily precipitation records from 20 meteorological stations for the period 1951–2012 for the analysis. Daily flow records for six gauge stations covering different time periods (Table 1) were subjected to a statistical analysis. Both meteorological and hydrological data were acquired from the IMGW-PIB. Whilst four flow gauge stations (Bondary, Narew, Suraża, and Strękowa Góra) are located on the Narew River and are used for direct dam impact assessment, the remainig two (Narewka and Fasty, located on the Narew tributaries and not influenced by the dam) constitute control stations to eliminate potentially wrong inferences due to the climate impact on the flow regime. To assess the significance level of the precipitation trend, the non-parametric Pettitt's test [37] was applied, with a significance level of $p \leq 0.05$. This test provides the assessment of the null hypothesis H_0, implying that the data are homogeneous throughout the period of observation. Pettitt's test was reported in the literature to be sufficient for detecting break points in a set of long-term observations [38,39]. In addition, flow-duration analysis was conducted, yielding the annual minimum; maximum; and 25, 50, and 75 percentile discharge for all gauge stations. Statistical analysis of flow data homogeneity based on Pettitt's test was conducted to indicate significant breakpoints and timescale trends to assess the impact of dam construction on the flow regime.

Additionally, flow-duration curves (FDCs) have been created for every gauge station to express the overall regime change in the analyzed time frame. The FDC is a plot that shows the percentage of time during that discharge in a stream is likely to equal or exceed some specified value of interest. In this study, the flow datasets for each gauge station were divided into two subsets (pre-dam, until 1988; and post-dam, since 1989) for which FDCs were created separately. Finally, temporal patterns of floods and droughts were analyzed. We applied standard thresholds of river discharge assessment, such as the median of the highest annual discharges (MHQ, proxy for a bankfull discharge) and average lowest discharge (ALQ), both from multi-year records, for flood and drought analysis, respectively. The study calculated the number of days with discharge higher than MHQ and lower than ALQ for selected gauge stations to reveal whether the occurrence and recurrence of floods and droughts, that is, extreme hydrological phenomena critical for the functioning and preservation of anastomosing river branches and related biocenoses, have maintained trends similar to pre-establishment of the dam or have changed. For a comprehensive view on the aspect of floods and drought distribution and frequency, we analyzed these phenomena in for all years considered (1951–2013) and separately for summer (May–October) and winter (November–April). In addition, we analyzed floods and droughts of the Narew River for the period from 1975 to 2013 because reference data for the SR were available. This approach allowed us to draw conclusions on the importance of the length of the river discharge dataset with respect to its usefulness for before–after control–impact studies of downstream effects of dams, which has already proven to be a critical issue in downstream effect analysis [40].

3. Results

3.1. Precipitation

The rainfall statistics (Figure 2) based on Pettitt's test clearly indicate a moderate increasing trend over the last 60 years with some wetter and drier periods. Over the period analysed, no statistically significant differences in monthly sums of precipitation were detected between the results from 20 stations analysed (statistics available in the Supplementary material, Section 1). Average annual precipitation sum within the Upper Narew catchment in the years 1951–2012 equalled 598 mm. The total annual precipitation was slightly higher in 1970–1980 than in 1951–1969 or 1981–2012. This pattern was significantly broken in 2010 due to an extremely wet year with flooding across Poland; however, the pattern appears to be continuing in recent years. This characteristic precipitation pattern is a crucial result, and in addition to dam construction, it needs to be taken into account as a potential key controlling factor with respect to further interpretation of flow regime changes.

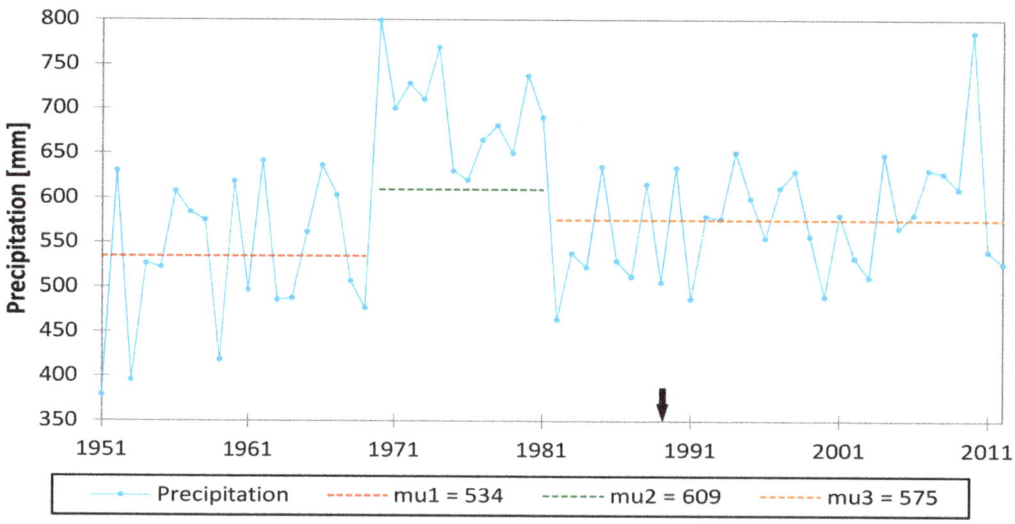

Figure 2. Total annual precipitation for the time period 1951–2012 in the Upper Narew Catchment. "mu" stands for the average value at each time period. The black arrow indicates the year the SR started operation.

3.2. Discharge

3.2.1. Minimum Discharge

The analysis of the annual minimum discharge pattern based on Pettitt's test indicated diverse results for each gauge station. No significant trend was detected in the analyzed period for the Fasty

(Figure 3A) and Narewka (Figure 3B) gauge stations located on the tributaries of the Narew River, where the influence of the SR on the flow regime was not an issue. In contrast, characteristic trends and breakpoints were noted for gauge stations located along the Narew River. However, due to different occurrence times, the detected trends were most likely driven by varied factors. For the Bondary gauge station (Figure 3C), located immediately downstream of the SR, one significant break point occurred in 1994, indicating an average increase of 60% between the separated time periods. Moving on further downstream of the Narew gauge station (Figure 3D), two significant breakpoints were detected (1971 and 1994). During the first time period (1951–1971), the annual minimum flow was 1.6 $m^3 \cdot s^{-1}$, increasing from 1972 to 1993 by 87%, and increasing yet again in 1994–2013 by 17%. The Sura˙z gauge station (Figure 3E) recorded trends identical to those at Narew but of lower magnitude in the corresponding time periods (an increase of 71% from 1972 to 1993 and of 14% from 1994 to 2013). Whilst the first increase of the annual minimum flow is most likely related to the precipitation increase, as the occurrence of both coincide in time, the latter is most likely due to reservoir operation.

The Str˛ekowa Góra gauge station (Figure 3F), which is located in the most downstream part of the catchment seems to reflect only precipitation-driven changes of the minimum flow pattern, which increased from 1972 to 1987 by 55% and decreased by 11% from 1988 to 2012. This indicates that the dam impact subsides at some point of the Narew River between the Suraz˙ and Str˛ekowa Góra gauge stations.

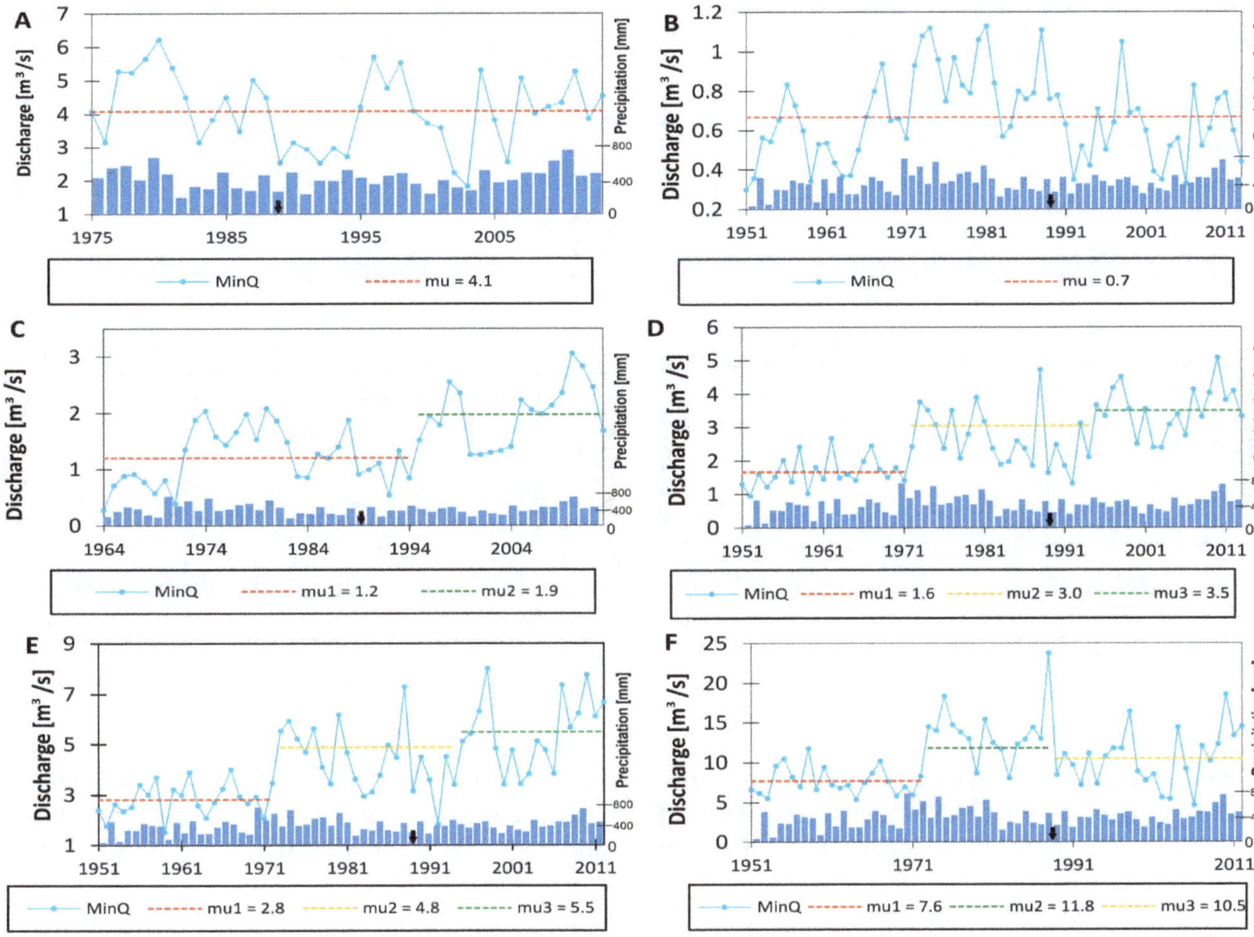

Figure 3. Annual minimum discharge for selected gauge stations of the Upper Narew Catchment: (**A**) Fasty, (**B**) Narewka, (**C**) Bondary, (**D**) Narew, (**E**) Suraż, and (**F**) Strękowa Góra. "mu" stands for the average value of each time period. The black arrow indicates the year the SR started operation.

3.2.2. First Quartile Discharge

The analysis of the first quartile flow, which represents the median of the lower half of the dataset indicates moderate changes. At the three gauge stations (Fasty, Narewka, and Bondary; Figure 4A–C, respectively), no significant trend was recognized during the analyzed period. In the case of minimum flow, a similar response was observed at the Narew (Figure 4D) and Suraż gauge stations (Figure 4E). An increase occurred in 1970, with rates reaching 71% and 60% at the Narew and Suraż gauge stations, respectively. Taking into account the time of appearance, this correlates with the break point of the precipitation pattern shift observed in 1970. It is similar to the response of the minimum flow pattern, with one major exception: it did not subside after 1980 when the precipitation decreased. Considering changes detected at the Strękowa Góra gauge station (Figure 4F), the flow trends precisely follow the precipitation pattern, increasing at first by 50% in 1972–1987, and finally decreasing by 11%.

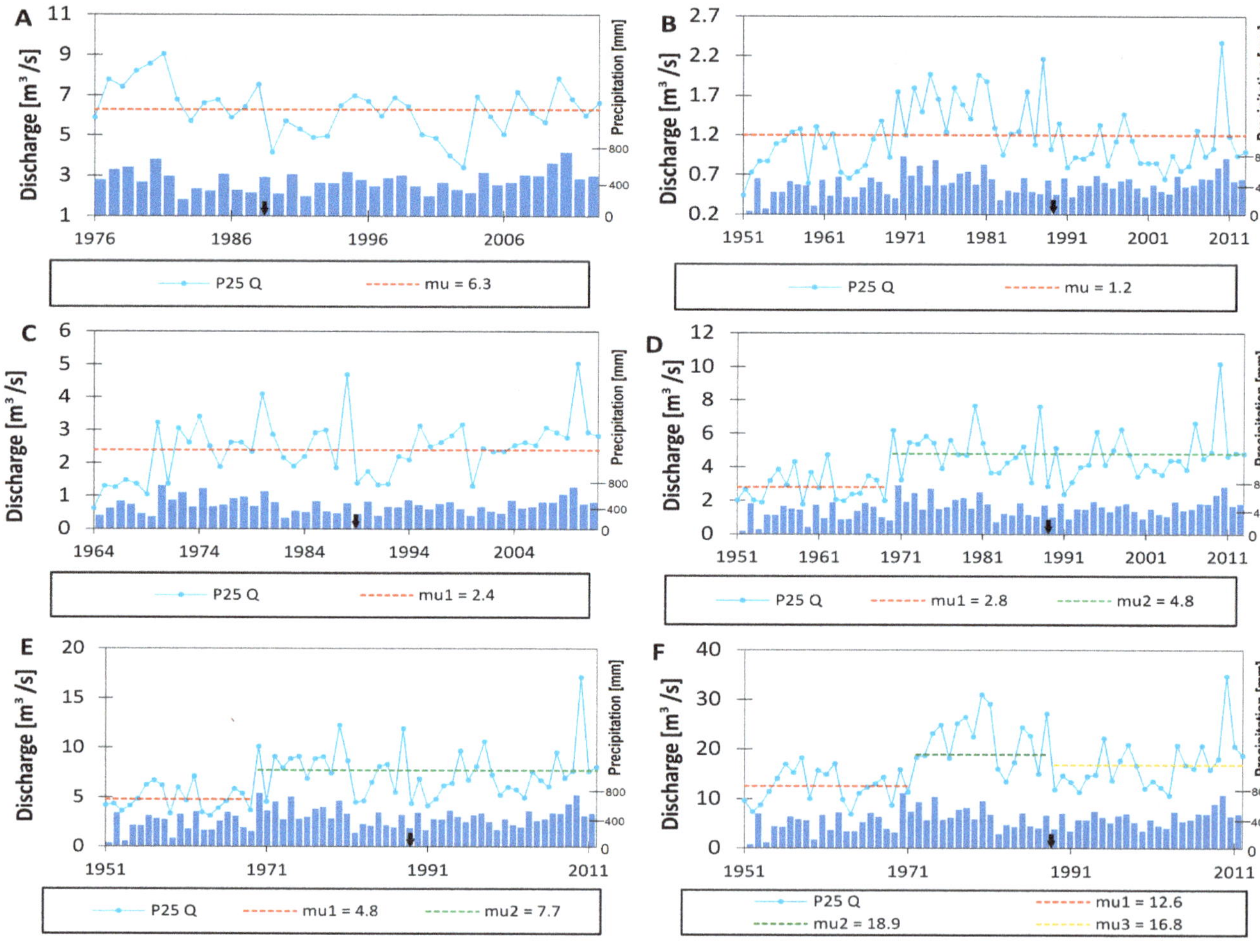

Figure 4. Annual first quartile discharge for selected gauge stations in the Upper Narew Catchment: (**A**) Fasty, (**B**) Narewka, (**C**) Bondary, (**D**) Narew, (**E**) Suraż, and (**F**) Strękowa Góra. "mu" stands for the average value of each time period. The black arrow indicates the year the SR started operation.

3.2.3. Second Quartile Discharge

The analysis of the second quartile flow, which simply represents the median of the observed dataset, indicated no significant trends for the two gauge stations (Fasty and Bondary; Figure 5A,C). Two characteristic patterns were observed at the remaining gauge stations. First, at the Narewka

(Figure 5B) and Strękowa Góra (Figure 5F) gauge stations, we observed two significant breakpoints of the trend reflecting variation in precipitation from 1970 to 1980. Second, at the Narew (Figure 5D) and Suraż (Figure 5E) gauge stations, we observed one significant trend shift in 1970. Regarding the rates of increase, the highest rate was noted at the Narew and Suraż gauge stations (64% and 60%, respectively) and a significantly lower one was detected at the Fasty and Strękowa Góra gauge stations (42% and 34%, respectively). The decreasing trends observed at the latter two gauge stations reached 10%.

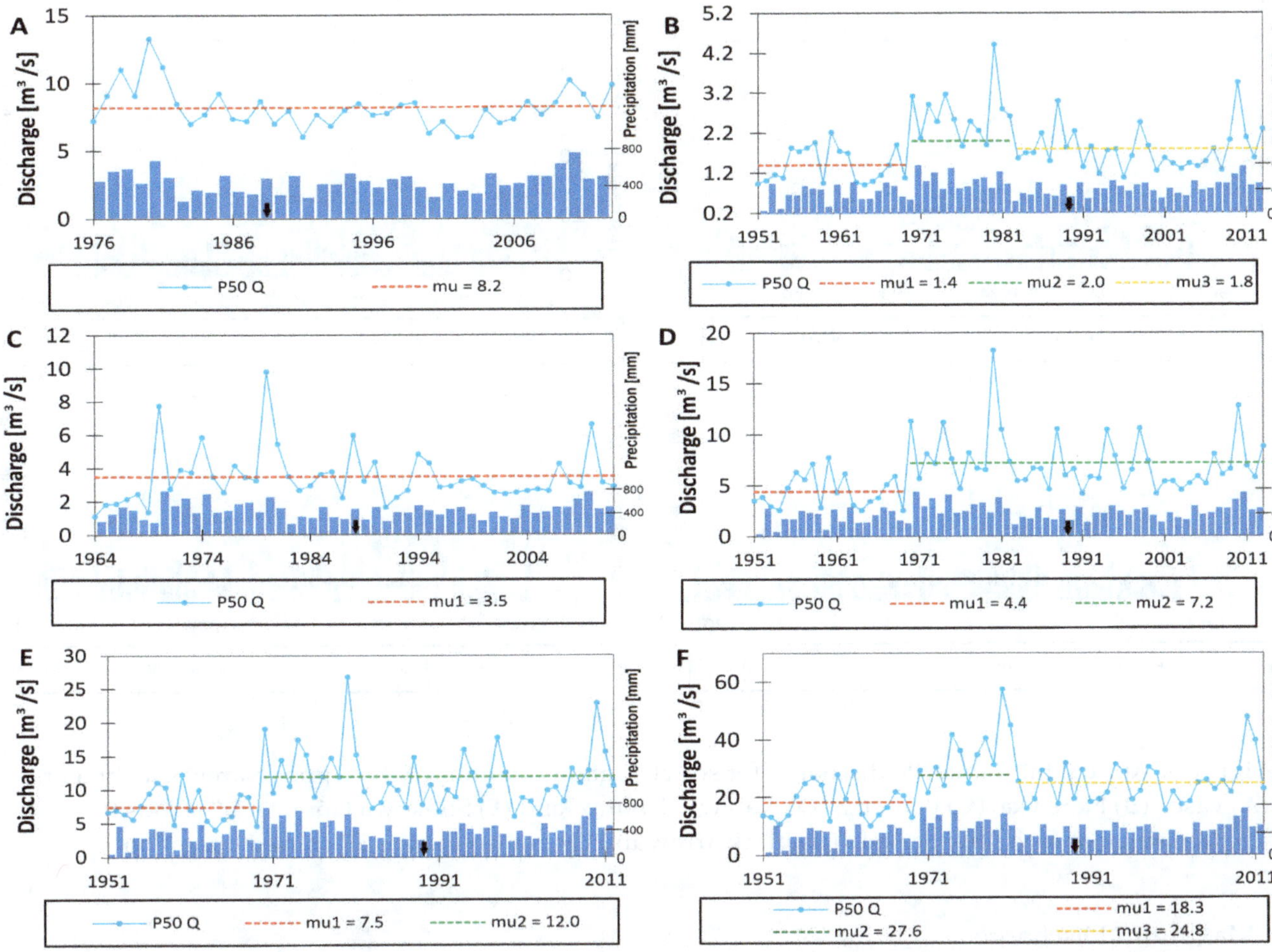

Figure 5. Annual second quartile discharge for selected gauge stations of the Upper Narew Catchment: (**A**) Fasty, (**B**) Narewka, (**C**) Bondary, (**D**) Narew, (**E**) Suraż, and (**F**) Strękowa Góra. "mu" stands for the average value of each time period. The black arrow indicates the year the SR started operation.

3.2.4. Third Quartile Discharge

The analysis of the third quartile flow, which represents the median of the higher half of the dataset, indicates diverse changes. No significant alteration of the trend during the analyzed period was detected at the three gauge stations (Fasty, Bondary, and Strękowa Góra; Figure 6A,C and F). Identical changes were observed at the Narewka (Figure 6B) and Narew (Figure 6D) gauge stations, accurately reflecting the precipitation pattern, which increases in 1970 (by 42%) and decreases in 1983 (by 15%). The flow pattern at the Suraż gauge station (Figure 6E) seems to have partial fluctuations in precipitation, because the increase date coincides with the time in both cases. However, there is a disagreement in the subsidence seen in total precipitation, which is not visible in the flow alteration noticed around 1980.

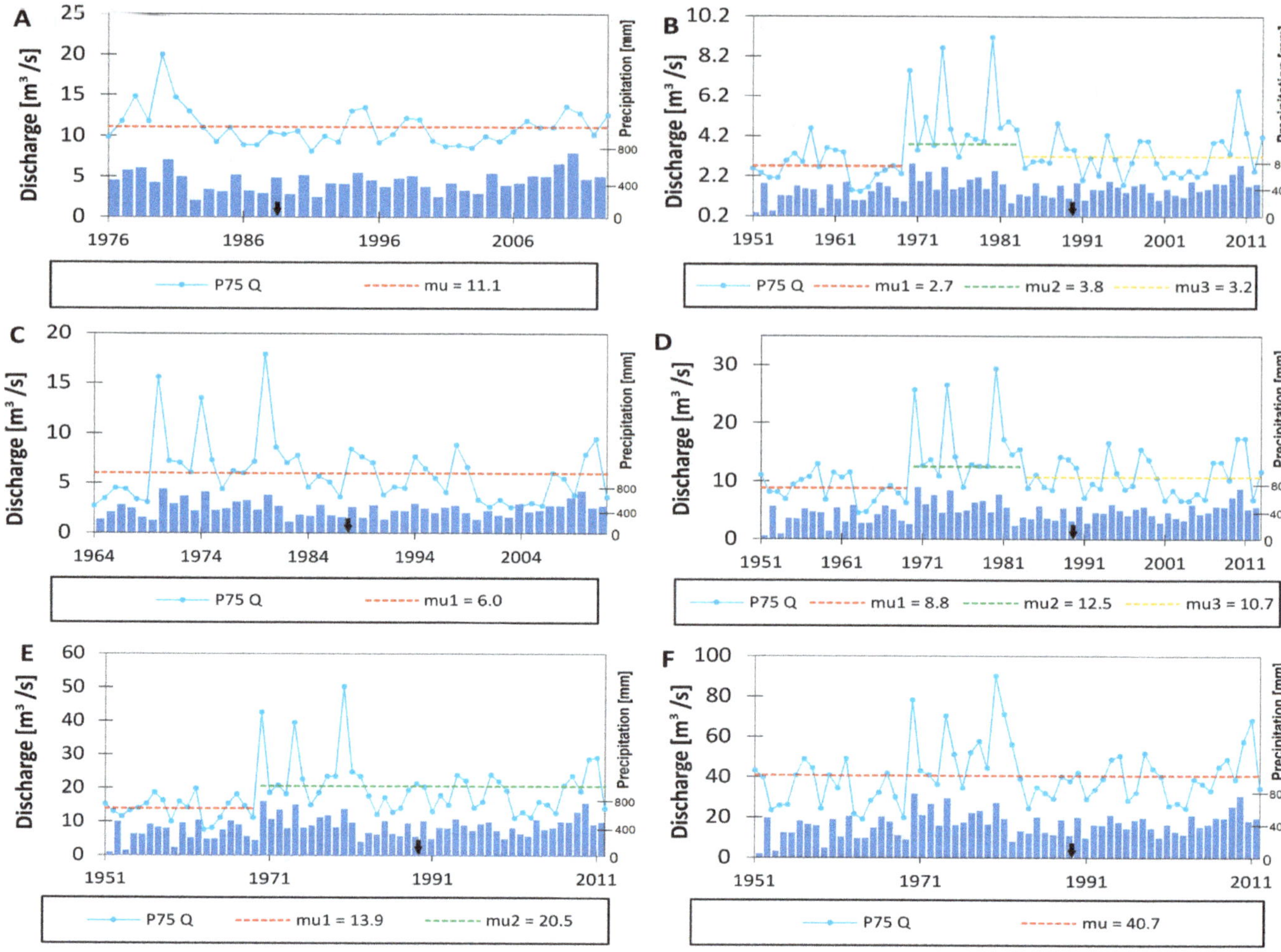

Figure 6. Annual third quartile discharge for selected gauge stations of the Upper Narew Catchment: (**A**) Fasty, (**B**) Narewka, (**C**) Bondary, (**D**) Narew, (**E**) Suraż, and (**F**) Strękowa Góra. "mu" stands for the average value of each time period. The black arrow indicates the year the SR started operation.

3.2.5. Maximum Discharge

The changes in the annual maximum flow trends in the analyzed period seem to be the most conspicuous. Among the investigated gauge stations, two indicated no significant trend changes (Fasty and Narewka; Figure 7A,B). However, an explicit decrease was noted at the remaining four stations. Although the direction of change is consistent, the time of occurrence differs. This proves that the driving force of such change is not homogenous at all stations. In particular, it is most probable that the shifts occurring in 1988 at the Bondary (Figure 7C), Narew (Figure 7D), and Suraż (Figure 7E) gauge stations are caused by the dam construction. A substantial decrease was detected in all cases, reaching 67%, 52% and 33% at the aforementioned gauge stations, respectively. Although the decrease (by 43%) was also noted at the Strękowa Góra gauge station (Figure 7F), the time of the breakpoint occurrence (1983) suggests that it is most likely driven by precipitation decrease.

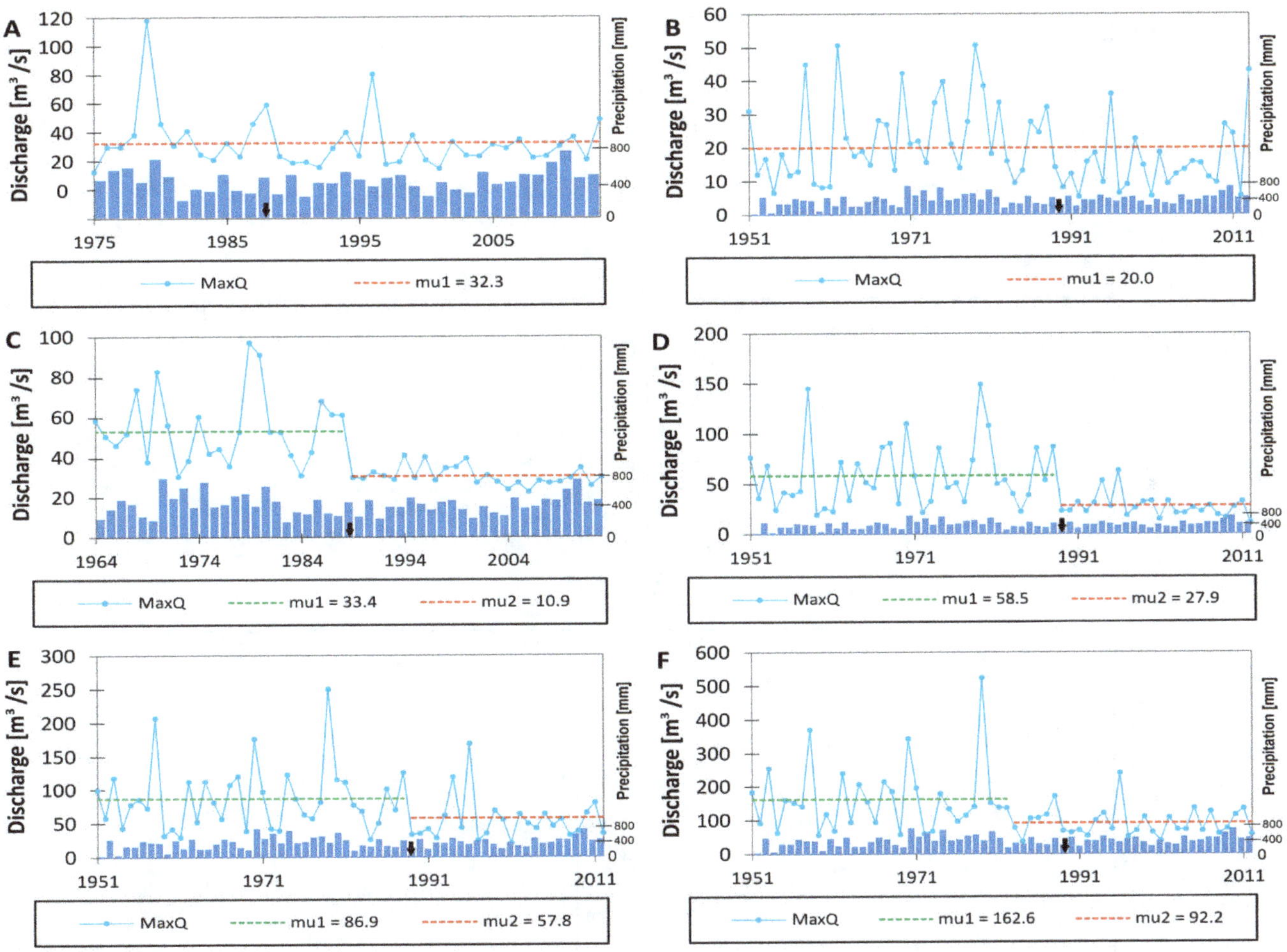

Figure 7. Annual maximum discharge for selected gauge stations of the Upper Narew Catchment: (**A**) Fasty, (**B**) Narewka, (**C**) Bondary, (**D**) Narew, (**E**) Suraż, and (**F**) Strękowa Góra. "mu" stands for the average value of each time period. The black arrow indicates the year the SR started operation.

3.3. Flow-Duration Curve

The results indicating the change in the FDC prove the undeniable impact of reservoir construction on the flow regime of the Narew River. The alterations at the Fasty (Figure 8A) and Narewka (Figure 8B) gauge stations exclusively reflect climate-driven changes because both are located at the Narew tributaries, which are not impacted by the dam construction. Figure 8A,B show that the shape of the curve remains similar during pre- and post-dam periods; only a regular shift (overall decrease) in magnitude is detected in all exceedance probability intervals. In contrast, such a shift does not occur for gauge stations located directly on the Narew River (Figure 8C–F). Instead, a flattening of the curve is observed at all gauge stations during the post-dam period, leading to varying intersections of the curves at different exceedence probability points. The response is homogenous at all gauges, indicating that the daily streamflow decreases in the lower percentiles (i.e., higher flow) and increases in the higher percentiles (i.e., lower flow). Moving from upstream (Bondary) to downstream (Strękowa Góra), the magnitude of change seems to decline gradually, whereas the curve intersections shift to lower percentiles.

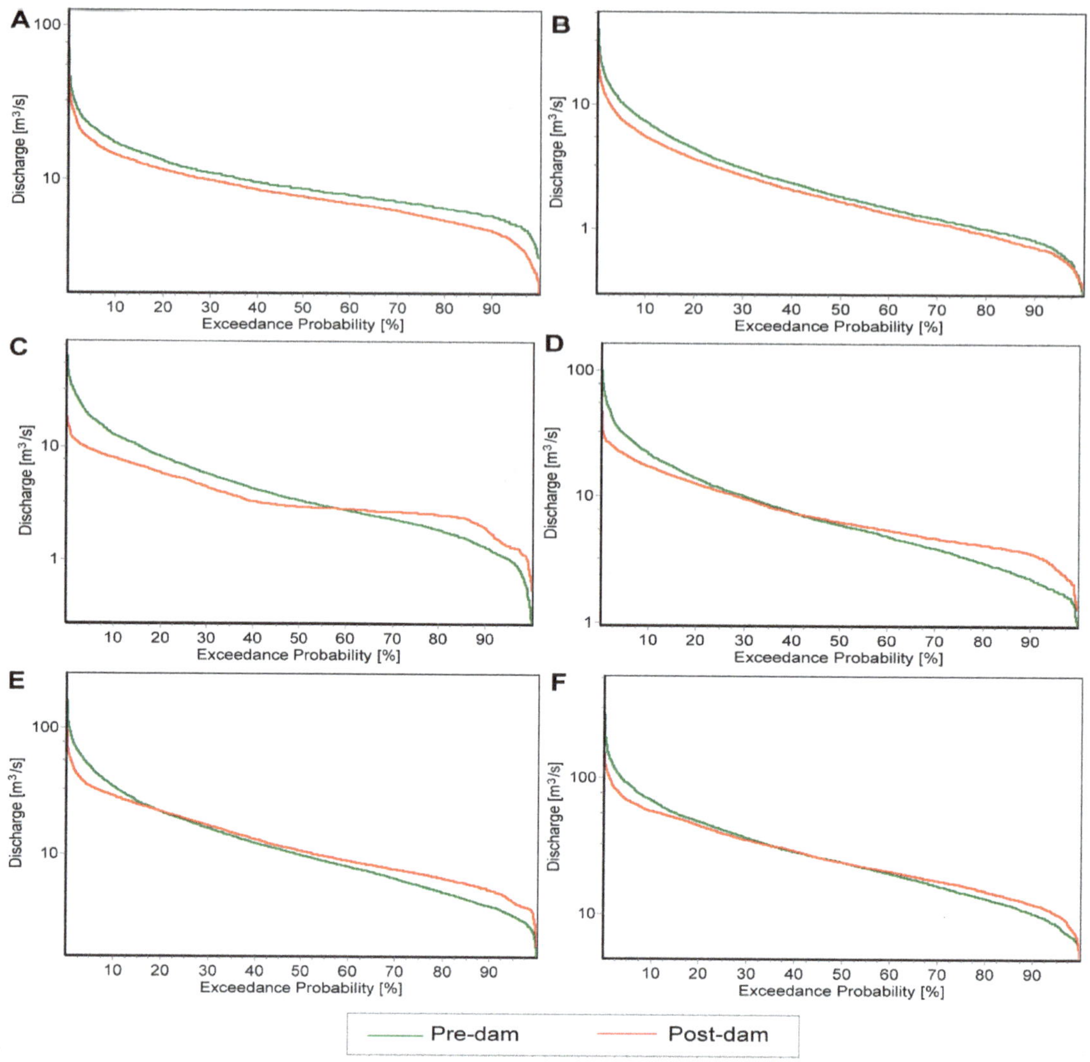

Figure 8. The changes in FDCs at selected gauge stations of the Upper Narew Catchment: (**A**) Fasty, (**B**) Narewka, (**C**) Bondary, (**D**) Narew, (**E**) Suraż, and (**F**) Strękowa Góra. The discharge is presented on a logarithmic scale.

3.4. Ocurrence and Recurrence of Floods and Droughts in Different Seasons

The analysis of temporal trends of floods and droughts revealed significant changes in the ocurrence and recurrence of these extreme phenomena (Figures 9 and 10). In the period "before" the SR was established, the frequency of years without flooding (daily average discharge higher than MHQ) reached only 0.1 (floods did not occur in 4 out of 39 analyzed years). In the period "after" the SR was established, the frequency of years without flooding and higher than the applied threshold reached 0.3 (floods did not occur in 7 out of the 23 analyzed years). The longer inundation periods observed in the Narew River Valley in the past, mainly in March and April, which were assessed using records from the Suraż gauge station, are most likely related to the unmanaged flow regime, which is affected by snowmelts and thaws. Although the annual recurrence of floods tends to decrease on the regional scale (a statistically significant decrease was recorded both for Narew in Suraż and the reference catchment Narewka in Narewka; Figure 10G,T), the decrease of the spring thaw flood recurrence and duration is statistically significant only for Narew (Figure 10I,L vs. Figure 10V,Z). This observation leads to the conclusion that the SR has a significant and considerably high influence on the recurrence and duration of floods for the Narew River: (1) the contemporary recurrence of Narew floods on the annual basis is 34% lower than before the SR was established; and (2) the contemporary duration of spring thaw floods declines by 56%, from an average of 26 days to 15 days before and after the SR started to operate, respectively.

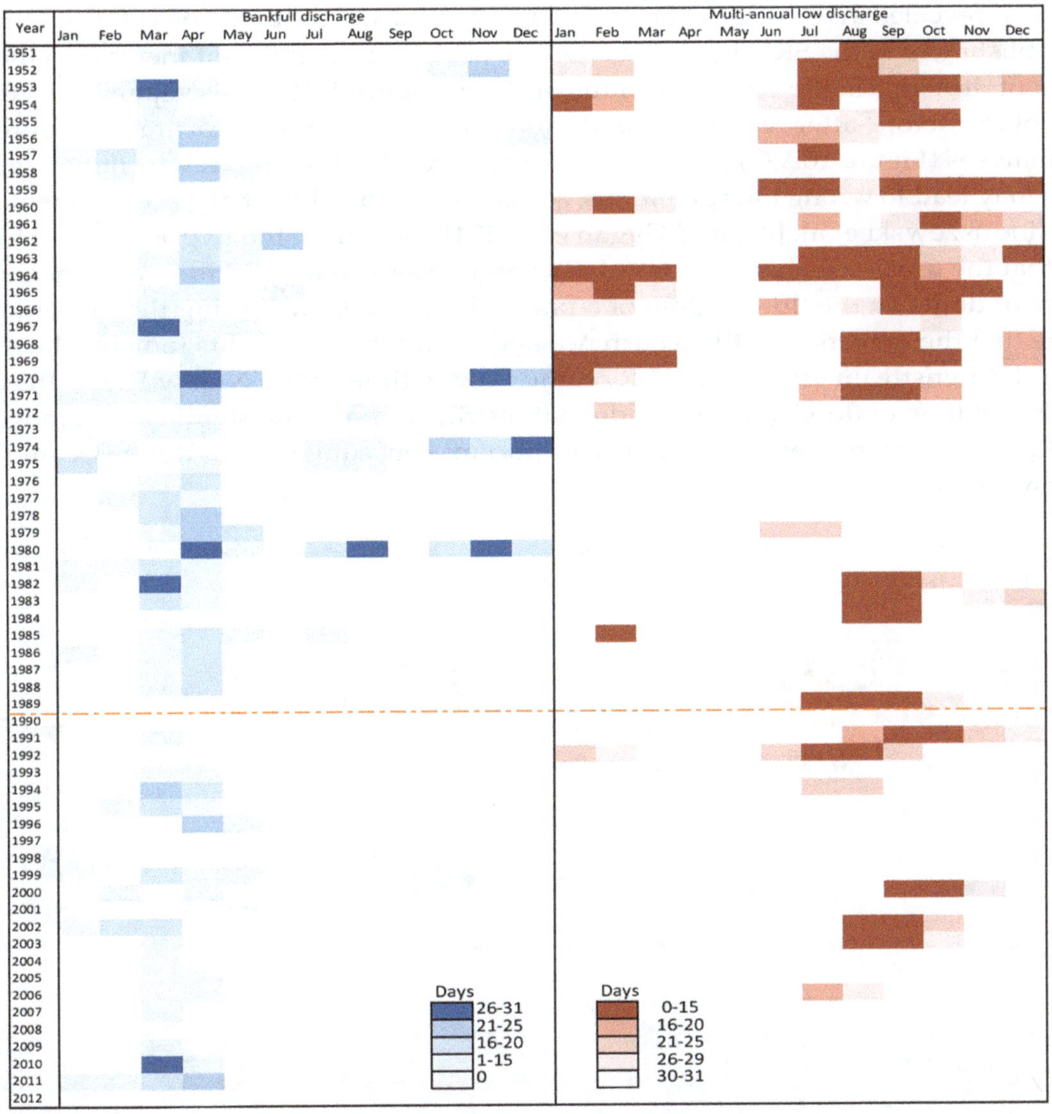

Figure 9. Matrix presenting the occurrence and duration of floods (blue) and droughts (red) of the Narew River at the Suraż gauge station.

The detection of a significant difference of the flood duration (decline) in Narewka (reference catchment) between the whole year (Figure 10T) and summer (V-X; Figure 10U) leads to the conclusion that the operation of the SR does not significantly reduce summer flooding. In terms of flooding, the flow regime of the other reference catchment (Supraśl River, Fasty gauge station) does not show significant changes throughout the analyzed years (Figure 10Y–Z). Despite the fact that the recurrence and duration of flooding generally decreases with respect to the whole year from 1951–2013, for both summer–autumn and winter–spring phenomena, the before–after comparison is not statistically significant (p values range from 0.188 to 0.608).

The SR has a completely different influence on the low flow of the Narew River. On the regional scale, the duration and recurrence of droughts lower than the applied threshold tends to significantly increase (e.g., in the reference catchments of Narewka and Supraśl; Figure 10M–O and P–S, respectively). In contrast, both the short- (1975–2013) and long-term (1951–2013) horizon recurrence and duration of droughts decrease in the case of the Narew downstream SR, similar to short-term observations presented by Romanowicz and Osuch [28]. A statistically significant decline of the drought frequency was recorded in summer (May–October; Figure 10B,E), winter, and in the long term (whole year; Figure 10A,C), which is an important downstream effect of the SR. Although no statistical significance was revealed in the case of Narewka when comparing the durations of droughts before and after the establishment of the SR (Figure 10M–O), we observed a significant increase in the recurrence and duration of droughts in the case of the Supraśl River (Figure 10P,R). Discrepancies in the statistical significance of the before–after regime of floods and droughts between short-term (Figure 10D,J) and long-term analyses (Figure 10A,G) prove that using short sets of data to reveal flow regime changes due to dams may lead to wrong interpretations, as also shown by Huh et al. [40]. Different conclusions drawn by Mioduszewski et al. [27] and Cygan et al. [24] confirmed the hypothesis of Huh et al. [40] who stated that the use of too short a series of discharge records may lead to wrong conclusions about the influence of dams on the flow regime of a particular river. Considering the presented data, one can conclude that the reduction of the recurrence and duration of droughts remains the most notable and unilateral downstream effect of the SR on the flow regime of the Narew River. In combination with the interpretation of floods, we conclude that the SR supports the significant homogenization of the Narew River's discharge, which likely has important consequences for downstream riverine and riparian ecosystems.

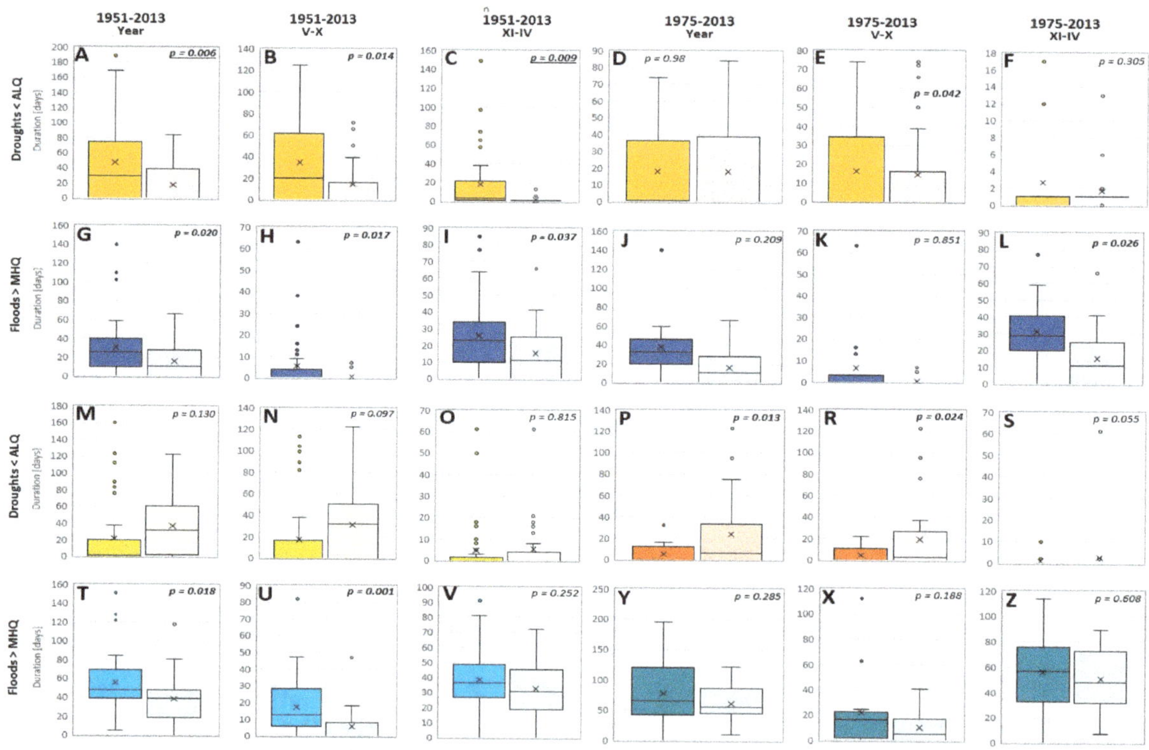

Figure 10. Comparison of the "before" (1951–1988/1975–1988) and "after" (1989–2013) durations of droughts (Q < ALQ; orange/yellow) and floods (Q > MHQ; blue/dark blue) for Narew–Suraż (**A–L**), Narewka–Narewka (MNO–TUV), and Supraśl–Fasty (PRS–YXZ). (**A–C**,**G–I**,**M–O**,**T–V**) represent data from 1951 to 2013. (**D–F**,**J–L**,**P–S**,**Y–Z**) reflect data from 1975 to 2013.

4. Discussion

4.1. *Hydrological Aspects*

The results show that the SR altered the flow regime of the Narew River. Although the influence of the SR operation on the average and median flows is not significant compared with studies of Mioduszewski et al. [27] and Cygan et al. [24], and as presented by Kiczko et al. [26]—strongly depends on the river stretch, we revealed the influence of the SR on the extreme flow of the Narew River. Marcinkowski et al. [23] investigated seasonal changes in water flow and indicated that average flows are lower in April but higher in February and March, which is determined by dam operations, in which water is released from the reservoir in advance of the spring thaw to prevent flooding in municipalities close to the reservoir (see Supplementary material, Section 2).

A temporal precipitation increase recorded in the past (1970–1980) was taken into account in addition to the construction of the SR, as one of the most crucial factors altering the river flow regime, which was also observed by Mioduszewski et al. [27]. The impact of precipitation fluctuations is clearly visible in Figures 5B,F and 6B,D, which show that the temporal increase of the median flow precisely follows the precipitation pattern.

Marcinkowski et al. [41], by means of a hydrological model and nine General Circulation Model–Regional Climate Model (GCM–RCM) runs, evaluated pure effect of climate on water resources for the Upper Narew catchment for the 2020–2050 time horizon. They postulated that the median of projected changes in water yield, i.e., the portion of precipitation that reaches the stream, indicate an average annual increase of 9%. Notably, they observed the most pronounced increase in winter, and a substantially lower increase in other seasons. Given the fact, that the SR controls the release of water, storing most of it during the spring season, the increased flow caused by climate change might be significantly suppressed.

The differences observed for the recurrence and duration of droughts between the Narew River and reference catchments of Narewka and Supraśl may result from land-use changes [28]. However, the aspects of land use in Narew and the analyzed reference catchments are controlled by similar drivers originating from environmental policies (e.g., the Common Agricultural Policy of the EU, responsible for subsidising grassland farming in valuable riparian wetlands). Hence, completely antagonistic trends of the recurrence and duration of droughts of the Narew River compared with control catchments, also reported by Romanowicz and Osuch [28], tend to be a clear flow-regime change attributed to the SR's operation. This downstream effect of the SR can be considered positive, both from environmental and socioeconomic perspectives. Higher water levels during droughts are likely to prevent desiccation, which may reduce CO_2 emissions from drained wetlands in some areas of the valley. However, so far, no links have been found between increasing river water levels during droughts and rising groundwater levels in the valley [27,42], indicating that it may be difficult to mitigate general groundwater decline in the area with an appropriate water spill control from the SR. Hence, the role of the SR in improving downstream habitat quality by decreasing the frequency of deep droughts and increasing the water levels in the river appears to be negligible. On the other hand, the lack of long droughts in recent years could also be considered as less stressful for the riverine ecosystem. One might suspect that if this stress factor occurs repeatedly over time, biota would have to adapt to this phenomenon, and species with higher resistance to this stress factor would become more abundant in the ecosystem. However, so far, not much is known about the role of droughts as an ecosystem stress factor influencing the specific and unique biodiversity of river systems.

The proven and evident reduction of the spring thaw flood duration and frequency that can be attributed to the operation of the SR in the analyzed years along with recent milder winters and less snow accumulation than before [42] can contradict the environmental effects of wetland restoration. Bush encroachment in open areas of Narew Valley wetlands, which was reported to be successfully mitigated by birch removal [43], can continue because lowered and shortened inundation allows the re-establishment of young birch stands and feedbacks with evapotranspiration [44]. Hence, shrub removal, although so far successful and important, appears in this case to be a measure tackling the result of the process rather than affecting the process of bush encroachment itself.

4.2. Hydromorphological Aspects

Results of the reconnection of abandoned side arms and anabranches of the Narew River in the NNP area appear to be contradicted by the operation of the SR as well. Although sidearm reconnection projects implemented in the Narew Valley were reported to have positive environmental effects [45], one could suspect that facing a continuing trend of high flow reduction attributed to the function of the SR, managing appropriate levels of sidearm connectivity to the main river channel in the future will remain a challenging task. In addition to negative responses of biota to river maintenance work performed in the Narew River System [46], flow-regime changes remain an important issue to be addressed in future river management strategies and also within the NNP.

The NNP, located downstream of the SR, has been facing the loss of anabranches in recent decades [23]. It has been recognized that the highly variable flood-prone flow regime characterized by the occurrence of seasonal high discharges is one of the most crucial factors for global anastomosing rivers to persist [47,48]. Frequent or high-magnitude flooding is assumed as a precondition for avulsion and the eventual formation of new channels. The SR-induced decrease of the magnitude and duration of floods can, therefore, be deemed as one of the factors, among others recognized by Marcinkowski et al. [23], speeding up the gradual extinction of the valuable anastomosing character of the Narew River. It is critical to address this issue in any amendments to contemporary water management instructions governing the flow regime of the Narew River.

4.3. Ecological Aspects

Another crucial factor responsible for anabranch loss was reported by Marcinkowski et al. [23] and relates to the uncontrolled expansion of Common Reed (*Phragmites australis*) in the NNP in recent decades. As reported by Próchnicki [49], the analysis of aerial imagery revealed a two-fold increase of the reed share in the valley vegetation between 1987 and 1997. It was justified by the overall shift in management strategies and cessation of floodplain mowing, which was very popular until 1980 in this region. However, as recognized in other studies, it could also be determined based on water-level changes. These water-level changes could reflect intensification of drainage and land reclamation directly within the river valley, but—noteworthy—could also result from the SR operation. Stromberg et al. [50] stated that aquatic plant species have been observed to increase in association with riverine alterations such as river channelization, stabilized water levels, reduced frequency of inundation, and altered timing of water and sediment flow. Moreover, Galatowitsch et al. [51] showed that the intensive invasion of *Phragmites australis* in the Platte River (Nebraska, USA) was caused by changes of the natural flow regime after the catchment has been highly altered by water development for irrigation and hydropower production. They stated that rapid expansion of reed followed the significant decrease of the flood flow. Therefore, it is highly probable that the invasion of Common Reed in the NNP, which coincides with the construction of the SR, might be caused by SR-induced flow regime changes, in addition to positive correlations of reed expansion and the cessation of mowing. Reporting downstream effects of the SR on the flow regime of the Narew River, and considering a number of environmental issues examined in the Narew Valley in recent years, similarly to Romanowicz and Osuch [28] we stress that further research is critically needed to reveal the influence of the SR-induced flow regime change on biota and biocoenosis of this area in a site-specific context. Based on the knowledge of the response of riparian environments to reservoir-influenced flow regime changes [1,3,4,6,7,50,51], we stress that keeping the status quo with respect to the regulation of river discharge by the SR can eventually result in irreversible changes of the downstream environment of the Narew River and its valley, such as the deterioration of the main objects of environmental conservation of the NNP related to a functioning anastomosing river channel. Considering the results of Piniewski et al. [52], and given the findings presented in this paper, we also stress that the SR-induced flow regime changes of Narew are likely to be more unilateral and stronger than the ones resulting from prospective climatic changes.

Now that 25 years have passed since the first spill of water was released from the fully filled SR, and given that data-based quantification of the influence of the SR on the Narew River flow regime allows drawing comprehensive and statistically relevant conclusions, the environmental management of the area should anticipate SR-induced flow regime changes of the Narew River.

Although not many studies present acceptable solutions allowing functioning dam-reservoir systems and sustaining resilient anastomosing rivers downstream, we hope that our findings will allow other researchers to view environmental factors of the Narew River and its valley with a different perspective. This perspective should emphasize flow-regime alteration as potentially a critical aspect, followed by the response of ecosystems.

4.4. Implications for Management

In view of our results, all considerations concerning instructions for water release and water management in lowland reservoirs, in particular those relating to the SR, gain new meaning and should be revisited. The observations of Kiczko and Napiórkowski [53] and Kiczko et al. [54], addressing the need for anticipation of environmental issues in water management measures implemented at the reservoir, should be used to provide answers to new management questions, such as 'How should water release from the reservoir be managed in order to keep the level of habitat degradation risk as low as possible?' and 'Is optimization of water release possible when limitations originate from both environmental and agricultural requirements?'.

Finally, a possible negative influence of the SR on downstream environments should be revisited in terms of gains and losses. Although the reservoir itself provides new habitats for birds, remains an important local attraction for recreational fishing, and provides some hydropower energy, it is necessary to calculate environmental and socioeconomic benefits and trade-offs of the SR. This calculation should anticipate actions related to environmental restoration in the Narew Valley and declining areas of valuable habitats downstream, because they might result from the negative influence of the downstream effects of the SR. Revealing the full set of benefits and losses associated with the function of the dam is expected to support (economically) decisions on the maintenance of the dam or on its removal in the future. Observing quick and positive responses of river systems to dam removals [5] fosters hope that there is still a chance to preserve Europe's unique anastomosing system before its complete degradation, which appears to result partly from the operation of the SR, by managing the landscape of the Narew Valley.

5. Conclusions

(1) The SR impact on the flow regime differs among gauge stations and indicators. The most significant and direct change was observed for extreme flow (minimum and maximum) of three subsequent gauges (Bondary, Narew, and Suraż). In each of them, a substantial decrease of high flow and an increase of low flow were detected. The quartile discharge alteration was more blurred and seems to be rather climate change-affected, reflecting precipitation pattern fluctuations.

(2) The SR has a significant and considerable influence on the recurrence and duration of floods of the Narew River: (1) the contemporary recurrence of Narew floods on an annual basis is 34% lower than before the SR was established; and (2) the contemporary duration of spring thaw floods has declined by 56%, from an average of 26 days before to 15 days after the SR has started to operate.

(3) An overall and significant increase in low flow was noted along the course of the Narew River downstream of the SR (in particular, at the Suraż gauge station), while a significant increase in the frequency and duration of droughts was detected in reference catchments of Narewka and Supraśl. However, the role of the SR in improving the quality of downstream riparian habitats tends to be insignificant, because the continuously reported trend of groundwater-level decline in the valley seems not to be reversed once the SR has been established.

(4) The significant change in floods and the flood–drought balance revealed in this study and attributed to the function of the SR poses a great threat with respect to the extinction of anabranches depending on the regular occurrence of high flow. This conclusion needs to be accounted for in the future management of the most valuable anastomosing stretch of the river in the NNP.

(5) The flow regime of rivers in control catchments (Narewka and Supraśl) reflects exclusively temporal changes of precipitation. The flow regime of the Narew River downstream of the SR from 1951 to 2013 presents different dynamics. Flow regime alterations of the Narew River observed for neighboring gauges—that is, the decrease of the duration and extent of spring flooding, increase of low flow, and altered flow durations—can be attributed to the operation of the SR. This observation is contradictory to several previous hydrological studies [24,27], which might have drawn wrong conclusions on the small (or even lack of) influence of the SR on the flow regime of the Narew based on the analysis of too short a data series. However, our observations on the influence of the SR on the flow regime of the Narew River confirm some of the previous suspicions and hypotheses of hydrologists working with limited sets of data [28] and environmentalists tackling biocoenosis of the area [42,55,56].

(6) With respect to the presented conclusions, some changes of riparian and riverine ecosystems of the Narew River and its valley attributed to climate change, land-use modification, or a switch in the role of riparian vegetation in water consumption, and reported in numerous scientific studies, require revision in terms of considering the SR as the dominant factor inducing the flow regime of the Upper Narew River.

Supplementary Materials
Figure S1: Monthly precipitation sum for 20 gauging stations in the Upper Narew catchment, Figure S2: Flow regime in Suraż gauging station 1951–2012. A—1st quartile monthly discharge, B—median monthly discharge, C—3rd quartile monthly discharge, Table S1: T test results of statistical significance of monthly precipitation sum between meteorological stations in the Upper Narew catchment.

Acknowledgments: We would like to thank three anonymous reviewers for their comments and suggestions, which helped to improve the manuscript. The study presented in this paper was financed by the National Science Centre, Poland, under grant number 2015/19/N/ST10/01629.

Author Contributions: Both authors equally contributed to all steps of the data analysis and manuscript preparation.

References

1. Dynesius, M.; Nilsson, C. Fragmentation and flow regulation of river systems in the northern third of the world. *Science* **1994**, *266*, 753–762. [CrossRef] [PubMed]
2. Graf, W.L. Dam nation: A geographic census of American dams and their large-scale hydrologic impacts. *Water Resour. Res.* **1999**, *35*, 1305–1311. [CrossRef]
3. Graf, W.L. Downstream hydrologic and geomorphic effects of large dams on American rivers. *Geomorphology* **2006**, *79*, 336–360. [CrossRef]
4. Magilligan, F.J.; Nislow, K.H. Changes in hydrologic regime by dams. *Geomorphology* **2005**, *71*, 61–78. [CrossRef]
5. O'Connor, J.E.; Duda, J.J.; Grant, G.E. 1000 dams down and counting. *Science* **2015**, *348*, 496–497. [CrossRef] [PubMed]
6. Bejarano, M.D.; Sordo-Ward, A.; Alonso, C.; Nilsson, C. Characterizing effects of hydropower plants on sub-daily flow regimes. *J. Hydrol.* **2017**, *550*, 186–200. [CrossRef]
7. Kibler, K.M.; Alipour, M. Flow alteration signatures of diversion hydropower: An analysis of 32 rivers in Southwestern China. *Ecohydrology* **2017**, *10*, e1842. [CrossRef]
8. Maheu, A.; St-Hilaire, A.; Caissie, D.; El-Jabi, N. Understanding the thermal regime of rivers influenced by small and medium size dams in Eastern Canada. *River Res. Appl.* **2016**, *32*, 2032–2044. [CrossRef]
9. Sanyal, J. Predicting possible effects of dams on downstream river bed changes of a Himalayan river with morphodynamic modelling. *Quat. Int.* **2017**, *453*, 48–62. [CrossRef]
10. Williams, G.P.; Wolman, M.G. *Downstream Effects of Dams on Alluvial Rivers*; Geological Survey Professional Paper 1286; USGS: Washington, DC, USA, 1984.
11. Ahn, J.M.; Jung, K.Y.; Shin, D. Effects of coordinated operation of weirs and reservoirs on the water quality of the Geum River. *Water* **2017**, *9*, 423.
12. Cooper, A.R.; Infante, D.M.; Wehrly, K.E.; Wang, L.; Brenden, T.O. Identifying indicators and quantifying large-scale effects of dams on fishes. *Ecol. Indic.* **2016**, *61*, 646–657. [CrossRef]

13. Song, Y.; Cheng, F.; Murphy, B.R.; Xie, S. Downstream effects of the Three Gorges Dam on larval dispersal, spatial distribution and growth of the four major Chinese carps call for reprioritizing conservation measures. *Can. J. Fish. Aquat. Sci.* **2017**. [CrossRef]
14. Valentin, S.; Wasson, J.G.; Philippe, M. Effects of hydropower peaking on epilithon and invertebrate community trophic structure. *River Res. Appl.* **1995**, *10*, 105–119. [CrossRef]
15. Chen, Q.; Zhang, X.; Chen, Y.; Qiongfang, L.; Qiu, L.; Liu, M. Downstream effects of a hydropeaking dam on ecohydrological conditions at subdaily to monthly time scales. *Ecol. Eng.* **2015**, *77*, 40–50. [CrossRef]
16. Han, J.; Zhang, W.; Fan, Y.; Yu, M. Interacting effects of multiple factors on the morphological evolution of the meandering reaches downstream the Three Gorges Dam. *J. Geogr. Sci.* **2017**, *27*, 1268–1278. [CrossRef]
17. Owusu, K.; Obour, P.B.; Nkansah, M.A. Downstream effects of dams on livelihoods of river-dependent communities: The case of Ghana's Kpong Dam. *Geogr. Tidsskr. Dan. J. Geogr.* **2016**, *117*, 1–10. [CrossRef]
18. Jiang, L.; Ban, X.; Wang, X.; Cai, X. Assessment of hydrologic alteration caused by the Three Gorges Dam in the middle and lower reaches of Yangtze River, China. *Water* **2014**, *6*, 1419–1434. [CrossRef]
19. Lee, J.E.; Heo, J.-H.; Lee, J.; Kim, N.W. Assessment of flood frequency alteration by dam construction via SWAT Simulation. *Water* **2017**, *9*, 264. [CrossRef]
20. Zhang, X.; Dong, Z.; Gupta, H.; Wu, G.; Li, D. Impact of the Three Gorges Dam on the hydrology and ecology of the Yangtze River. *Water* **2016**, *8*, 590. [CrossRef]
21. Mbaka, J.G.; Mwaniki, M.W. A global review of the downstream effect of small impoundments on stream habitat conditions and macroinvertebrates. *Environ. Rev.* **2015**, *23*, 257–263. [CrossRef]
22. Gradziński, R.; Baryła, J.; Doktor, M.; Gmur, D.; Gradziński, M.; Kędzior, A.; Paszkowski, M.; Soja, R.; Zieliński, T.; Żurek, S. Vegetation-controlled modern anastamosing system of the upper Narew River (NE Poland) and its sediments. *Sediment. Geol.* **2003**, *157*, 253–276. [CrossRef]
23. Marcinkowski, P.; Grabowski, R.C.; Okruszko, T. Controls on anastomosis in lowland river systems: Towards process-based solutions to habitat conservation. *Sci. Total Environ.* **2017**, *609*, 1544–1555. [CrossRef] [PubMed]
24. Cygan, B.; Niedbała, J.; Piekarski, M.K. Wpływ Zbiornika Siemianówka na kształtowanie się charakterystyk hydrologicznych rzeki Narwi. In Proceedings of the Conference Materials Zagospodarowanie Zlewni Bugu i Narwi w Ramach Zrównoważonego Rozwoju 2003, Popowo, Poland, 23–24 May 2003. (In Polish)
25. Jekatierynczuk-Rudczyk, E.; Górniak, A. Influence of Siemianówka Reservoir on Narew River below dam. In *Ecosystem of Siemianówka Reservoir in 1990–2004 and its Restoration*; Górniak, A., Ed.; Department of Hydrobiology, University of Białystok: Białystok, Poland, 2006; pp. 193–199. (In Polish)
26. Kiczko, A.; Romanowicz, R.J.; Osuch, M. Impact of water management policy on flow conditions in wetland areas. *Phys. Chem. Earth* **2011**, *36*, 638–645. [CrossRef]
27. Mioduszewski, W.; Gajewski, G.; Biesiada, M. Zróżnicowanie stosunków wodnych w dolinie Narwi w granicach Narwiańskiego Parku Narodowego. *Water Environ. Rural Areas* **2004**, *11*, 39–50. (In Polish)
28. Romanowicz, R.J.; Osuch, M. Assessment of land use and water management induced changes in flow regime of the Upper Narew. *Phys. Chem. Earth* **2011**, *36*, 662–672. [CrossRef]
29. Grabowska, M.; Ejsmont-Karabin, J.; Karpowicz, M. Reservoir-river relationships in lowland, shallow, eutrophic systems: An impact of zooplankton from hypertrophic reservoir on river zooplankton. *Pol. J. Ecol.* **2013**, *61*, 759–768.
30. Grabowska, M.; Mazur-Marzec, H. The effect of cyanobacterial blooms in the Siemianówka Dam Reservoir on the phytoplankton structure in the Narew River. *Oceanol. Hydrobiol. Stud.* **2011**, *40*, 19–26. [CrossRef]
31. Karpowicz, M. Influence of eutrophic lowland reservoir on Crustacean zooplankton assemblages in river valley oxbow lakes. *Pol. J. Environ. Stud.* **2014**, *23*, 2055–2061.
32. Karpowicz, M. Microcrustacean (*Cladocera, Copepoda*) source-sink dynamics in a lowland river ecosystem with a dam reservoir. *Oceanol. Hydrobiol. Stud.* **2016**, *45*, 297–303. [CrossRef]
33. Banaszuk, P.; Wysocka-Czubaszek, A. Phosphorus dynamics and fluxes in a lowland river: The Narew anastomosing river system, NE Poland. *Ecol. Eng.* **2005**, *25*, 429–441. [CrossRef]
34. Banaszuk, H.; Banaszuk, P.; Gradziński, R.; Kamocki, A.K.; Mioduszewski, W.; Okruszko, T.; Próchnicki, P.; Szewczyk, M. *Przyroda Podlasia: Narwiański Park Narodowy*; Ekonomia Podlasia: Podlasie, Poland, 2004; ISBN 83-87231-07-X. (In Polish)
35. BIPROMEL. *Siemianówka Reservoir–Water Management Rules*; Technical Report; Bipromel: Warszawa, Poland, 1999. (In Polish)
36. Sokołowski, J. *Monografia Zbiornika Wodnego Siemianówka*; WZMIUW: Warszawa, Poland, 1999. (In Polish)

37. Pettitt, A.N. A non-parametric approach to the change-point problem. *Appl. Stat.* **1979**, *28*, 126–135. [CrossRef]
38. Rybski, D.; Neumann, J. A Review on the Pettitt Test. In *In Extremis*; Kropp, J., Schellnhuber, H.J., Eds.; Springer: Berlin/Heidelberg, Germany, 2011; pp. 202–213, ISBN 978-3-642-14863-7.
39. Tan, X.; Gan, T.Y.; Shao, D. Effects of persistence and large-scale climate anomalies on trends and change points in extreme precipitation of Canada. *J. Hydrol.* **2017**, *550*, 453–465. [CrossRef]
40. Huh, S.; Dickey, D.A.; Meador, M.R.; Ruhl, K.E. Temporal analysis of frequency and duration of low and high streamflow: Years of record needed to characterise streamflow variability. *J. Hydrol.* **2005**, *310*, 78–94. [CrossRef]
41. Marcinkowski, P.; Piniewski, M.; Kardel, I.; Szcześniak, M.; Benestad, M.; Srinivasan, R.; Ignar, S.; Okruszko, T. Effect of Climate Change on Hydrology, Sediment and Nutrient Losses in Two Lowland Catchments in Poland. *Water* **2017**, *9*, 156. [CrossRef]
42. Banaszuk, P.; Kamocki, A. Effects of climatic fluctuations and land-use changes on the hydrology of temperate fluviogenous mire. *Ecol. Eng.* **2008**, *32*, 133–146. [CrossRef]
43. Kamocki, A.; Kołos, A.; Banaszuk, P. Can we effectively stop the expansion of trees on wetlands? Results of a birch removal experiment. *Wetl. Ecol. Manag.* **2016**, *25*, 359–367. [CrossRef]
44. Grygoruk, M.; Batelaan, O.; Mirosław-Świątek, D.; Szatyłowicz, J.; Okruszko, T. Evapotranspiration of bush encroachments on a temperate mire meadow—A nonlinear function of landscape composition and groundwater flow. *Ecol. Eng.* **2014**, *73*, 598–609. [CrossRef]
45. Deoniziak, K.; Hermaniuk, A.; Wereszczuk, A. Effects of wetland restoration on the amphibian community in the Narew River Valley (Northeast Poland). *Salamandra* **2017**, *53*, 50–58.
46. Grygoruk, M.; Frąk, M.; Chmielewski, A. Agricultural rivers at risk: Dredging results in a loss of macroinvertebrates. Preliminary observation from the Narew Catchment, Poland. *Water* **2015**, *7*, 4511–4522. [CrossRef]
47. Nanson, G.C.; Knighton, A.D. Anabranching rivers: Their cause. character and classification. *Earth Surf. Process. Landf.* **1996**, *21*, 217–239. [CrossRef]
48. Schumann, R.R. Morphology of Red Creek, Wyoming, an arid-region anastomosing channel system. *Earth Surf. Process. Landf.* **1989**, *14*, 277–288. [CrossRef]
49. Próchnicki, P. The expansion of common reed (phragmites australis (cav.) trin. ex steud.) in the anastomosing river valley after cessation of agriculture use (Narew River valley, NE Poland). *Pol. J. Ecol.* **2005**, *53*, 353–364.
50. Stromberg, J.C.; Lite, S.J.; Marler, R.; Paradzick, C.; Shafroth, P.B.; Shorrock, D.; White, J.M.; White, M.S. Altered stream-flow regimes and invasive plant species: The Tamarix case. *Glob. Ecol. Biogeogr.* **2007**, *16*, 381–393. [CrossRef]
51. Galatowitsch, S.M.; Larson, D.L.; Larson, J.L. Factors affecting post-control reinvasion by seed of an invasive species, Phragmites australis, in the central Platte River, Nebraska. *Biol. Invasions* **2016**, *18*, 2505–2516. [CrossRef]
52. Piniewski, M.; Laize, C.L.R.; Acreman, M.; Okruszko, T.; Schneider, C. Effects of climate change on environmental flow indicators in the Narew Basin, Poland. *J. Environ. Qual.* **2011**, *43*, 155–167. [CrossRef] [PubMed]
53. Kiczko, A.; Napiórkowski, J. Aspiration-Reservation Decision Support System fo Siemianówka Reservoir. In *Modelling of Hydrological Processes in the Narew Catchment*; Springer: Berlin, Germany, 2011; pp. 111–121.
54. Kiczko, A.; Romanowicz, R.; Napiórkowski, J.; Piotrowski, A. Integration of reservoir management and flow routing model—Upper Narew Case Study. *Publ. Inst. Geophys. Pol. Acad. Sci.* **2008**, *E-9*, 41–56.
55. Pugacewicz, E. Zmiany w awifaunie lęgowej doliny Górnej Narwi w latach 1986–2007. *Dubelt* **2012**, *4*, 1–41. (In Polish)
56. Szewczyk, M.; Dembek, W.; Kamocki, A. Response of Riparian vegetation to the decrease of flooding: Narew National Park, Poland. In Proceedings of the International Conference 'Towards Natural Flood Reduction Strategies', Warsaw, Poland, 6–13 September 2003. Available online: http://www.academia.edu/download/36018017/3_91.pdf (accessed on 2 August 2017).

4

Research on Cascade Reservoirs' Short-Term Optimal Operation under the Effect of Reverse Regulation

Changming Ji [1], Hongjie Yu [1,*], Jiajie Wu [1], Xiaoran Yan [1] and Rui Li [2]

1 School of Renewable Energy, North China Electric Power University, Beijing 102206, China; cmji@ncepu.edu.cn (C.J.); 1121550126@ncepu.edu.cn (J.W.); yanxiaoran1014@163.com (X.Y.)
2 Overseas Exam Service Center, Shijiazhuang Information Engineering Vocational College, Shijiazhuang 050035, China; liruijoyce@163.com
* Correspondence: yhjyeah@163.com

Abstract: Currently research on joint operation of a large reservoir and its re-regulating reservoir focuses on either water quantity regulation or water head regulation. The accuracy of relevant models is in need of improvement if the influence of factors such as water flow hysteresis and the aftereffect of tail water level variation are taken into consideration. In this paper, given the actual production of Pankou-Xiaoxuan cascade hydropower stations that combines two operation modes ('electricity to water' and 'water to electricity'), a coupling model of their short-term optimal operation is developed, which considers Xiaoxuan reservoir's regulating effect on Pankou reservoir's outflow volume and water head. Factors such as water flow hysteresis and the aftereffect of tail water level variation are also considered to enhance the model's accuracy. The Backward Propagation (BP) neural network is employed for precise calculation of the downstream reservoir's inflow and the upstream reservoir's tail water level. Besides, we put forth Accompanying Progressive Optimality Algorithm (APOA) to solve the coupling model with aftereffect. An example is given to verify the scientificity of the proposed model and the advantages of APOA. Through analysis of the model calculation results, the optimal operation rules of the cascade reservoirs are obtained in terms of water quantity regulation and water head regulation, which can provide scientific reference for cascade reservoirs' optimal operation.

Keywords: reverse regulation; coupling model; aftereffect; accompanying progressive optimality algorithm

---

1. Introduction

In the downstream of a large reservoir, the construction of a small re-regulating reservoir can not only ensure the upstream hydropower station's peak-regulating capability in the power grid, but also exert a positive effect on downstream water supply, ecology, shipping, etc., spawning significant comprehensive utilization benefits. Therefore, this kind of development mode has been widely adopted in water conservancy construction across the world, e.g., the Three Gorges-Gezhouba cascade reservoirs in China [1], the Xiaolangdi-Xixiayuan cascade reservoirs in China [2], the Bigge-Ahausen cascade reservoirs in Germany [3], the Srinagarindra-Tha Thung Na cascade reservoirs in Thailand, and so forth [4].

The downstream reservoir's reverse regulation effect on the upstream reservoir is mainly embodied in two facets: (1) water quantity regulation, i.e., using the downstream reservoir's storage capacity to optimize the upstream reservoir's outflow; (2) water head regulation, i.e., controlling the downstream reservoir's water level to optimize the overall output given its backwater effect on the upstream reservoir's tail water level. To date, many scholars have studied the joint operation of a large reservoir and its re-regulating reservoir. For example, Richter, B. D. proposed that the re-regulating reservoir can readjust the upstream reservoir's outflow to eliminate the unnatural fluctuations caused by hydropower operation, so that water discharge to the downstream channel can move in a way much closer to natural flow, thus protecting the downstream ecological environment [5]. Bai, T. et al. established a model of Xiaolangdi-Xixiayuan cascade hydropower stations' joint peak-regulating operation, which used Xixiayuan reservoir to regulate Xiaolangdi reservoir's outflow volume, thus alleviating the conflict between water regulation and power dispatching during Xiaolangdi hydropower station's operation process [6]. Taking the Three Gorges-Gezhouba cascade reservoirs as an example, Cai, Z. et al. analyzed the law of the influence of Gezhouba reservoir's water level on the cascade hydropower generation during different operation states of the Three Gorges reservoir, and obtained the control table of optimal water levels [7]. However, the above studies have only considered either water quantity regulation or water head regulation. For instance, Bai, T. did not consider Xixiayuan reservoir's influence on Xiaolangdi reservoir's tail water level in calculation. For simplification, Cai, Z. used the 'runoff operation' hypothesis on Gezhouba reservoir, whose water level stayed unchanged during calculation. So far, there has been no research that considers the combination of water quantity regulation and water head regulation; hence the above models somewhat deviate from the actual operation of cascade reservoirs, with limitations in their conclusions.

Additionally, in previous studies, to solve the cascade reservoirs' short-term optimal operation model, the calculation of such variables as downstream reservoir's inflow and upstream reservoir's tail water level was often simplified (e.g., the upstream reservoir's outflow was directly used as the downstream reservoir's inflow [8], the upstream reservoir's tail water level was calculated via the stage-discharge relation method [9], etc.). These simplifications neglect the impact of factors such as water flow hysteresis and the aftereffect of tail water level variation, which makes model variables deviate from reality and reduces model accuracy. Yet, in the study of reverse regulation, downstream reservoir's inflow and upstream reservoir's tail water level are the key variables during calculation, as they can reflect the connection of water quantity and water head between the upstream and the downstream reservoirs. Therefore, it is important to accurately calculate these two variables. Research about water flow hysteresis is focused mainly on flow travel time, which is generally treated as a fixed value in primary studies [10]. In recent years, some scholars have proposed that flow travel time varies dynamically with the change of upstream reservoir outflow [11]. Although this improvement takes into consideration the influence of reservoir outflow's rate on its travel speed, it is still regarded as a steady flow without considering flow attenuation during travel. Consequently, there is still a certain deviation between the study results and the actual situation. Apart from the stage-discharge relation method, research on the variation rule of tail water level mostly adopts hydraulic methods that determine reservoir tail water level by constructing and solving a hydrodynamic model of open-channel unsteady flow [12,13]. However, hydraulic models are unsuitable for practical production due to their stringent requirements for boundary condition data and slow calculation speed. Accordingly, to seek a universal and efficient method for further exploration and accurate simulation of cascade reservoirs' operation process with various complex factors is of great significance in improving model accuracy.

Short-term optimal operation is mostly applied to reservoirs' practical production. Therefore, besides model accuracy, there is also high demand for solving methods of the model in terms of computation time, calculation results, and so on. Among the many methods to work out reservoir optimal operation models, Dynamic Programming (DP) is widely used thanks to its advantages such as global convergence and high stability [14,15]. However, DP is restricted by the 'curse of dimensionality' [16], which means that the amount of calculation increases exponentially with the growing number of reservoirs. Moreover, the no-aftereffect condition is no longer satisfied in cascade reservoirs' short-term optimal operation when factors such as water flow hysteresis and the aftereffect of tail water level variation are taken into account, which makes DP inapplicable. In this regard, Ji, C. et al. tried to solve the model of cascade reservoirs' short-term optimal operation with aftereffect by using Multi-Stage Dynamic Programming (MSDP) [17]. This method has global convergence, but its problem of 'dimensionality curse' is even more serious, which renders it inapplicable to practical production.

Mei, Y. proposed Multi-Dimensional Dynamic Programming Approximation Algorithm [18], which has a small amount of calculation, but may not produce the actual global optimal solution due to approximation in the calculation process. Although modern intelligent algorithms, such as Genetic Algorithm [19,20], Particle Swarm Optimization Algorithm [21,22], and Ant Colony Optimization Algorithm [23,24], can solve problems with aftereffect in a short period of time, they tend to get stuck in the local optimal solution. Until now, there has been no mature method that can fully meet the requirements of actual production, so presently it is necessary to develop a method that can accurately figure out the model's global optimal solution in a relatively short time.

In view of the above problems in the study of short-term optimal operation of cascade reservoirs under the effect of reverse regulation, this paper combines the two operation modes ('electricity to water' and 'water to electricity') of Pankou-Xiaoxuan cascade hydropower stations to build a coupling model of their short-term optimal operation, which considers both Xiaoxuan reservoir's regulating effect on Pankou reservoir's outflow and the former's backwater effect on the latter's tail water level. Through the BP neural network, factors such as water flow hysteresis and the aftereffect of tail water level variation are studied. Furthermore, Accompanying Progressive Optimality Algorithm (APOA) is put forward to solve the proposed coupling model, so as to provide scientific guidance for the short-term optimal operation of cascade reservoirs under the effect of reverse regulation.

2. Coupling Model

2.1. Objective Function

The subject of this paper is Pankou-Xiaoxuan cascade hydropower stations located in Hubei province in China. With an installed capacity of 500 MW, the Pankou hydropower station is an important peak-regulating power station in the province. Its regular power operation is controlled by Hubei Provincial Power Dispatching Center (a typical 'electricity to water' mode). With an installed capacity of 50 MW, Xiaoxuan hydropower station, as the re-regulating hydropower station for Pankou, mainly regulates Pankou's outflow for power generation (a typical 'water to electricity' mode), which is controlled by Han Jiang Hydropower Development Co., Ltd. (hereafter referred to as the Company, Shiyan, China). Although the two hydropower stations have different dispatching superordinates, they both belong to the Company as their decision maker. The Company hopes that through Xianxuan reservoir's reverse regulation over Pankou reservoir, the overall power generation efficiency of the whole cascade hydropower stations can be improved. Long-term production practice has proved that Xiaoxuan reservoir's water level has a certain backwater effect on Pankou reservoir's tail water level. On one hand, while maintaining a high water level of Xiaoxuan reservoir is apparently beneficial to its own power generation, it will raise the tail water level of Pankou reservoir, thus lowering its productive water head and affecting its power generation efficiency, which means consuming more water to complete the same power generation quota. On the other hand, Xiaoxuan reservoir can operate at a low water level to enhance the power generation efficiency of Pankou hydropower station, but at the cost of losing some of its own power generation benefits. Plus, when Pankou hydropower station is in operation to regulate peak load, its outflow is so uneven that Xiaoxuan reservoir has to re-regulate it to meet the demand for downstream ecological flow. Hence, for the economic operation of the cascade hydropower stations, it is of crucial importance to determine the operation mode of Xiaoxuan hydropower station that can reversely regulate Pankou reservoir's outflow and water head in a scientific manner.

Against the backdrop of coupling the two operation modes ('electricity to water' and 'water to electricity'), the aim of the cascade hydropower stations is to balance Pankou's power generation efficiency with Xiaoxuan's generated energy. Accordingly, from the angle of the whole cascade hydropower stations' total energy, two optimization criteria—minimum energy consumption and maximum generated energy—are integrated to develop the objective function in the coupling model of Pankou-Xiaoxuan cascade reservoirs' short-term optimal operation [25,26], as shown by Equation (1).

Unlike the traditional objective function that involves water quantity, this one calculates water energy, with Pankou's power generation efficiency reflected by its hydroenergy consumption for power generation (meaning that the less hydroenergy consumption, the higher its power generation efficiency). In this way, the operation goals of both Pankou and Xiaoxuan are unified in the form of energy, which avoids the emergence of a multi-objective problem. In addition, as the two optimization criteria conflict with each other, the objective function is the subtraction of the former from the latter, finally seeking the maximum power generation benefits of the whole cascade hydropower stations:

$$E = \max \sum_{t=1}^{T} \left[N_{2,t}(q_{2,t}, H_{2,t})\Delta t - 3600\lambda_{1,t} q_{1,t}(N_{1,t}, H_{1,t})\Delta t \right] \tag{1}$$

where E is the remainder of Xiaoxuan's generated energy after deducting Pankou's hydroenergy consumption in the operation period, unit: kWh; T is the number of calculation periods over the entire operation period; $N_{2,t}(q_{2,t}, H_{2,t})$ is Xiaoxuan's power output when its power generation flow and water head are $q_{2,t}$ and $H_{2,t}$ respectively in period t, unit: kW, and is obtained according to the power characteristic curve of its generator set; Δt is the length of the calculation period, unit: h, and $\Delta t = 0.25$ h; $\lambda_{1,t}$ is the energy efficiency coefficient of Pankou hydropower station in period t, unit: kWh/m^3, and its physical meaning is the amount of energy contained in every cubic meter of Pankou reservoir's water in period t; $q_{1,t}(N_{1,t}, H_{1,t})$ is Pankou's power generation discharge when its output and water head are $N_{1,t}$ and $H_{1,t}$ respectively in period t, unit: m^3/s, and is also obtained according to the power characteristic curve of its generator set.

2.2. Constraint Conditions

Now that the coupling model combines the two operation modes, the constraint conditions must contain particular constraints specific to these two modes, as shown by Equations (2) and (3), while other regular constraints are expressed by Equations (4)–(10).

(1) Output command constraint

Because Pankou hydropower station operates in the 'electricity to water' mode, its output in each period is already determined:

$$N_{1,t} = N_t \tag{2}$$

where $N_{i,t}$ is the output of hydropower station i in period t, unit: kW (i = 1 for Pankou, i = 2 for Xiaoxuan and the same hereafter); N_t is the output command from the superior dispatching center for Pankou hydropower station in period t, unit: kW.

(2) Water level constraint at the end of the operation period

Because Xiaoxuan hydropower station operates in the 'water to electricity' mode, its water level at the end of the operation period is already determined:

$$Z_{2,T+1} = Z_{2,\text{end}} \tag{3}$$

where $Z_{i,t}$ is the water level of reservoir i at moment t, unit: m; $Z_{2,\text{end}}$ is Xiaoxuan reservoir's controlled water level at the end of the operation period, unit: m.

(3) The upper and lower bounds of water level:

$$Z_i^{\min} \leq Z_{i,t} \leq Z_i^{\max} \tag{4}$$

where $Z_i^{\max}$, $Z_i^{\min}$ are the upper and lower bounds of water level of reservoir i, respectively, unit: m.

(4) The upper and lower bounds of output:

$$N_i^{\min} \leq N_{i,t} \leq N_i^{\max} \tag{5}$$

where $N_i^{\max}$, $N_i^{\min}$ are the upper and lower bounds of output of hydropower station i, respectively, unit: kW.

(5) The upper and lower bounds of flow rate:

$$q_i^{\min} \leq q_{i,t} \leq q_i^{\max} \tag{6}$$

where $q_{i,t}$ is the power generation discharge of hydropower station i in period t, unit: m^3/s, and in this paper, it equals to reservoir outflow because water abandonment is not taken into consideration; $q_i^{\max}$, $q_i^{\min}$ are the upper and lower bounds of power generation discharge of hydropower station i, respectively, unit: m^3/s.

(6) Water balance constraint:

$$V_{i,t+1} = V_{i,t} + 3600(Q_{i,t} - q_{i,t})\Delta t \tag{7}$$

where $V_{i,t}$, $V_{i,t+1}$ are the storage capacity of reservoir i at respectively the beginning and the end of period t (i.e., moment t and moment t + 1), unit: m^3; $Q_{i,t}$ is the inflow of reservoir i in period t, unit: m^3/s.

(7) Vibration zone constraint:

$$N_{i,j} \notin [N_{i,j}^{\min}, N_{i,j}^{\max}] \tag{8}$$

where $N_{i,j}$ is the output of generator unit j in hydropower station i, unit: kW; $N_{i,j}^{\min}$, $N_{i,j}^{\max}$ are the upper and lower bounds of the vibration zone of generator unit j in hydropower station i, unit: kW.

(8) Ecological flow constraint

To meet the demand for downstream ecological flow [27], Xiaoxuan reservoir's outflow should not be less than 16.7 m^3/s:

$$q_{2,t} \geq q_e \tag{9}$$

where q_e is the demand for downstream ecological flow, unit: m^3/s, and q_e = 16.7 m^3/s.

(9) Auxiliary power constraint

To prevent the occurrence of inverse power transmission, when Pankou hydropower station halts production, it is Xiaoxuan hydropower station that undertakes the auxiliary power supply mission for the running of both stations:

$$N_{2,t} \geq N_s \ when \ N_{1,t} = 0 \tag{10}$$

where N_s is the auxiliary power need of the cascade hydropower stations, unit: kW, and N_s = 2000 kW.

The ecological flow constraint and the auxiliary power constraint can well reflect Xiaoxuan reservoir's regulation effect on Pankou reservoir's discharge volume. Since the completion of Xiaoxuan reservoir, Pankou reservoir no longer bears the task of downstream ecological water supply, and the task of auxiliary power supply has been transferred to Xiaoxuan hydropower station. Therefore, when Pankou hydropower station does not undertake peak-regulating tasks, all its generator units can stop running instead of operating in the low efficiency zone just to meet the constraints of ecological flow and auxiliary power supply.

3. Calculation of the Model's Key Variables

Between the upstream and the downstream reservoirs, there is a certain hydraulic connection in terms of water quantity and water head [28], as shown in Figure 1. Water quantity connection refers to the upstream reservoir's outflow turning into the downstream reservoir's inflow after travelling through the interval river channel. Water head connection refers to that the downstream reservoir's backwater raises the water level of the interval channel, causing a jacking effect on the upstream reservoir's tail water level. Due to these two kinds of connections, the re-regulating reservoir can adjust the upstream reservoir's discharge volume and water head. Thus, the downstream reservoir's inflow and the upstream reservoir's tail water level, which can respectively reflect these two types of connections, become the key variables in the coupling model of Pankou-Xiaoxuan cascade reservoirs' short-term optimal operation under the effect of reverse regulation. Computation of the key variables has a critical influence on the model's accuracy in that an error in their calculations will cause the model's results to deviate from reality, losing its instructive significance to practical production. Thus, it is very important to calculate them accurately.

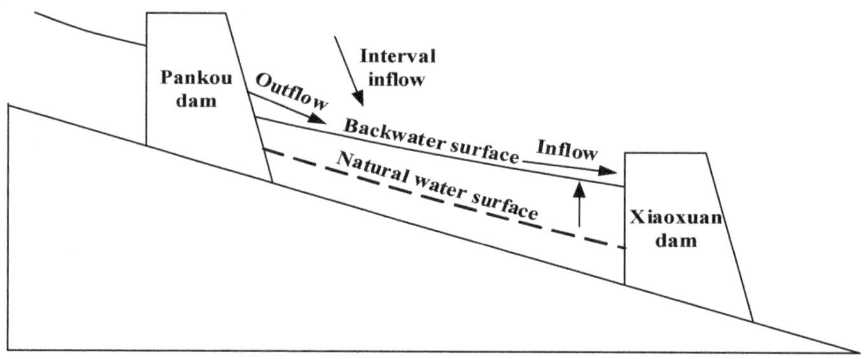

Figure 1. Schematic diagram of the cascade reservoirs' hydraulic connection.

3.1. The Downstream Reservoir's Inflow Considering Water Flow Hysteresis and Interval Inflow

In the cascade reservoirs, the downstream reservoir's inflow is composed of the upstream reservoir's outflow and the interval inflow. The upstream reservoir's outflow arrives at the downstream reservoir with water flow hysteresis that mainly involves two parts. (1) It takes some time for the water flow to travel from the upstream reservoir to the downstream reservoir. (2) During this travel process, the upstream reservoir's outflow, which is an unsteady flow, goes through not only displacement but also attenuation to some degree [29]. Figure 2 illustrates the comparison between Pankou reservoir's outflow and Xiaoxuan reservoir's inflow during a certain power generation process, from which it can be seen that in contrast to Pankou reservoir's outflow, Xiaoxuan reservoir's inflow somewhat lags behind and gets smoother. Water flow hysteresis makes it a complicated hydraulics problem as to when and how Pankou reservoir's outflow will reach Xiaoxuan reservoir.

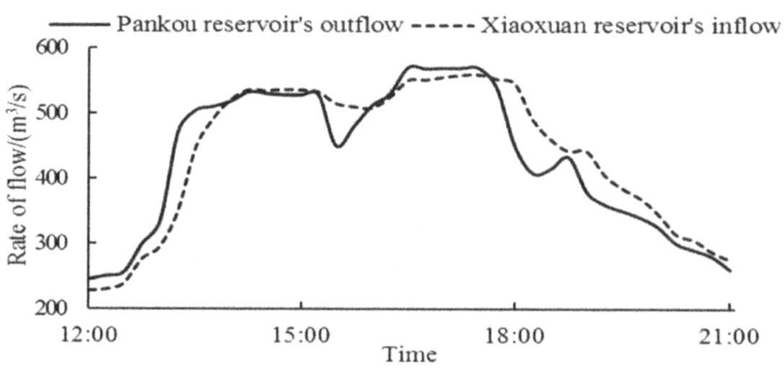

Figure 2. Pankou's outflow and Xiaoxuan's inflow.

The dam sites of Pankou and Xiaoxuan are only 10.4 km apart, without any tributary in between, and the interval catchment area is fairly small, so the interval inflow is often neglected in practical production. However, long-term monitoring data show that on average Xiaoxuan reservoir's inflow is slightly greater than Pankou reservoir's outflow, so the existence of the interval inflow cannot be ignored in order to improve the model's accuracy. The interval inflow between Pankou and Xiaoxuan is mainly formed by runoff generation and confluence in the interval basin, which largely depends on the basin's rainfall [30]. When the interval inflow is taken into consideration, the calculation of Xiaoxuan reservoir's inflow will become more complicated in that not only the evolution of water current in the river but also the hydrologic cycle from precipitation to runoff is involved. Since the physical mechanisms of these natural processes are not fully understood, it is difficult to accurately simulate them. Hence in this paper, with masses of data, the BP neural network is used to figure out the relationship between Xiaoxuan reservoir's inflow and Pankou reservoir's outflow as well as the interval basin's rainfall, so that the accurate value of Xiaoxuan reservoir's inflow can be worked out [31].

The BP neural network is a multilayer feedforward network trained by the error backward propagation algorithm. As one of the most widely-used neural networks, it can approximate any nonlinear function with any precision. Structurally, the BP neural network is divided into one input layer, one output layer, and several hidden layers [32]. In the calculation of Xiaoxuan reservoir's inflow, the output layer is represented by Xiaoxuan reservoir's inflow in the current period ($Q_{2,t}$). As for the input layer, Pankou reservoir's outflow ($q_{1,t}$~$q_{1,t-4}$) as well as the interval basin's precipitation (P_t~P_{t-4}) in the current and the previous four periods are chosen as the primary data and screened by the correlation coefficient method. It is demonstrated in Table 1 that the correlation coefficient between Pankou reservoir's outflow and Xiaoxuan reservoir's inflow is smaller in the earlier period. For $q_{1,t}$, $q_{1,t-1}$ and $q_{1,t-2}$, their correlation coefficients with $Q_{2,t}$ are similar, whereas the correlation coefficients of $q_{1,t-3}$ and $q_{1,t-4}$ with $Q_{2,t}$ are noticeably smaller. Therefore, $q_{1,t}$, $q_{1,t-1}$ and $q_{1,t-2}$ are selected as the input layer data. Compared with Pankou reservoir's outflow, the interval basin's precipitation has a much smaller correlation coefficient, which changes little over time. Considering that the interval inflow accounts for only a small portion of Xiaoxuan reservoir's inflow, the precipitation P_{t-1}, with the largest correlation coefficient with $Q_{2,t}$ in this regard, is selected as the input layer data. In 1989, Cybenko, G. and Hornik, K. proved that the three-layer network (with one input layer, one output layer, and one hidden layer) can simulate any complex nonlinear problems [33,34]. Therefore, one hidden layer is set up whose node numbers are determined according to Equation (11) [35]:

$$h = \sqrt{u + v} + a \tag{11}$$

where h is the number of nodes in the hidden layer; u is the number of nodes in the input layer; v is the number of nodes in the output layer; a is an adjustment coefficient between 0 and 10, and according to tentative calculation when $a = 4$ the neural network yields the best training results in this study.

After the structure of the neural network is established, the historical data of 2012–2016 are input into the neural network for training. The neuron transfer function is a hyperbolic tangent function, and after 10,000 times of training, the network is saved for later calculation of Xiaoxuan reservoir's inflow. Figure 3 shows the calculated Xiaoxuan reservoir's inflow during the power generation process mentioned before. Compared with the actual inflow, the inflow produced by the BP neural network has a relative error of 1.11%, which is more accurate than that obtained by the translation method (relative error: 3.84%), which only considers the travel time (i.e., Pankou reservoir's outflow merely translates into Xiaoxuan reservoir's inflow).

Table 1. Correlation coefficients between input layer data and the output layer.

Input Layer Data	$q_{1,t}$	$q_{1,t-1}$	$q_{1,t-2}$	$q_{1,t-3}$	$q_{1,t-4}$	P_t	P_{t-1}	P_{t-2}	P_{t-3}	P_{t-4}
Correlation Coefficient	0.91	0.89	0.88	0.83	0.79	0.19	0.20	0.19	0.19	0.18

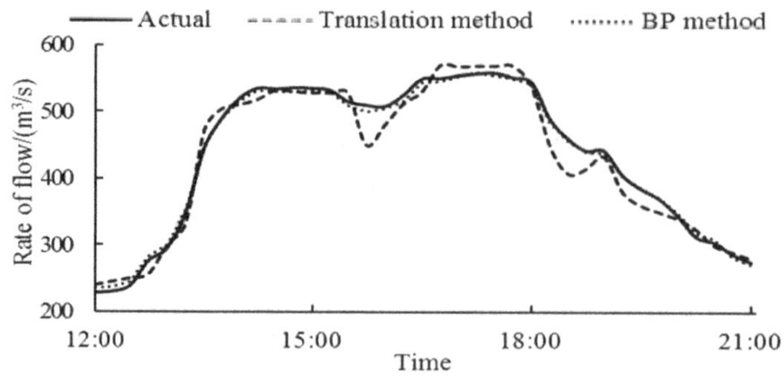

Figure 3. Calculation results of Xiaoxuan's inflow.

3.2. The Upstream Reservoir's Tail Water Level Considering the Influence of Dual Aftereffect Factors

During the operation of cascade reservoirs, under the backwater effect of the downstream reservoir, the upstream reservoir's tail water level is closely related to the downstream reservoir's water level. Different water levels of the downstream reservoir have different degrees of backwater effect on the upstream reservoir's tail water level. Figure 4 is the stage-discharge curve of Pankou reservoir corresponding to different water levels of Xiaoxuan reservoir. Long-term practical production shows that Pankou reservoir's tail water level obtained by the stage-discharge relation method still deviates from the actual value. This kind of deviation can be explained by two reasons. For one thing, the tail water level-discharge curve is based on the assumption that the reservoir outflow is a steady flow. However, in fact, it is an unsteady flow whose characteristics result in a certain aftereffect of tail water level variation [36], that is, the tail water level in the current period is related to that in the previous period. For another, it takes some time for Xiaoxuan reservoir's backwater to reach Pankou dam, meaning that Pankou reservoir's tail water level in the current period is not necessarily influenced by Xiaoxuan reservoir's water level in the current period but the previous several periods [37], which is another aftereffect of Pankou's tail water level variation. Under the influence of such dual aftereffect factors, Pankou's tail water level in the current period is affected by three kinds of variables: Pankou's outflow in the current period, Pankou's tail water level in the previous period, and Xiaoxuan's water level in the previous n (n is determined by the travel time of Xiaoxuan's backwater) periods. As the physical mechanism is very complicated, the BP neural network is again used for calculation.

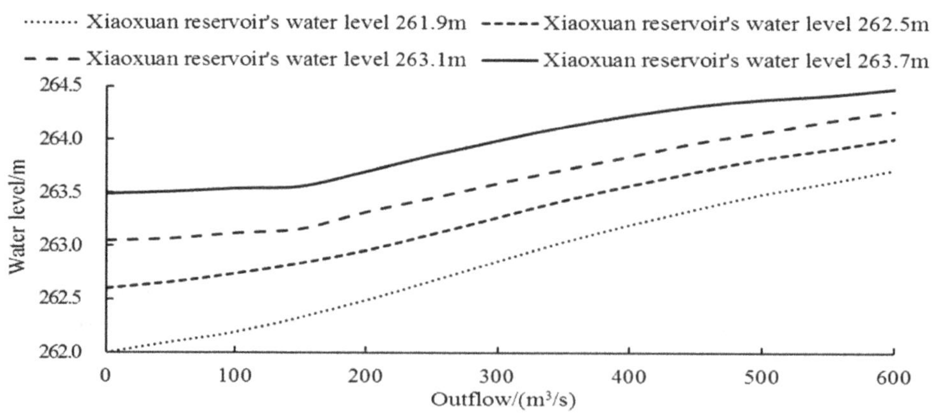

Figure 4. Pankou's stage-discharge curve.

Here, Pankou reservoir's tail water level in the current period ($DZ_{1,t}$) represents the output layer. As for the input layer data, aside from Pankou reservoir's outflow in the current period ($q_{1,t}$) and its tail water level in the previous period ($DZ_{1,t-1}$), it is also needed to decide which periods of Xiaoxuan reservoir's water level are concerned. Similarly, Xiaoxuan reservoir's water levels in the current and the previous four periods ($Z_{2,t}$~$Z_{2,t-4}$) are chosen as the primary data to be filtered by the correlation coefficient method. From Table 2 it can be seen that the correlation coefficient between Xiaoxuan's water level and Pankou's tail water level gradually decreases with time. For $Z_{2,t}$, $Z_{2,t-1}$ and $Z_{2,t-2}$, their correlation coefficients have little difference, while for $Z_{2,t-3}$ and $Z_{2,t-4}$ their correlation coefficients are apparently smaller, which indicates that the time it takes for Xiaoxuan reservoir's backwater to reach Pankou is between 0 and 2 periods. Thus, $Z_{2,t}$, $Z_{2,t-1}$ and $Z_{2,t-2}$ are selected as the input layer data. The configuration of the hidden layer and the training parameters is similar to that described in Section 3.1. After the training is completed, the network is saved for later calculation of Pankou reservoir's tail water level. Figure 5 shows the resulted tail water level variation of Pankou reservoir in the same power generation process mentioned before. Compared with the actual variation, the results of the BP neural network only produce a calculation error of 0.001 m, which is notably smaller than the calculation error (0.027 m) produced by the stage-discharge relation method, thus proving the advantage of the BP neural network.

Table 2. Correlation coefficients between input layer data and the output layer.

Input layer Data	$Z_{2,t}$	$Z_{2,t-1}$	$Z_{2,t-2}$	$Z_{2,t-3}$	$Z_{2,t-4}$
Correlation Coefficient	0.74	0.73	0.73	0.70	0.67

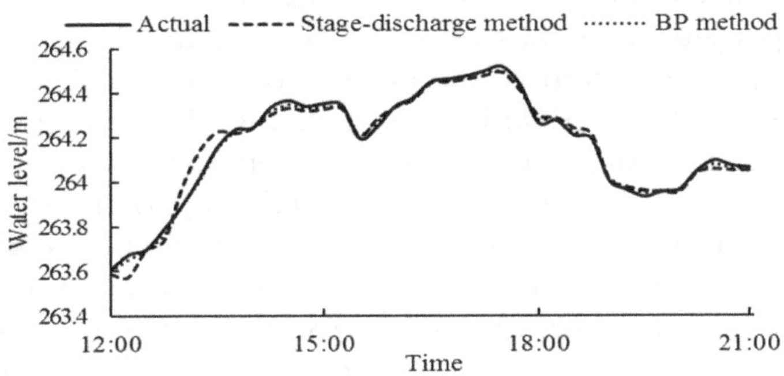

Figure 5. Calculation results of Pankou's tail water level.

4. Model Solution

When factors such as water flow hysteresis and the aftereffects of tail water level variation are taken into account, the decision variable (power generation flow) of the current period for the cascade reservoirs is not only related to the state (reservoir water level) of this period but also the state of the previous several periods. Since the 'no aftereffect' condition is not satisfied, the proposed model of short-term optimal operation cannot be solved by DP. Although MSDP can deal with the problem with aftereffect, it is restricted by a serious 'curse of dimensionality' [17]. The subject of this study is taken as an example for further explanation. The short-term optimal operation of Pankou-Xiaoxuan cascade reservoirs is an issue about the joint operation of two reservoirs. Nonetheless, as Pankou hydropower station operates in the 'electricity to water' model, once the operation state of Xiaoxuan hydropower station is determined, Pankou's operation state will also be settled (by in-plant economic operation calculation), which reduces the issue to an essentially one-dimensional problem. In this one-dimensional scenario, suppose that the number of discrete points of the reservoir water level is M and the number of interrelated periods is N (i.e., the state of the former $N-1$ periods influences the decision of the current period), then in each multi-stage calculation of MSDP there are M^{N+1} combinations of Xiaoxuan's water level and the total amount of calculation is $T \cdot M^{N+1}$ with T times of multi-stage calculation. Since the value of $T \cdot M^{N+1}$ grows exponentially with the increase in N, even in a one-dimensional circumstance the dimensionality curse still emerges due to the interconnection of N periods. As the number of dimensions increases to L ($L > 1$), the total amount of calculation will become $T \cdot (M^{N+1})^L$ with $(M^{N+1})^L$ combinations of water level at each multistage, which is a more serious problem of dimensionality curse. In view of this, we come up with an improved POA (i.e., APOA) to solve the model of the cascade reservoirs' short-term optimal operation with aftereffect.

4.1. Basic Principle of APOA

POA is an improved DP that simplifies a complicated multi-stage decision problem into a series of two-stage decision problems, considerably reducing the complexity and effectively mitigating the dimensionality curse [38]. In contrast to POA, APOA has two facets of improvements.

(1) When Xiaoxuan hydropower station is being optimized, Pankou reservoir's water level is taken as the variable that spatially accompanies Xiaoxuan reservoir's water level for calculation. In each discrete computation where Xiaoxuan's water levels in the current and the previous n periods are known, the magnitude of Xiaoxuan's backwater effect on Pankou's tail water level is definite. In addition, because Pankou's tail water level in the previous one period is already known, in the current period, Pankou's tail water level is only related to Pankou's outflow. Under the condition of a definite output of Pankou hydropower station, its reservoir outflow can be obtained through trial calculation. Its water level can then be figured out according to its inflow and outflow. Hence, for each discrete water level of Xiaoxuan, there is only one corresponding water level of Pankou and accordingly, one hydrograph of Pankou corresponding to each hydrograph of Xiaoxuan in the operation period.

(2) In each two-stage calculation, the variables of the preceding related $N - 1$ periods are taken as the temporally-accompanying variables of the current period for calculation. Considering the interrelation of N periods, the decision of the current two-stage calculation is affected by those decisions made in the previous $N - 1$ periods. APOA divides the problem with T periods into $T - 1$ sub problems, each of which includes two main calculation periods accompanied by $N - 1$ related periods. As shown in Figure 6, in the calculation of period t and period $t + 1$, the state and the decision of other periods (including the accompanying $N - 1$ related periods) are known, thereby effectively tackling the aftereffect problem. In each two-stage calculation, there are M combinations of Xiaoxuan's water level and $T - 1$ times of two-stage calculation for one iteration, so the amount of calculation is $(T - 1) \cdot M$. If the precision requirement is satisfied after K times of iteration, the total amount of calculation will be $K \cdot (T - 1) \cdot M$, evidently smaller than $T \cdot M^{N+1}$ produced by MSDP. As a result, APOA can effectively alleviate the problem of dimensionality curse.

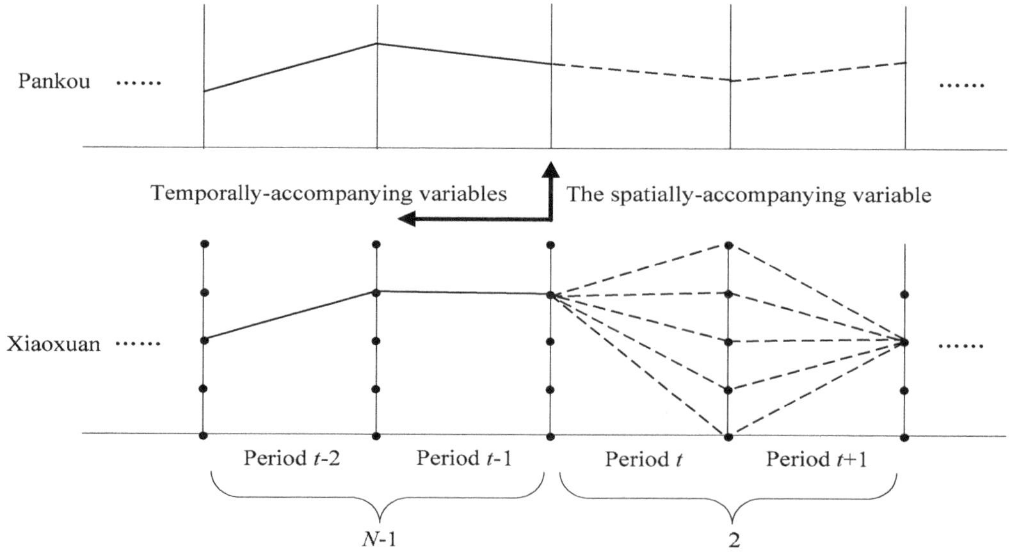

Figure 6. Schematic diagram of APOA.

4.2. Calculation Procedure of APOA

The calculation procedure of APOA is detailed as follows:

Step 1: Determine the number of associated periods. Use the correlation coefficient method to calculate the number of periods connected with the current period in terms of water quantity and water head. If the first m periods are related to the current period in terms of Xiaoxuan reservoir's inflow and the first n periods are related to the current period in terms of Pankou reservoir's tail water level, then the number of the related periods can be calculated by Equation (12), where $m = 2$, $n = 2$ and $N = 3$ in this article:

$$N = \max\{m+1, n+1\} \tag{12}$$

Step 2: Train the BP neural networks. The BP neural networks are trained by inputting training data, including the type of training data, starting and ending time, as well as the structure and parameters of the neural networks, which are detailed in Sections 3.1 and 3.2. After the training is completed, the neural network for calculating Xiaoxuan reservoir's inflow (net_1) and that for Pankou reservoir's tail water level (net_2) are obtained.

Step 3: Set the initial solution. The selection of initial solutions has a major influence on the convergence speed and computation time of APOA [39]. Long-term experience of the dispatchers is often embodied in the operation rule of the hydropower station, according to which a relatively optimal solution can be obtained. Thus, in this paper, such a solution is used as the initial solution. It should be noted that in the initial solution, the initial water level of Xiaoxuan reservoir ($Z_2^0 = \{Z_{2,1}^0, Z_{2,2}^0, \ldots, Z_{2,T+1}^0\}$) is the most critical variable while that of Pankou reservoir ($Z_1^0 = \{Z_{1,1}^0, Z_{1,2}^0, \ldots, Z_{1,T+1}^0\}$) is only the variable accompanying it.

Step 4: Optimize progressively. Carry out the optimization step by step from period 1 to period T. When the calculation progresses to period t and period $t + 1$ (accompanied by the two related periods of $t - 2$ and $t - 1$), fix Xiaoxuan reservoir's water levels at all moments except that at moment $t + 1$ and calculate all the discrete points of its water level at that moment. Xiaoxuan reservoir's inflow and Pankou reservoir's tail water level are calculated by the BP neural networks, as shown in Equation (13), with Pankou reservoir's water level acting as the variable accompanying Xiaoxuan's water level. After running through all the discrete points, select the optimal water level of Xiaoxuan reservoir ($Z_{2,t+1}^*$) and correspondingly the optimal water level of Pankou ($Z_{1,t+1}^*$) at moment $t + 1$. Make $Z_{2,t+1}^1 = Z_{2,t+1}^*$ and $Z_{1,t+1}^1 = Z_{1,t+1}^*$.

$$\begin{array}{c} Q_{2,t} = net_1(q_{1,t}, q_{1,t-1}, q_{1,t-2}, P_{t-1}), Q_{2,t+1} = net_1(q_{1,t+1}, q_{1,t}, q_{1,t-1}, P_t) \\ DZ_{1,t} = net_2(q_{1,t}, DZ_{1,t-1}, Z_{2,t}, Z_{2,t-1}, Z_{2,t-2}), DZ_{1,t+1} = net_2(q_{1,t+1}, DZ_{1,t}, Z_{2,t+1}, Z_{2,t}, Z_{2,t-1}) \end{array} \tag{13}$$

Step 5: Iterative calculation. After Step 4 is executed, the first iteration completes and produces a new solution $Z_2^1 = \{Z_{2,1}^1, Z_{2,2}^1, \ldots, Z_{2,T+1}^1\}$ accompanied by $Z_1^1 = \{Z_{1,1}^1, Z_{1,2}^1, \ldots, Z_{1,T+1}^1\}$. Compare Z_2^1 with Z_2^0 in terms of the value of the objective function. If the difference between their values is less than the calculation precision (ε), then the convergence condition is satisfied and Z_2^1 (accompanied by Z_1^1) is the final solution. Otherwise replace the initial solution with Z_2^1 and repeat Step 4 until the convergence condition is satisfied.

5. Case Study

The Du River, the largest tributary of the Hanjiang River, is located across Shaanxi province and Hubei province in China. Pankou-Xiaoxuan cascade hydropower stations, whose basic information is listed in Table 3, lie in the upper reaches of the Du River, as shown in Figure 7. Under the 'electricity to water mode, Pankou hydropower station generates electricity in accordance with the daily instructions (in the form of a 96-point load curve) from the superior dispatching center. In this paper, the operation period is one day with 96 calculation periods lasting 15 min each. Considering the different conditions among flood season, dry season, and transition season, 10 representative days are selected from each season of 2016–2017 for calculation. The results are analyzed in terms of model solving methods and the patterns of reverse regulation.

Table 3. Parameters of Pankou-Xiaoxuan cascade hydropower stations.

Items	Unit	Pankou	Xiaoxuan
Normal water level	M	355	264
Dead water level	M	330	261.3
Regulation volume	10^8 m^3	11.2	0.0678
Regulation performance	-	annual regulation	daily regulation
Installed capacity	MW	500	50
Operation mode	-	'electricity to water'	'water to electricity'

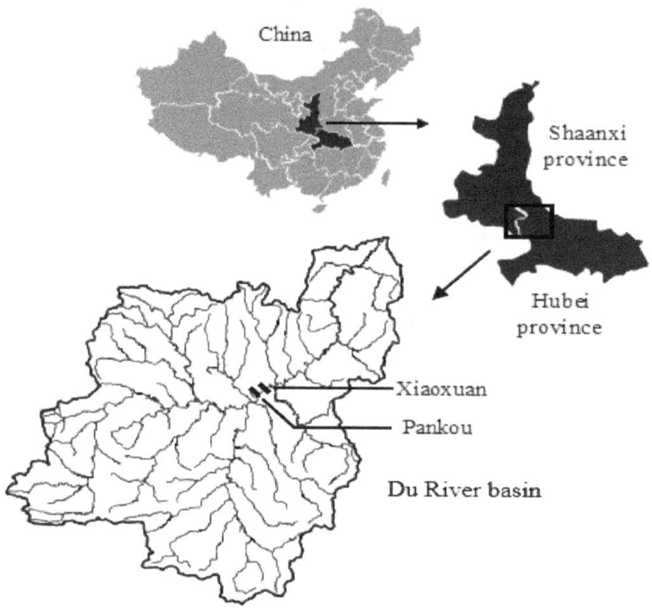

Figure 7. The location of Pankou-Xiaoxuan cascade reservoirs.

5.1. Comparative Analysis of Model Solving Methods

Take 10 December 2016 as an example to develop the coupling model of Pankou-Xiaoxuan cascade reservoirs' short-term optimal operation and use DP, POA, MSDP and APOA, respectively, to work it out. While water flow hysteresis and the aftereffect of tail water level variation are taken into account in MSDP and APOA by using BP neural network to calculate Xiaoxuan reservoir's inflow and Pankou reservoir's tail water level, the two factors are not considered in DP and POA where Pankou's outflow is directly taken as Xiaoxuan's inflow and Pankou's tail water level is obtained by the stage-discharge relation method. The calculated results of each method are statistically summarized in Table 4, in which the optimization margin refers to the ratio of the energy increase after optimization (the difference between the two objective functions) to the actual generated energy.

Table 4. The statistics of calculated results.

Items	Unit	Actual	DP	POA	MSDP	APOA
Pankou's hydroenergy consumption	10^3 kWh	1518.5	1519.4	1519.4	1518.9	1518.9
Xiaoxuan's generated energy	10^3 kWh	267.7	297.5	297.5	307.4	307.4
Objective function	10^3 kWh	−1250.8	−1221.9	−1221.9	−1211.5	−1211.5
Optimization margin	%	-	1.61	1.61	2.19	2.19
Computation time	s	-	673.10	86.56	9.43×10^5	156.37

5.1.1. Comparative Analysis of the Objective Function Value of Each Method

MSDP has been theoretically proved to have global convergence, so on the premise of the same calculation conditions (taking into account the water flow hysteresis, the aftereffect of tail water level variation and the with same amount of discrete points), the objective function calculated by APOA is compared with that calculated by MSDP in order to assess the calculated results of APOA. It can be seen from Table 4 that for 10 December 2016, the objective function of APOA equals that of MSDP. Among the other 29 representative days, there are 20 days where their objective functions are equal and 9 days where their objective functions are very close. According to the results of the 30 representative days, the average optimization margin of APOA is 0.93%, only 0.03% less than that of MSDP, which demonstrates the advantage of APOA.

Similarly, the comparative analysis of POA and DP can also illustrate the advantage of POA. It is shown from Table 4 that the objective functions of DP and POA are different from those of MSDP and APOA. This is due to their different calculation conditions. Thus, there is no point in comparing the objective functions between the former two and the latter two. Instead, we make comparisons in terms of the length of computation time and the calculation accuracy.

5.1.2. Analysis of Computation Time

In this case, the amount of calculation is $T{\cdot}M^2$ for DP and $T{\cdot}M^{N+1}$ for MSDP with the aftereffect factors taken into consideration. With the number of discrete points $M = 560$ and the number of related periods $N = 3$, the amount of calculation of MSDP is 313,600 times that of DP. However, Table 4 shows that the computation time of MSDP is about 1400 times that of DP, far fewer than 313,600 times. This is because MSDP imposes constraint on the feasible domain and leaves out those discrete points that do not satisfy the constraints, saving a great deal of calculation time. Nevertheless, its computation time of 262 h still cannot meet the efficiency requirement in practical production.

As an improved algorithm for alleviating the dimensionality curse of DP, POA is characterized by an obviously smaller amount of calculation compared with DP, as demonstrated by the statistical results from Table 4. Unlike the contrast between DP and MSDP, APOA has the same form of calculation amount as POA ($K{\cdot}(T-1){\cdot}M$) despite considering aftereffect factors. No matter how the number of iterations (K) may change, the change in the amount of calculation is linear rather than exponential, so the amount of calculation for APOA and POA is in the same order of magnitude. Table 4 shows that the calculation time of APOA is 156.37 s, which is 1.8 times that of POA but only 1/6000 that of MSDP. This means that with the same calculation results, APOA substantially shortens the computation time in comparison with MSDP, which can meet the efficiency requirement in actual production.

5.1.3. Analysis of Calculation Accuracy

The accuracy of calculation depends on how close the calculation result is to the actual process. As is shown in Figure 8a, the outflow graphs of Pankou reservoir obtained by APOA and POA basically coincide, with merely a slight difference that can be seen from the partially enlarged detail (the reason will be analyzed in detail in Section 5.2). While the difference of Pankou's outflow processes obtained by the two methods is small, the calculated tail water level graphs differ distinctly, as is displayed in Figure 8b. For POA, the shape of Pankou's tail water level graph resembles that of the reservoir's outflow graph, with steep rises and falls. However, Pankou's tail water level variation is a continuous and slow process. The stage-discharge relation method used in POA only takes into account the impact of Pankou's outflow and Xiaoxuan's water level in the current period and neglects the aftereffect of tail water level variation, making the calculation results inconsistent with the reality. On the contrary, the BP neural network used in APOA considers various factors that influence Pankou's tail water level and produces a relatively smooth graph, which tallies more with the actual situation. Therefore, APOA has higher calculation accuracy than POA with respect to Pankou's tail water level.

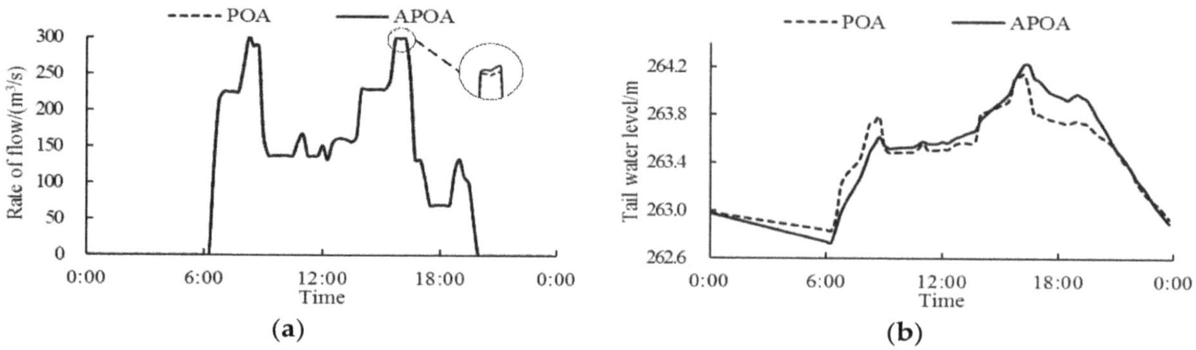

Figure 8. Comparison of Pankou's calculated results between POA and APOA. (**a**) Pankou's outflow process; (**b**) Pankou's tail water level process.

Next, we analyze from the angle of Xiaoxuan reservoir's inflow. As the BP neural network adopted in APOA takes water flow hysteresis into consideration, the corresponding inflow of Xiaoxuan reservoir, as shown in Figure 9a, slightly lags behind and is smoother compared to POA. Due to water flow hysteresis, Xiaoxuan's operation strategy obtained by APOA, as signified by Xiaoxuan's water level process in Figure 9b and its output process in Figure 9c, also lags behind that obtained by POA, especially around the periods when Pankou starts power generation. On that day, it is Pankou hydropower station's downtime until 6:30 when power generation begins, spurring an increase in its outflow. In the calculation of POA, Pankou's outflow is directly used as Xiaoxuan's inflow, causing Xiaoxuan to receive inflow at the exact moment and gradually increase its output as well as water level. Yet actually, when Pankou starts operation, its generator units are at the climbing stage with a relatively small outflow that travels a long time (about two periods) to reach Xiaoxuan. Accordingly, during the two periods of 6:30–6:45 and 6:45–7:00 Xiaoxuan's inflow is too little to keep the reservoir water level rising while discharging water for power generation. Because of the low accuracy in calculating Xiaoxuan reservoir's inflow, the operation scheme produced by POA is against the natural law and impractical in actual production.

The calculation results of APOA, however, show that Xiaoxuan reservoir's water level does not rise until 7:00 when Pankou's outflow from the previous two periods has already got to Xiaoxuan. Therefore, Xiaoxuan does follow the rule of water balance, discharging water for power generation and meanwhile raising reservoir water level. To sum up, owing to the excellent performance of the BP neural network in the calculation of Xiaoxuan reservoir's inflow, the accuracy of APOA is higher than that of POA.

5.2. *Analysis of the Reverse Regulation Rule*

The optimal operation scheme obtained by APOA is compared to the actual operation scheme in order to find out the reverse regulation rule and get the optimal operation mode of the cascade reservoirs. Figure 10 is a comparison chart of the optimal and the actual operation schemes on 10 December 2016. During the actual operation, Pankou generates electricity as instructed, with its water level and output process shown in Figure 10a, while Xiaoxuan's water level and output process are shown in Figure 10b. Between 0:00 and 6:30, the downtime of Pankou hydropower station, Xiaoxuan hydropower station keeps operating with about 2 MW of output to meet the requirements for auxiliary power and ecological flow, and its water level drops steadily and slowly. After 6:30, as Pankou begins power generation, Xiaoxuan also increases its output until Pankou halts operation at 20:00. Then Xiaoxuan's operation returns to about 2 MW of output until the end of the operation period.

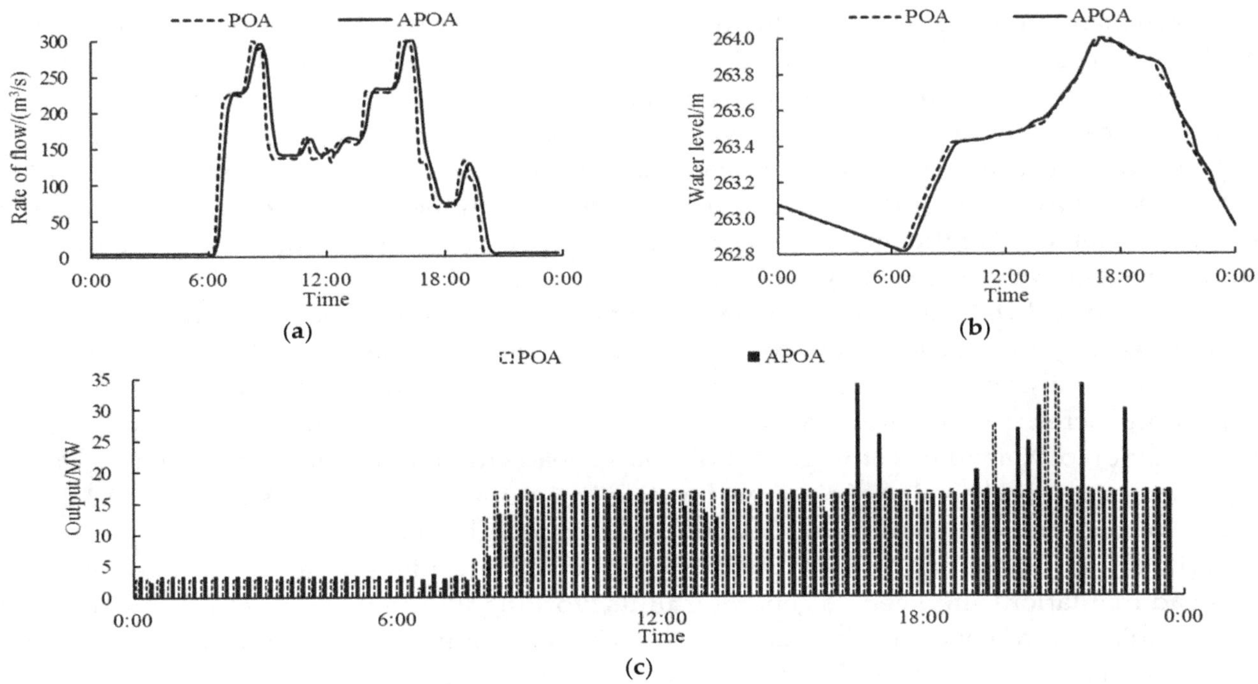

Figure 9. Comparison of Xiaoxuan's calculated results between POA and APOA. (**a**) Xiaoxuan's inflow process; (**b**) Xiaoxuan's water level process; (**c**) Xiaoxuan's output process.

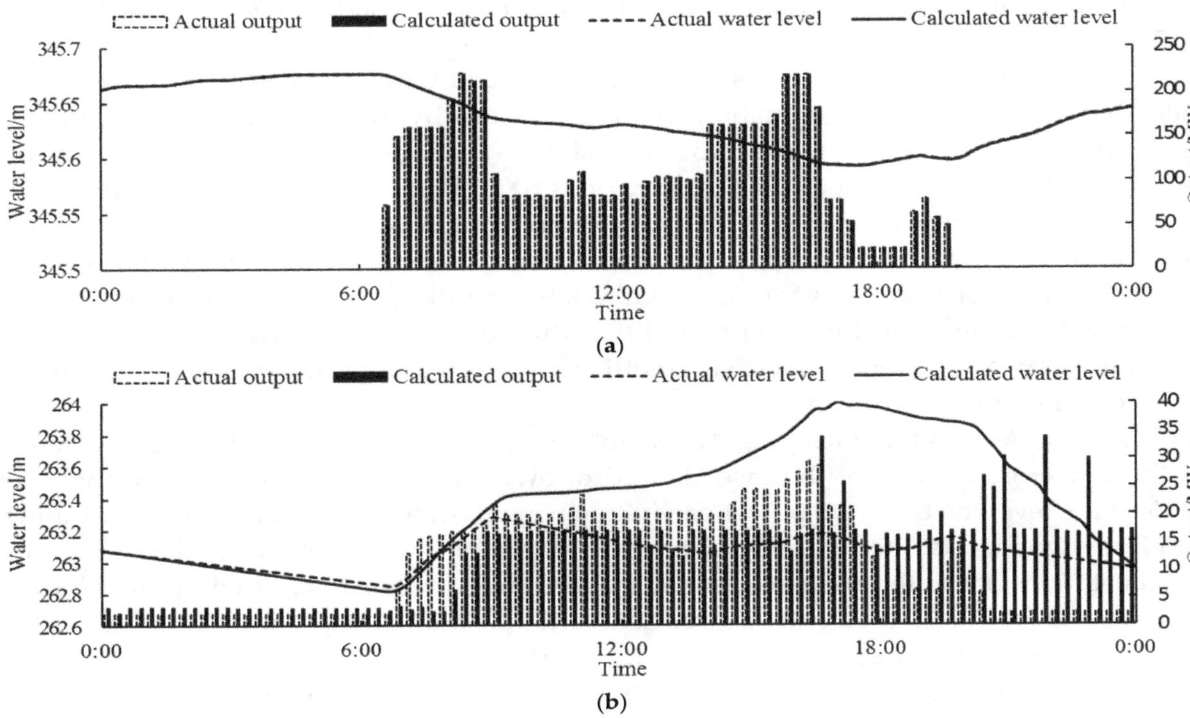

Figure 10. Comparison between the optimal and the actual operation schemes. (**a**) Pankou; (**b**) Xiaoxuan.

In contrast to the actual operation scheme, the optimal operation scheme produces a markedly different operation strategy for Xiaoxuan. For one thing, after Pankou commences power generation, Xiaoxuan's water level steadily rises and nearly reaches the normal water level at 17:00, and keeping a high water level yields greater output with the same flow rate. For another, after Pankou halts operation at 20:00, Xiaoxuan's power generation remains efficient with about 16 MW of output as its water level is still high. As a result, Xiaoxuan's generated energy in the optimal operation scheme is 307.4 thousand kWh, an increase of 39.7 thousand kWh from the actual situation (267.7 thousand kWh).

As for Pankou, the elevated water level of Xiaoxuan increases the backwater effect on its tail water level and affects its power generation efficiency. However, it can be seen from Figure 10a that the calculated hydrograph of Pankou virtually coincides with the actual one. In addition, it can be obtained from Table 4 that the hydroenergy consumption of Pankou in the optimal operation scheme is 1518.9 thousand kWh, only 0.4 thousand kWh more than the actual situation (1518.5 thousand kWh), meaning that the backwater effect of Xiaoxuan on Pankou's power generation efficiency is insignificant.

As far as the whole cascade hydropower stations are concerned, the elevated water level of Xiaoxuan creates a growth in generated energy that is remarkably greater than the hydroenergy consumption increase of Pankou caused by its fall in power generation efficiency. Moreover, the optimization of Xiaoxuan's storage and discharge leads to more efficient utilization of Pankou's outflow. Consequently, the value of objective function in the optimal operation scheme is 39.3 thousand kWh greater than the actual value, with an optimization margin of 2.19%.

A primary conclusion can be drawn from these calculated results—to raise Xiaoxuan's water level (water head regulation) and keep its generator units working within the high-efficiency zone (water quantity regulation) can boost the power generation benefits of the cascade hydropower stations.

Thus, to make the above conclusion more convincing, the other 29 representative days are also analyzed. As the above analyses have involved the representative day in the dry season, Figure 11 only displays the results of 1 July 2017 (flood season) and 12 June 2017 (transition season). Besides, it can be seen from Figure 10a that in the optimal scheme and the actual scheme, Pankou's operation processes are basically the same, and the focus of this article is Xiaoxuan's re-regulating effect on Pankou. Therefore, only Xiaoxuan's operation schemes are displayed in Figure 11. It can be seen from Figures 10b and 11 that whatever the season is, the calculated hydrographs of Xiaoxuan are above the actual ones and that in the optimal operation scheme, Xiaoxuan hydropower station's output is around 16 MW, 32 MW or 48 MW. When such output is assigned to its generator units, the generator units operate in the high-efficiency zone (the rated output of Xiaoxuan's generator unit is 16.7 MW, near which the power generation efficiency is relatively high), which also verifies the above conclusion.

It should be noted that during Pankou's downtime, if Xiaoxuan discharges water by efficiently operating one generator unit, its water level will decline continually because of its small reservoir capacity. If the downtime lasts long, to prevent the situation of a too low water level with no water for use, Xiaoxuan has to generate electricity inefficiently with a flow that just satisfies the needs of ecological flow and auxiliary power.

It is easy to understand that keeping Xiaoxuan's generators operating in the high-efficiency zone benefits the power generation of the cascade hydropower stations. As for the reason why raising Xiaoxuan's water level can boost the power generation benefits of the cascade hydropower stations, it is discussed from two aspects: the efficiency of the generator units and the total water head of the cascade hydropower stations. From the angle of generators' efficiency, while Pankou hydropower station with a high water head is affected mainly by power generation flow and insignificantly by a minor change in water head, Xiaoxuan hydropower station with a low water head is sensitive to changes in water head. Suppose the impact of Xiaoxuan's water level variation on the cascade hydropower stations' total water head is neglected (i.e., when Xiaoxuan's water level is raised by 1 m, Pankou's tail water level will also rise by 1 m correspondingly and the cascade hydropower stations' total water head stays unchanged), then we design two operation schemes as shown in Table 5. On the basis of Scheme 1, Scheme 2 raises Xiaoxuan's water level by 2 m, thus increasing the water head of Xiaoxuan from 12 to 14 m while decreasing that of Pankou from 80 to 78 m. The calculation results show that with the same power generation flow of 100 m^3/s, Xiaoxuan sees an increase of 0.0198 in power generation efficiency, much greater than the reduction of 0.0063 that Pankou suffers. Therefore, to raise Xiaoxuan's water level by 2 m means an increase of 0.0021 in the overall power generation efficiency.

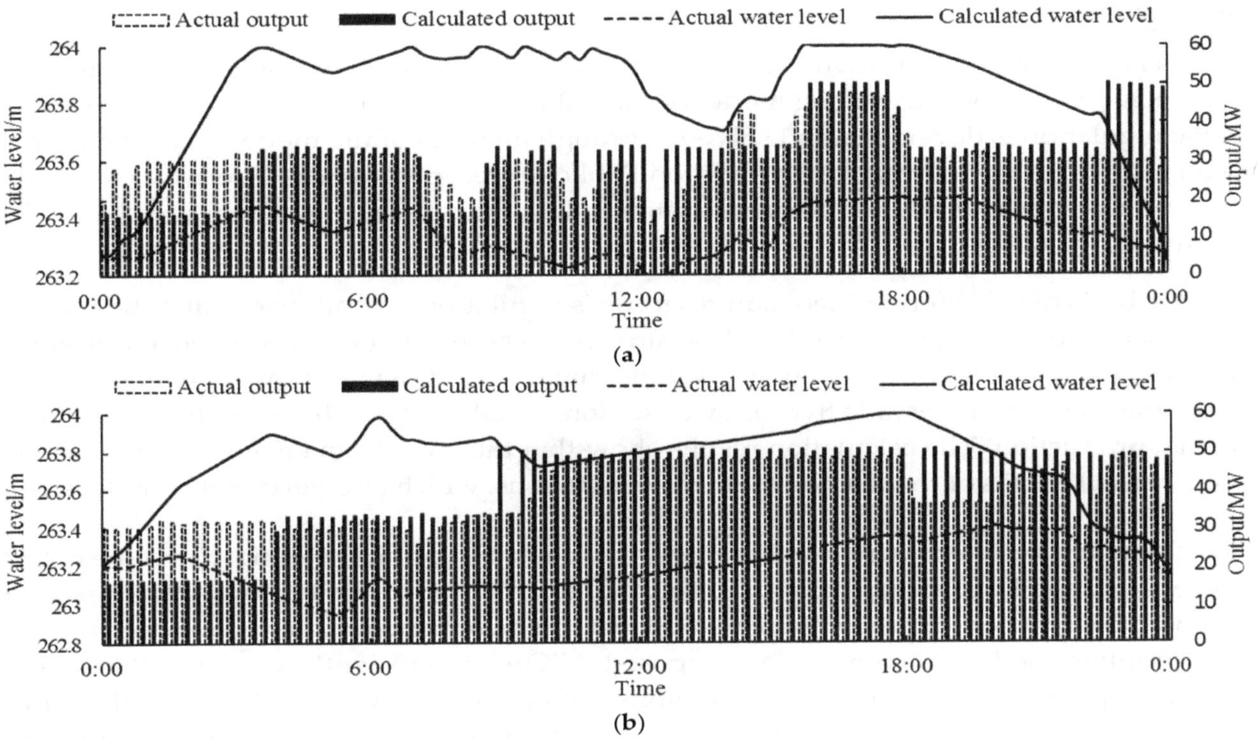

Figure 11. Comparison between optimal and actual operation schemes of Xiaoxuan, (**a**) 1 July; (**b**) 12 June.

In fact, the water level variation of Xiaoxuan reservoir has some influence on the total water head of the cascade hydropower stations. As is shown in Figure 1, the farther the backwater is from Xiaoxuan dam, the closer the backwater surface gets to the natural water surface. It is found from statistical analysis of historical data that an increase of 1 m in Xiaoxuan reservoir's water level will raise Pankou reservoir's tail water level by about 0.6 m. Therefore, Scheme 1 should actually become Scheme 3 after Xiaoxuan's water level is raised by 2 m, as shown in Table 5. From the angle of the cascade hydropower stations' total water head, Scheme 3 uses 0.8 m more of water head than Scheme 1 and further elevates the overall efficiency compared to Scheme 2.

Table 5. Comparison of different operation schemes $\eta = N/(9.81qH)$.

Items	Unit	Scheme 1	Scheme 2	Scheme 3
H_1	m	80	78	78.8
H_2	m	12	14	14
q_1	m^3/s	100	100	100
q_2	m^3/s	100	100	100
η_1	-	0.6467	0.6404	0.6429
η_2	-	0.8503	0.8701	0.8701
η	-	0.6732	0.6753	0.6771

In summary, when Xiaoxuan hydropower station raises its operation water level, even though Pankou's water head will fall due to the backwater effect, the impact on the efficiency of Pankou's generator units is unnoticeable. Besides, the increment in Xiaoxuan's water head can greatly enhance the efficiency of its generator units and thus considerably increase its generated energy. In contrast, the improvement of Xiaoxuan's power generation benefits is more significant. Therefore, it is beneficial for the cascade hydropower stations to raise Xiaoxuan's operation water level.

6. Conclusions

In this paper, Pankou-Xiaoxuan cascade reservoirs are taken as an example and integrated with their practical production, a coupling model of their short-term optimal operation under the effect of reverse regulation is developed and solved. Through analysis of the model's calculation results, Xiaoxuan reservoir's reverse regulation effect on Pankou reservoir is studied from two facets (water quantity and water head) and the optimal operation rules of the cascade reservoirs are obtained, with the following conclusions:

1. On the basis of considering Xiaoxuan reservoir's regulation on both water quantity and water head of Pankou reservoir, the model takes into account both Pankou's power generation efficiency and Xiaoxuan's generated energy to seek the maximum of overall power generation benefits from the angle of the cascade hydropower stations' total energy, which fits the requirements of actual production. The calculation results show that the model can effectively enhance power generation benefits of the cascade hydropower stations, which also verifies the model's validity.

2. The BP neural network has excellent performance in exploring water flow hysteresis and the aftereffect of tail water level variation, so that the accurate values of downstream reservoir's inflow and upstream reservoir's tail water level can be obtained, which significantly improves the coupling model's accuracy. The proposed APOA can efficiently work out the short-term optimal operation model of cascade reservoirs with aftereffect. With the merits and accuracy of its calculation results demonstrated, APOA is proved to meet the demand of actual production.

3. As for the rule of reverse regulation, from the aspect of water quantity regulation, Xiaoxuan reservoir should strategically store and discharge the inflow from Pankou reservoir and try to discharge flow in the mode where its generator units are in the high-efficiency zone, so that this portion of water can be utilized more efficiently; from the aspect of water head regulation, the increase in Xiaoxuan's generated energy brought by raising its operation water level is greater than Pankou's hydroenergy loss caused by the fall in its power generation efficiency. Therefore, to raise Xiaoxuan's operation water level is beneficial to power generation of the whole cascade hydropower stations.

It should be noted that we need to balance water quantity regulation with water head regulation. Excessive elevation of Xiaoxuan's operation water level may result in lack of space for water quantity regulation and even water abandonment of Xiaoxuan. Thus, it will be the focus of future research to study how high Xiaoxuan's operation water level should be raised, considering the uncertainty of Pankou's output.

Author Contributions: C.J. conceived and designed the experiments; H.Y. and X.Y. performed the experiments; H.Y. and J.W. analyzed the data; H.Y., J.W. and R.L. wrote the paper.

Acknowledgments: This study was financially supported by the National Key Research and Development Program of China (2016YFC0402208) and the National Natural Science Foundation of China (51709105). The authors are grateful to the reviewers for their comments and valuable suggestions.

References

1. Mo, L.; Lu, P.; Wang, C.; Zhou, J. Short-term hydro generation scheduling of Three Gorges-Gezhouba cascaded hydropower plants using hybrid MACS-ADE approach. *Energy Convers. Manag.* **2013**, *76*, 260–273. [CrossRef]
2. Liao, X.; Li, Y.; Zhu, X.; Zhao, K.; Bai, T.; Gao, K. Research on peak regulation capacity of Xiaolangdi-Xixiayuan cascade hydropower station. *Int. J. Hydroelectr. Energy* **2011**, *29*, 29–31.
3. Dickel, K.; Klein, P.; Nielinger, A. Rehabilitation of the weir at Ahausen reservoir. *Hydropower Dam* **2001**, *4*, 94–97.

4. Tospornsampan, J.; Kita, I.; Ishii, M.; Kitamura, Y. Optimization of a multiple reservoir system operation using a combination of genetic algorithm and discrete differential dynamic programming: A case study in Mae Klong system, Thailand. *Paddy Water Environ.* **2005**, *3*, 29–38. [CrossRef]
5. Richter, B.D.; Thomas, G.A. Restoring environmental flows by modifying dam operations. *Ecol. Soc.* **2007**, *12*, 181–194. [CrossRef]
6. Bai, T.; Chang, J.; Fang, J.; Huang, Q.; Zhu, X.; Liao, X. Study on short-term peak load of Xiaolangdi and Xixiayuan cascade hydropower stations. *J. Hydroelectr. Eng.* **2012**, *31*, 83–88.
7. Cai, Z.; Zhang, A.; Zhang, J. Optimized control on water level between Three Gorges dam and Gezhouba dam and its benefit analysis. *J. Hydroelectr. Eng.* **2010**, *29*, 10–17.
8. Shawwash, Z.K.; Siu, T.K.; Russell, S.O.D. The B.C. hydro short term hydro scheduling optimization model. *IEEE Trans. Power Syst.* **2000**, *15*, 1125–1131. [CrossRef]
9. Habert, J.; Ricci, S.; Papa, E.L.; Thual, O.; Piacentini, A.; Goutal, N.; Jonville, G.; Rochoux, M. Reduction of the uncertainties in the water level-discharge relation of a 1D hydraulic model in the context of operational flood forecasting. *J. Hydrol.* **2016**, *532*, 52–64. [CrossRef]
10. Li, A.; Wang, L.; Lin, W.; Ji, C. Application of immune particle swarm optimization algorithm to short-term optimal dispatch of cascade hydropower stations. *J. Hydraul. Eng.* **2008**, *39*, 426–432.
11. Ge, X.; Zhang, L.; Shu, J.; Xu, N. Short-term hydropower optimal scheduling considering the optimization of water time delay. *Electr. Power Syst. Res.* **2014**, *110*, 188–197. [CrossRef]
12. Shang, Y.; Liu, R.; Li, T.; Zhang, C.; Wang, G. Transient flow control for an artificial open channel based on finite difference method. *Sci. China Technol. Sci.* **2011**, *54*, 781–792. [CrossRef]
13. Wang, N.; Wang, C.; Du, S.; Le, H.; Li, J. Application of dynamic relationship between water level and discharge in hydrodynamic flood forecasting model. *Int. J. Hydroelectr. Energy* **2014**, *32*, 68–72.
14. Bellman, R. Dynamic programming. *Science* **1966**, *153*, 34–37. [CrossRef] [PubMed]
15. Saadat, M.; Asghari, K. Reliability improved stochastic dynamic programming for reservoir operation optimization. *Water Resour. Manag.* **2017**, *31*, 1795–1807. [CrossRef]
16. Chu, W.; Yang, T.; Gao, X. Comment on "High-dimensional posterior exploration of hydrologic models using multiple-try DREAM (ZS) and high-performance computing" by Eric Laloy and Jasper A. Vrugt. *Water Resour. Res.* **2014**, *50*, 2775–2780. [CrossRef]
17. Ji, C.; Li, C.; Wang, B.; Liu, M.; Wang, L. Multi-stage dynamic programming method for short-term cascade reservoirs optimal operation with flow attenuation. *Water Resour. Manag.* **2017**, *31*, 4571–4586. [CrossRef]
18. Mei, Y. Dynamic programming model without Markovin property of cascade reservoirs operation and its application. *Adv. Water Sci.* **2000**, *11*, 194–198.
19. Prasanchum, H.; Kangrang, A. Optimal reservoir rule curves under climatic and land use changes for Lampao Dam using Genetic Algorithm. *KSCE J. Civ. Eng.* **2017**, *22*, 351–364. [CrossRef]
20. Chang, L.C. Guiding rational reservoir flood operation using penalty-type genetic algorithm. *J. Hydrol.* **2008**, *354*, 65–74. [CrossRef]
21. Kumar, D.N.; Reddy, M.J. Multipurpose reservoir operation using particle swarm optimization. *J. Water Resour. Plan. Manag.* **2007**, *2006*, 192–201. [CrossRef]
22. Xie, W.; Ji, C.; Wu, Y.; Li, X. Particle swarm optimization based on cultural algorithm for flood optimal scheduling of hydropower reservoir. *J. Hydraul. Eng.* **2010**, *41*, 452–457.
23. Afshar, M.H.; Ketabchi, H.; Rasa, E. Elitist continuous ant colony optimization algorithm: Application to reservoir operation problems. *Int. J. Civ. Eng.* **2006**, *4*, 274–285.
24. Jalali, M.R.; Afshar, A.; Marino, M.A. Multi-reservoir operation by adaptive pheromone re-initiated ant colony optimization algorithm. *Int. J. Civ. Eng.* **2007**, *5*, 284–301.
25. Chen, J.; Wu, Y.; Hu, B.; Cheng, T. Improved mathematical model of minimum energy consumption for load distribution of cascade hydropower plants. *Automat. Electr. Power Syst.* **2017**, *41*, 155–160.
26. Jiang, Z.; Qin, H.; Ji, C.; Feng, Z.; Zhou, J. Two dimension reduction methods for multi-dimensional dynamic programming and its application in cascade reservoirs operation optimization. *Water* **2017**, *9*, 634. [CrossRef]
27. Suen, J.P.; Eheart, J.W. Reservoir management to balance ecosystem and human needs: Incorporating the paradigm of the ecological flow regime. *Water Resour. Res.* **2016**, *42*, 178–196. [CrossRef]
28. Yuan, B. Short-Term Optimization Dispatching of Cascaded Hydropower Daily-Optimized Dispatching for YILIHE Cascaded Hydropower. Master's Thesis, Kunming University of Science and Technology, Kunming, China, 2012.

29. Barbetta, S.; Moramarco, T.; Perumal, M. A Muskingum-based methodology for river discharge estimation and rating curve development under significant lateral inflow conditions. *J. Hydrol.* **2017**, *554*, 216–232. [CrossRef]
30. Papazafiriou, Z.G. Linear and nonlinear approaches for short-term runoff estimations in time-invariant open hydrologic systems. *J. Hydrol.* **1976**, *30*, 63–80. [CrossRef]
31. Othman, F.; Naseri, M. Reservoir inflow forecasting using artificial neural network. *Int. J. Phys. Sci.* **2011**, *6*, 433–440.
32. Coulibaly, P.; Anctil, F.; Bobée, B. Daily reservoir inflow forecasting using artificial neural network with stopped training approach. *J. Hydrol.* **2000**, *230*, 244–257. [CrossRef]
33. Cybenko, G. Approximation by superposition of sigmoidal function. *Math. Control Signal.* **1989**, *2*, 304–314. [CrossRef]
34. Hornik, K.; Stinchcombe, M.; White, H. Multilayer feedforward networks are universal approximators. *Neural Netw.* **1989**, *2*, 359–366. [CrossRef]
35. Shen, H.; Wang, Z.; Gao, C.; Qin, J.; Yao, F.; Xu, W. Determining the number of BP neural network hidden layer units. *J. Tianjin Univ. Technol.* **2008**, *24*, 13–15.
36. Liu, J.; Wang, J. Exploration of empirical calculation method about tailwater level in Gezhouba hydropower station. *Hydropower Autom. Dam Monit.* **2011**, *35*, 77–80.
37. Tang, H.; Chen, S.; Zhao, Y. Model and algorithm of short-term optimal scheduling of Three Gorges cascaded hydropower stations. *Int. J. Hydroelectr. Energy* **2008**, *26*, 133–136.
38. Lucas, N.J.D.; Perera, P.J. Short-term hydroelectric scheduling using the progressive optimality algorithm. *Water Resour. Res.* **1985**, *21*, 1456–1458. [CrossRef]
39. Zhong, P.; Zhang, J.; Xu, B.; Zhang, M. Daily optimal operation model of cascade reservoirs considering delay of flow propagation. *J. Hydroelectr. Eng.* **2012**, *31*, 34–38.

5

An Integrated Method for Accounting for Water Environmental Capacity of the River–Reservoir Combination System

Fen Zhao [1], Chunhui Li [1,*], Libin Chen [2] and Yuan Zhang [3,*]

1 Ministry of Education Key Lab of Water and Sand Science, School of Environment, Beijing Normal University, Beijing 100875, China; zhaofen@mail.bnu.edu.cn
2 College of Resources and Environment, Yangtze University, Wuhan 430100, China; lbchen@yangtzeu.edu.cn
3 Sate Key Laboratory of Environmental Criteria and Risk Assessment, Chinese Research Academy of Environmental Sciences, Beijing 100012, China
* Correspondence: chunhuili@bnu.edu.cn (C.L.); zhangyuan@craes.org.cn (Y.Z.) (C.L.)

Abstract: The security of drinking water is a serious issue in China and worldwide. As the backup source of drinking water for the Changde City in China, the Huangshi Reservoir suffers from the threat of eutrophication due to the water quality of the reservoir ecosystem being affected by the tributaries that carry Non-Point Source (NPS) pollutants. The calculation of the water environmental capacity (WEC) can provide a scientific basis for water pollution control, which refers to the maximum amount of pollutants that the water can accommodate. In this paper, according to the hydrological characteristics of the river–reservoir combination system, a one-dimensional (1-D) water quality model and the Environmental Fluid Dynamics Code (EFDC) model were chosen to calculate the water environmental capacity of each functional zone in this basin. The quantity control of pollution from the tributaries was conducted based on the combined results of the water environmental capacity calculation from the EFDC model and a one-dimensional (1-D) river water quality model. The results show that total water environmental capacity of the tributaries included a chemical oxygen demand (COD) of 421.97 tons; ammonia nitrogen (NH_3-N) of 40.99 tons; total nitrogen (TN) of 35.94 tons; and total phosphorus (TP) of 9.54 tons. The water environmental capacity of the Huangshi Reservoir region accounts for more than 93% of the total capacity. The reduction targets of the major pollutants in the Huangshi Reservoir and its four major input rivers, which are, namely, the Bamao River, the Longtan River, the Fanjiafang River, and the Dongtan River, have been determined to achieve the water quality objectives for the reservoir in 2020 and 2025. The results will be helpful for the local water quality management and will provide a valuable example for other similar water source reservoirs.

Keywords: drinking water resources; water environmental capacity (WEC); Environmental Fluid Dynamics Code (EFDC) model; the Huangshi Reservoir

---

1. Introduction

In recent years, instead of flood control and irrigation, drinking water supplementation has become the primary purpose of reservoirs in China [1]. The issue of drinking water security plays a decisive role in the national economy and social wellbeing in China [2,3]. With the enormous development of aquaculture, water pollution in reservoirs has become a serious problem with the water quality of drinking water source directly influencing people's health, which is related to the economic development and stability of the general social situation.

The water environmental capacity (WEC) refers to the maximum amount of pollutants that the water can accommodate under the designed hydrological conditions and the specified environmental objectives without destroying its own function [4,5]. Generally, the calculation of WEC can provide a scientific basis for water pollution control as it provides a baseline for the total pollutant reduction needed. Previous studies mainly focused on various methods of calculating WEC [6–18], including analytical methods, dynamic mechanism model methods, and the calculation of the water environment capacity in a specific area [11,14,16,19–23]. Most case studies mainly use a river [12,15,16,19,23], lake, and reservoir [6,14,20,22] as objects in China, but there has been less research on the WEC of a certain river–reservoir combination system.

In addition, there are many models and equations available to calculate the WEC of rivers, lakes, and reservoirs. For this study, it is important to select the proper water quality simulation model. In calculating WEC, the use of three-dimensional modeling for simulating the water ecosystems has been used extensively and successfully by many studies [24–26]. They can simulate and predict the overall water quality of the water bodies for evaluating the WEC. In this sense, they are superior to the traditional field monitoring methods, which are mainly based on the field data monitored at limited locations [27–31].

Many studies have indicated that the inflow discharges will affect the variations of pollutants or nutrients in reservoirs [3–5,32–34]. As the backup drinking water source reservoir for Changde City in China, the Huangshi Reservoir, with multiple tributaries, is currently suffering from the threat of eutrophication, with the influence of pollutants or nutrients from the tributaries of Huangshi Reservoir having not been taken into consideration. It is necessary to calculate the WEC of the river–reservoir combination system. The purpose of this study is to simulate the WEC and control the amount of pollution in a drinking water reservoir with multiple tributaries using a widely used model, the environmental fluid dynamic model (EFDC), coupled with a 1-D convection equation.

This research paper is arranged as follows: firstly, the EFDC model is applied to analyze the changing trends of the water quality in the Huangshi Reservoir and its tributaries. Secondly, the 1-D water quality model and the EFDC model were chosen to calculate the WEC of each functional zone of the river–reservoir combination system. Finally, the quantity control of pollution of the multiple tributaries is analyzed based on the combined results of the WEC calculation.

2. Study Area

2.1. Study Area

The Huangshi Reservoir is a large-scale water conservancy project, which mainly provides the benefits of irrigation, flood control, power generation, and a habitat for fish, among others. It is located in the Huangshi town, Hunan Province, China. The reservoir is mainly fed by the convergences of the Bamao River, the Longtan River, the Fanjiafang River, the Liujiaxi River, and the Dongtan River. The boundary of the reservoir and the tributaries is shown in Figure 1, with the main characteristics of the tributaries shown in Table 1. The weather is relatively moderate with an annual average temperature of 16.8 °C. The annual average precipitation is about 1465.2 mm, the annual mean evaporation is about 1284.2 mm, and the relative humidity is 77%. The total storage capacity of the reservoir is 6.02×10^9 m^3, and the effective storage capacity is 3.38×10^9 m^3. The watershed has a total area of 494.34 km^2, with forests being the major land use type that accounts for 85.12% of the total basin area.

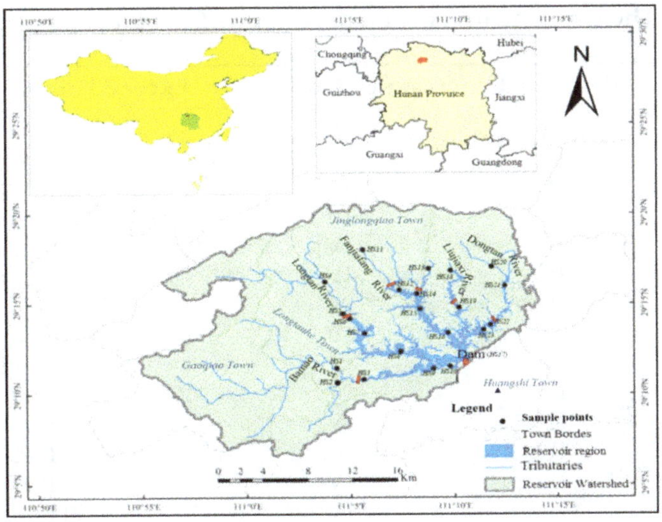

Figure 1. The location of the Huangshi Reservoir and the sample points of water quality.

Table 1. The features of the tributaries.

Rivers	Control Area (km^2)	Length of Reaches (km)	Water Quality Objectives (River Water Quality Standards) *	Flow (m^3/s)
Bamao River	102.9	10	II	1.429
Longtan River	154.4	14.2	II	1.2769
Fanjiafang River	61.2	4	II	0.6726
Dongtan River	40.3	5.8	II	0.4089

* Note: for the values of Water quality objectives, see also Table 2.

Table 2. The surface water environmental quality standard of China [35] (units: mg/L).

Indicators	II	III	IV	V
COD	15	20	30	40
NH_3-N	0.5	1	1.5	2
TP	0.1 (lake and reservoir 0.025)	0.2 (lake and reservoir 0.05)	0.3 (lake and reservoir 0.1)	0.4 (lake and reservoir 0.2)
TN	0.5	1	1.5	2

2.2. *Water Quality Assessment in the Huangshi Reservoir Basin*

A total of 23 representative sampling points was set up in the Huangshi Reservoir area for the assessment of the water quality and hydrobiology (shown in Figure 1), from which two points were monitored in the Bamao River (HS1, HS2), two points were monitored in the Longtan River (HS4, HS5), two points were monitored in the Dongtan River (HS20, HS21), and only one point was monitored in both the Fanjiafang River (HS13) and the Liujiaxi River (HS18). The rest of the 15 points were monitored in the reservoir region. According to the monitored water quality data of the sampling points in the years of 2013–2015 for the surface water environmental quality standard of China (GB3838–2002), the water quality of the Huangshi Reservoir is in accordance with the requirements of drinking water quality. Namely, the water quality of Huangshi Reservoir is classified as degree II and III (shown in Table 2). However, the main water quality indexes, including the chemical oxygen demand (COD), the ammonia nitrogen (NH_3-N), the total nitrogen (TN), and the total phosphorus (TP), exceed the environmental quality standards (shown in Tables 3 and 4), with the Longtan River exceeding the standards most compared to with the other four tributaries.

Table 3. The water quality evaluation result of the Huangshi Reservoir.

Sections	Standard	Water Quality	Exceed Factors and Multiples *
HS7	II	III	TN (0.92)
HS8	II	V	TN (3.96)
HS9	II	IV	TN (1.32)
HS16	II	IV	TN (1.18), TP (0.88)
HS17	II	IV	TN (1.70)
HS19	II	IV	TN (0.28), TP (2.36)
HS3	II	V	TN (2.86), TP (0.04)
HS6	II	IV	TN (0.88)
HS10	II	V	TN (2.38)
HS12	II	V	TN (2.06), TP (0.76)
HS14	II	IV	TN (1.54), TP (0.48)
HS15	II	V	TN (1.46), TP (4.68)
HS22	II	IV	TN (1.42), TP (0.92)
HS23	II	IV	TN (1.08), TP (0.08)

* Note: exceed multiples = (measured value − standard value)/standard value.

Table 4. The water quality evaluation result of the tributaries.

Tributaries	Sections	Standard	Water Quality	Exceed Factors and Multiples *
Bamao River	HS1		V	TN (0.90)
	HS2		IV	TN (0.49)
Longtan River	HS4		V	TN (0.90)
	HS5		V	TN (0.88), TP (1.54)
Fanjiafang River	HS11	III	V	TN (0.70)
	HS13		V	TN (0.71), TP (0.60)
Liujiaxi River	HS18		V	TN (0.86)
Dongtan River	HS20		V	TN (0.79)
	HS21		V	TN (0.53)

* Note: exceed multiples = (measured value − standard value)/standard value.

The monitored data of the sampling points set up in the Huangshi Reservoir (Figure 1), the water quality based on the assessment of the hydrobiology with the species, and the quantity measures are shown in Figure 2.

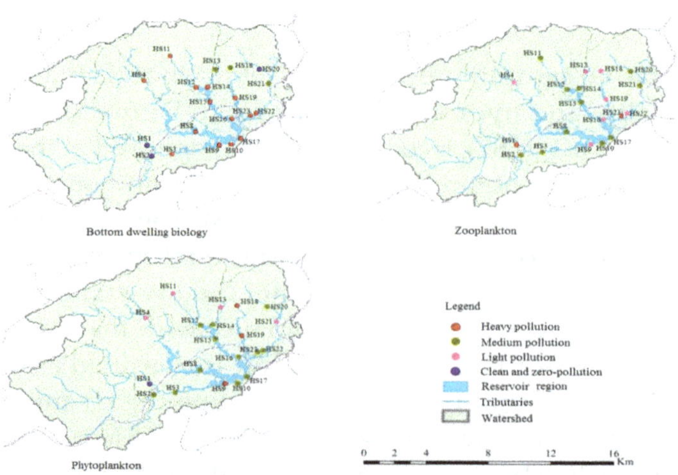

Figure 2. The assessment results of water quality.

The tributaries were seriously affected by agricultural non-point source (NPS) pollution and living pollution, while the water quality in the vicinity of the reservoir was worse. The water quality of the reservoir is increasingly worsening, with the levels of nitrogen and phosphorus nutrient salts exceeding the standard.

3. Methods

3.1. Water Environmental Capacity of the River-Reservoir Combination System

For fulfilling the purpose of water pollutant control in a drinking water reservoir with multiple tributaries, the paper selected 2014 as the status quo year, and 2020 and 2025 as the forecast years.

In order to simulate the WEC and control the amount of pollution of the Huangshi Reservoir with multiple tributaries in 2020 and 2025, we developed an integrated method to evaluate the WEC of the river–reservoir combination system, which consisted of the coupling of the 1-D model and the EFDC model. The technical route is as follows (Figure 3).

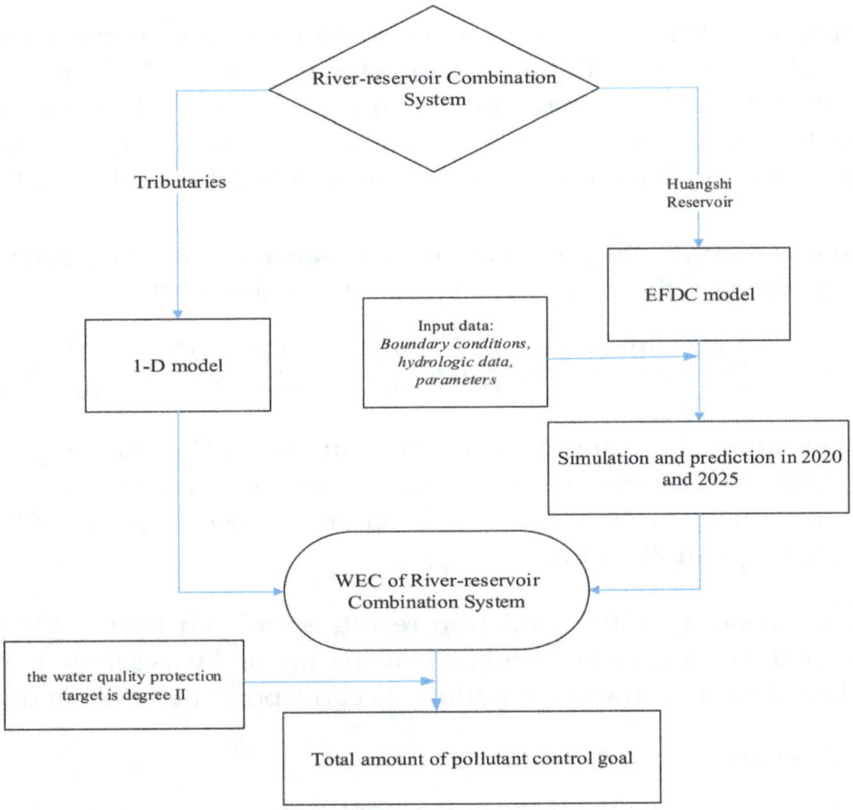

Figure 3. The technical route.

3.2. Water Environmental Capacity Model of the Tributaries

The equation for calculating the WEC was derived based on a 1-D convection-diffusion equation, with the adopted model expressed as follows [5]:

$$M = (C_S - C_0 exp(-kL/u))exp(kL/2u)Q_r \quad (1)$$

where M is the water environmental capacity per unit time (t); C_s is the pollutant concentration of the tail section in mg/L; C_0 is the pollutant concentration of the initial section in mg/L; k is the pollutant degradation coefficient in 1/s; and L is the length of the reach. A plurality of sewage outlets in the reach can be generalized as a centralized outfall and the outfall was located at the middle point of the reach. Therefore, it is equivalent to a concentrated point source, of which the self-purification length is half the length of the river reach. This means that the self-purification length is $L/2$ when the length of the river reach is L. In addition, u is the average flow velocity under the design flow in m/s; and Q_r is the design flow.

In this paper, the climatic characteristics have been considered [36] and the degradation coefficients of COD, NH_3-N, TN, and TP are 0.5, 0.25, 0.03, and 0.011/day, respectively.

3.3. *Water Environmental Capacity Model of the Huangshi Reservoir*

In this study, the EFDC model developed by Hamrick [37] was applied to the Huangshi Reservoir. This model is an integrated modeling system that has been successfully applied to many surface water systems, including rivers, estuaries, lakes, and coastal bays [12,38–40]. The EFDC model includes four major sub-models: (1) a hydrodynamic model; (2) a sediment transport model; (3) a toxic model; and (4) a water quality model. This simulation is mainly based on the EFDC hydrodynamic, convective diffusion, and first-order degradation modules of the water quality model.

The water quality module not only considers the wind direction, wind speed, and evaporation effect on the flow field and the pollutant transport but also the influence of the distribution characteristics of different types of aquatic plants and waves on the bottom stress. Its numerical simulation can estimate C, N, P, and other forms of nutrients, as well as a variety of algae. It is a strong model of water quality and movement and can accurately reflect the level of pollution in water.

In the water quality module, the governing mass-balance equation for each water quality state variable and the temperature of the water was expressed as follows [41]:

$$\frac{\partial C}{\partial t}+\frac{\partial(uC)}{\partial x}+\frac{\partial(vC)}{\partial y}+\frac{\partial(wC)}{\partial z}=\frac{\partial}{\partial x}\left(K_x\frac{\partial C}{\partial x}\right)+\frac{\partial}{\partial y}\left(K_y\frac{\partial C}{\partial y}\right)+\frac{\partial}{\partial z}\left(K_z\frac{\partial C}{\partial z}\right)+S_c \quad (2)$$

where C is the concentration of a water quality state variable or C is the temperature of the water; u, v and w are the velocity components in the x, y, and z directions, respectively; K_x, K_y and K_z are the turbulent diffusion coefficients in the x, y, and z directions, respectively; and SC is the internal and external sources and sinks per unit volume.

By combining the water quality simulation results simulated by the EFDC model with the environmental target of the Huangshi Reservoir, an inverse method was selected to calculate the WEC, which is able to deduce the concentration of pollutants corresponding to water quality targets.

3.3.1. Hydrological Parameters

The protection of the water quality of the Huangshi Reservoir is strict. The lowest water monthly average flow rate and the lowest water level condition in 2013 and 2014 were chosen to calculate the environmental capacity of the reservoir.

3.3.2. Boundary Conditions

Flow boundary conditions: There are six boundary conditions in the simulated water area, which are formed by the inflow of the five rivers using the flow boundary, and the dam using the water surface elevation.

Meteorological boundary conditions: The data were obtained from the Meteorological Data Sharing Service Network of China, including rainfall, evaporation, solar radiation, cloud coverage, atmospheric pressure, air temperature, air humidity, wind speed, wind direction, and so on.

Water quality boundary conditions: The concentrations of COD, NH_3-N, TN, and TP in the five main input rivers were used.

3.3.3. Parameters of Model

Due to the lack of longtime serial hydrological data and water quality monitoring data of inflows and outflows of the Huangshi Reservoir, the parameters of the hydrodynamic model of the Huangshi Reservoir referenced some classical cases of the hydrodynamic model in South China. The degradation coefficients of COD, NH_3-N, TN, and TP of the Huangshi Reservoir referenced the research report of comprehensive improvement of the water environment and ecological protection plan in the Shanxi Reservoir area. The degradation coefficients of COD, NH_3-N, TN, and TP are 0.03 /day, 0.03/day, 0.03/day, and 0.01/day, respectively. The parameters of the water quality model are shown in Table 5.

Table 5. The parameters of the water quality model.

Parameters	Roughness	Eddy Viscosity Coefficient	Diffusion Coefficient	COD	NH_3-N	TN	TP
Coefficient/Degradation coefficient	0.02 m	1 m^2/s	0.2	0.03/day	0.03/day	0.03/day	0.01/day

3.3.4. Model Verification

In this study, four main water quality variables were simulated, namely COD, NH_3-N, TN, and TP. The observed data from 2013 to 2014 were used for the verification of the water quality model.

The error statistics of the model verification are listed in Table 6. It is clear that the simulated values agree well with the measured values, which demonstrates that the established water quality model is able to simulate the process of water quality in the Huangshi Reservoir.

Table 6. The error statistics of the model verification (units: mg/L).

Characteristic Values	COD	NH_3-N	TN	TP
Compare numbers	12	12	12	12
Average observed	7.78	0.12	1.20	0.15
Average simulated	6.92	0.10	1.15	0.10
Average Error	−0.85	0.01	−0.04	0.02
Relative Error	20.29	28.10	12.99	157.42
Absolute Error	1.68	0.02	0.14	0.03
Root Mean Square Error	2.18	0.03	0.18	0.03
Relative Root Mean Square Error	47.28	70.38	61.56	139.98
Nash-Sutcliffe coefficient	−2.05	−11.89	−4.05	−50.35

4. Results

4.1. Water Quality Simulation and Prediction

The Huangshi Reservoir is the backup source of drinking water for Changde City. There are no industrial districts in the area and water quality is affected mainly by agriculture NPS pollutants, domestic sewage, and aquaculture pollutants.

Based on the pollutant loads discharged into the river in the status quo year of 2014 (shown in Table 7), the increase of the population and the development of urbanization, and the scale of eco-agriculture and the increase of aquaculture, the pollutant loads in the tributaries of the Huangshi Reservoir in 2020 and 2025 have been estimated according to the Report of Huangshi Reservoir protection planning from the Chinese Academy of Environmental Sciences (shown in Table 8). The values have also been converted into concentrations of pollutants in the tributaries of the Huangshi Reservoir, which is shown in Table 9.

Table 7. The pollutant quantity inlets into the river in 2014 (units: tons/year).

Area	COD	NH_3-N	TN	TP
	Inflow	Inflow	Inflow	Inflow
Bamao River	148.66	17.96	27.04	4.859
Longtan River	108.67	13.87	19.65	2.678
Fanjiafang River	116.8	14.6	21.36	3.34
Dongtan River	89.292	12.23	16.23	1.573
Reservoir region	197.64	97.54	179.03	44.375
Total amount	661.062	156.2	263.31	56.825

Table 8. The predicted values of the pollutant loads in the tributaries of the Huangshi Reservoir in 2020 and 2025 (units: tons/year).

Year	Pollutants	Bamao River	Longtan River	Fanjiafang River	Dongtan River
2020	COD	764.73	482.54	428.13	239.27
	NH_3-N	70.56	45.65	39.14	22.43
	TN	136.2	86.16	75.46	41.72
	TP	59.98	38.576	32.24	16.446
2025	COD	981.21	604.68	549.45	298.1
	NH_3-N	86.48	54.60	48.21	27.08
	TN	172.16	106.32	95.67	51.42
	TP	72.42	44.424	39.21	19.544

Table 9. The predicted values of the pollutant concentrations in the tributaries of the Huangshi Reservoir in 2020 and 2025 (units: tons/years).

Year	Pollutants	Bamao River	Longtan River	Fanjiafang River	Dongtan River
2020	COD	16.97	11.98	20.18	18.55
	NH_3-N	1.57	1.13	1.85	1.74
	TN	3.02	2.14	3.56	3.24
	TP	1.33	0.96	1.52	1.28
2025	COD	21.77	15.02	25.90	23.12
	NH_3-N	1.92	1.36	2.27	2.10
	TN	3.82	2.64	4.51	3.99
	TP	1.61	1.10	1.85	1.52

The hydraulic conditions adopted the average level of the status year and have no prediction for the future hydraulic conditions, that is, the hydraulic conditions almost stay constant in the model. The annual average wind speed used in Huangshi Reservoir was 2 m/s, the dominant wind direction was northeast, and the annual average evaporation was 1160 mm. The hydrological calculation of WEC was based on the driest monthly flow in 2013 and 2014. The pollutant loads estimated in 2020 and 2025 are allocated on a monthly basis.

The water quality of the Huangshi Reservoir without reduction measures simulated the predicted values of the inflow for COD, NH_3-N, TN, and TP in 2020 and 2025. The results are displayed in Figures 4 and 5, which respectively show the following:

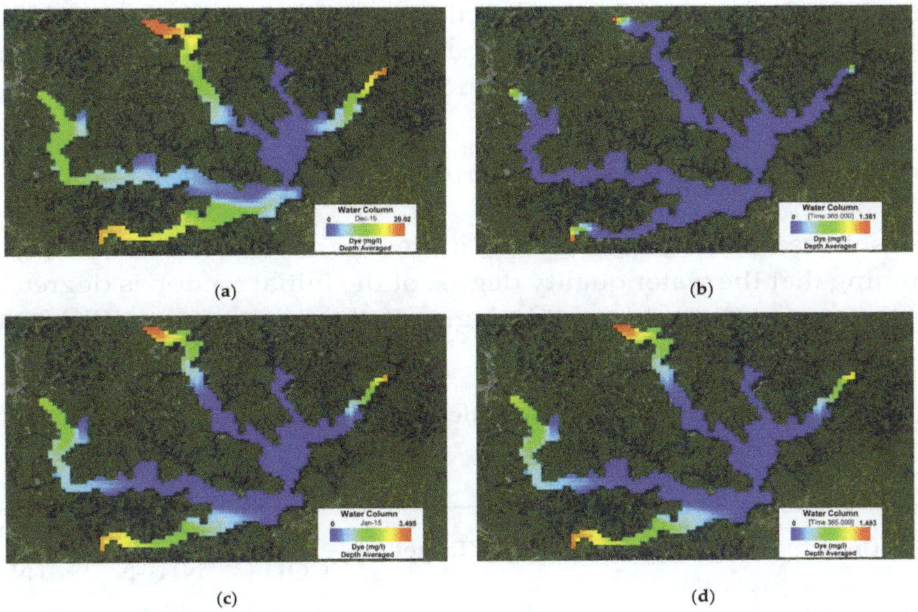

Figure 4. The simulated concentration distribution of COD, NH_3-N, TN, and TP in the Huangshi Reservoir in 2020. ((**a**) COD, (**b**) NH_3-N, (**c**) TN and (**d**) TP).

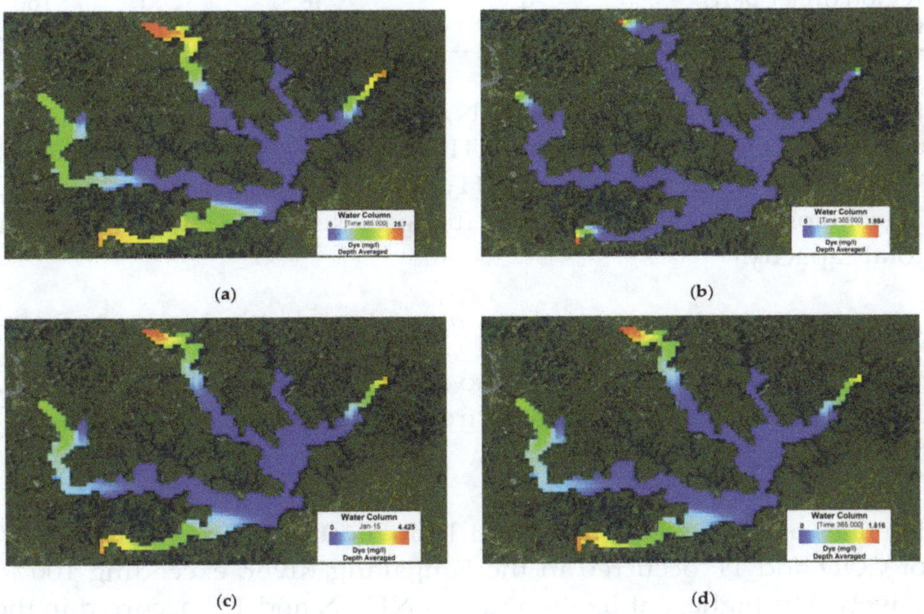

Figure 5. The simulated concentration distribution of COD, NH_3-N, TN, and TP in Huangshi Reservoir in 2025. ((**a**) COD, (**b**) NH_3-N, (**c**) TN and (**d**) TP).

(a) In 2020, the affected areas are the incoming parts of the tributaries, based on the concentration distribution of COD and NH_3-N. The main reservoir area is less affected. TN has a bigger area of the effect. In addition, the influence of TP from the tributaries on the water quality of the reservoir is confined to the inner part of the tributaries in the reservoir.

(b) In 2025, the influences of the tributaries on the water quality of the reservoir are confined to the inner parts of the tributaries in the reservoir, but they have bigger areas of effect on the concentrations of COD, NH_3-N, TN, and TP compared to those in 2020.

4.2. *Water Environmental Capacity of the River–Reservoir Combination System*

As the Huangshi Reservoir is a water source protection zone, the water quality protection target is degree II. Assuming that the water quality degree of the initial section is degree I when the water quality degree of the reservoir control point is degree II, the results of the WEC for the tributaries of each section are shown in Table 10.

Table 10. The water environmental capacities of the Huangshi Reservoir and tributaries (units: tons/year).

Sections	Water Quality Target	Water Environmental Capacity			
		COD	NH_3-N	TN	TP
Bamao River	II	185.46	15.37	13.57	3.60
Longtan River	II	131.32	13.87	12.11	3.22
Fanjiafang River	II	58.56	7.33	6.38	1.69
Dongtan River	II	46.63	4.42	3.88	1.03
Sum of rivers		421.97	40.99	35.94	9.54
Reservoir region (inner part)	II	4850	200	160	38
Total amount		5271.97	240.99	195.94	47.54

The highest WEC values of the COD, NH_3-N, TN, and TP were in the Bamao River, while the lowest WEC values of the COD, NH_3-N, TN, and TP were in the Dongtan River. The total WEC of the tributaries included a COD of 421.97 tons; an NH_3-N of 40.99 tons; a TN of 35.94 tons; and a TP of 9.54 tons. Thus, it is obvious that the WEC of the Huangshi Reservoir region mainly accounts for more than 93% of the total capacity.

4.3. *Total Amount of Pollutant Control Goal Based on the Water Environmental Capacity*

Based on the pollutant loads discharged into the river in 2014, the utilized peculiarities of the reservoir and tributaries, as well as the water environmental capabilities, were estimated and shown in Table 11.

The utilization rates of COD, NH_3-N, TN, and TP in most tributaries were over 100%. The highest utilization rate of COD and TP occurred in the Fanjiafang River, exceeding 100% by 2 times and 1.98 times, respectively. The highest utilization rate of NH_3-N and TN occurred in the Dongtan River, exceeding 100% by 2.77 times and 4.18 times, respectively.

The utilization rates of COD and NH_3-N in the reservoir region were 13% and 65%, respectively, while there was still a certain WEC. However, the maximum utilization rate of TN and TP exceeded 100% by 1.34 times and 1.2 times, respectively. These were mainly caused by the fish cage culture and the internal release of pollutants. Furthermore, this rate had a significant relationship with the high degree of eutrophication in the reservoir region.

Table 11. The water environmental capacity utilizations of the Huangshi Reservoir in 2014 (units: tons/year).

Area	COD			NH_3-N			TN			TP		
	Inflow	Capacity	Utilization Ratio (%)	Inflow	Capacity	Utilization Ratio (%)	Inflow	Capacity	Utilization Ratio (%)	Inflow	Capacity	Utilization Ratio (%)
Bamao River	148.66	185.46	80	17.96	15.37	117	27.04	13.57	199	4.859	3.6	135
Longtan River	108.67	131.32	83	13.87	13.87	100	19.65	12.11	162	2.678	3.22	83
Fanjiafang River	116.8	58.56	199	14.6	7.33	199	21.36	6.38	335	3.34	1.69	198
Dongtan River	89.292	46.63	191	12.23	4.42	277	16.23	3.88	418	1.573	1.03	153
Reservoir region	197.64	4850	4	97.54	200	49	179.03	160	112	44.375	38	117
Total amount	661.062	5271.97	13	156.2	240.99	65	263.31	195.94	134	56.825	47.54	120

According to the pollutant inflows into the reservoir and the WEC results, the pollutant control target can be determined based on the WEC. The water pollutant reductions of the Huangshi Reservoir region as well as the Bamao River, Longtan River, Fanjiafang River, and Dongtan River—which are the 4 major tributaries into the reservoir—have been determined for 2020 and 2025 (Tables 12 and 13).

Table 12. The pollutants reduction targets of the Huangshi Reservoir in 2020 (units: tons/year).

Area	COD	NH_3-N	TN	TP
	Reductions	Reductions	Reductions	Reductions
Bamao River	10.39	6.59	20.93	3.358
Longtan River	7.18	2.73	12.16	0.457
Fanjiafang River	92.74	9.89	20.55	3.062
Dongtan River	63.55	9.93	15.61	1.055
Reservoir region	−4560.48 *	−58.02 *	84.81	23.233
Total amount	−4386.62	−28.88	154.06	31.165

* Note: the symbol of "−"delegates' available capacity without reduction.

Table 13. The pollutants reduction targets of the Huangshi Reservoir in 2025 (units: tons/year).

Area	COD	NH_3-N	TN	TP
	Reductions	Reductions	Reductions	Reductions
Bamao River	86.64	13.78	33.19	6.17
Longtan River	61.22	8.19	20.68	1.898
Fanjiafang River	150.44	16.06	29.96	4.961
Dongtan River	103.48	14.23	21.89	1.827
Reservoir region	−4445.73 *	−2.79 *	175.45	44.122
Total amount	−4043.95	49.47	281.17	58.978

* Note: the symbol of "−" delegates' available capacity without reduction.

The pollutants in the tributaries of the Huangshi Reservoir will need a large number of cuts both in 2020 and 2025, with respect to the degree II national surface water quality standards. The largest COD reduction will occur in the Fanjiafang River, with approximately 61.3% and 71.9% of the inflow needing to be cut to meet the water quality objectives, respectively. Tributaries will need a total reduction of 173.8 to 401.8 tons per year, with the residual WEC for COD being approximately 4560.5 tons per year in the reservoir region.

The largest NH_3-N reduction will occur in the Dongtan River and the Fanjiafang River in 2020 and 2025. More than 69.2% of the pollutant inflow will need to be cut to meet the water quality objectives in 2020, which corresponds to an increase of 7.1% in 2025. Tributaries will need a total reduction of 29.14 and 52.2 tons per year, respectively, with the residual WEC for NH_3-N needing to be approximately 4560.5 and 2.79 tons per year in the reservoir region, respectively.

The largest TN reduction will occur in the Dongtan River in 2020 as well as in the Dongtan River and the Fanjiafang River in 2025. More than 80.1% and 74.59% of the inflow will need to be cut to meet the water quality objectives, respectively. The TN of the entire basin will need to be cut by 44.02% and 58.9%, respectively, which corresponds to an increase of 14.88% in 2025.

The largest TP reduction will occur in the Dongtan River and the Fanjiafang River in 2020 and 2025. More than 64.4% of the inflow will need to be cut to meet the water quality objectives, which corresponds to an increase of 10.19% in 2025. The inflow into the reservoir will be significantly larger than the WEC of the reservoir. The TP of the reservoir area will need to be reduced by 37.9% and 53.73%, respectively, with the entire basin needing to be cut by 39.6% and 55.37%, respectively.

Water pollution control is still an arduous task in all four tributaries and the reservoir region. The pollutants significantly exceed the state standards for drinking water sources in China, especially in the reservoir region, which will induce or aggravate the risk of water eutrophication.

5. Conclusions

This study constructed an integrated method to simulate the WEC of the Huangshi river–reservoir combination system, which coupled a 1-D convection-dispersion model with the EFDC model. It was further applied to study the quantity control of pollution of the input rivers based on the results of the WEC calculation and the degree II national surface water quality standards.

The highest WEC is in the Bamao River, while the lowest WEC is in the Dongtan River. Summarizing the WEC of the tributaries, the COD is 421.97 tons, NH_3-N is 40.99 tons, TN is 35.94 tons, and TP is 9.54 tons.

The utilization rates of COD, NH_3-N, TN, and TP in most of the tributaries are over 100%. The utilization rates of COD and NH_3-N in the reservoir region are 13% and 65%, respectively, and there is still a certain WEC. However, the loads of TN and TP exceed the maximum utilization rates by 1.34 times and 1.2 times, respectively. This is mainly caused by the fish cage culture and the internal release of pollutants. Furthermore, this has a significant relationship with the high degree of eutrophication in the reservoir region.

The pollutants in the tributaries of the Huangshi Reservoir will need a large number of inflow cuts to meet the water quality objectives in 2020 and 2025. The pollutants significantly exceed the state standard, especially in the reservoir region. Therefore, the reduction and control of pollutants are the main objectives.

To exploit the WEC completely, rationally, and continuously, a balance point needs to be found between environmental concerns and the social and economic development. As the study area is the backup source of drinking water for Changde City, the calculation of WEC can enable decision makers to determine load reductions and allocations. Various measures of controlling the total amount of pollutants can be applied to realize the improvement of the water quality. This perspective can be referred as a beneficial supplement for a single study of rivers or lake.

Additionally, in calculating WEC, the uncertainty analysis should be taken into consideration. In a practical river—reservoir combination system there are many uncertainties in terms of variable information while calculating WEC. More variable information will result in a more flexible WEC.

Acknowledgments: This research was supported by the National Water Pollution Control and Treatment Science and Technology Major Project (2017ZX07301-001), National Key Research and Development Program (2017YFC0404505). We would like to extend special thanks to the editor and the anonymous reviewers for their valuable comments in greatly improving the quality of this paper.

Author Contributions: F.Z. conceived and designed the experiments; C.L. performed the experiments; L.C. and F.Z. analyzed the data; L.C. and Y.Z. contributed reagents/materials/analysis tools; F.Z. wrote the paper.

References

1. Liu, W.; Chen, W.; Hsu, M.H. Using a three-dimensional particle tracking model to estimate the residence time and age of water in a tidal estuary. *Comput. Geosci.* **2011**, *37*, 1148–1161. [CrossRef]
2. Su, Q.; Qin, H.; Fu, G. Environmental and ecological impacts of water supplement schemes in a heavily polluted estuary. *Sci. Total Environ.* **2014**, *472*, 704–711. [CrossRef] [PubMed]
3. Shen, Y.; Wang, J.; Zheng, B.; Zhen, H.; Feng, Y.; Wang, Z.; Yang, X. Modeling study of residence time and water age in Dahuofang Reservoir in China. *Sci. China Phys. Mech. Astron.* **2011**, *54*, 127–142. [CrossRef]
4. Zhang, Y. The Development of Basic Concept of Water Environmental Capacity. *Res. Environ. Sci.* **1992**, *5*, 59–61. (In Chinese)
5. Yang, Z.; Cheng, C.; Tan, X.; Cheng, R.; Zhong, M.A. Analysis of water environmental capacity of Guanting Reservoir and its upstream basin. *J. Arid N.A. Res. Environ.* **2015**, *29*, 164–166. (In Chinese)
6. Gao, G.Q. Water environmental carrying capacity calculation and protection measures on Hefang Reservoir. *Appl. Mech. Mater.* **2011**, *1446*, 2537–2540. [CrossRef]
7. Han, L.X.; Yan, F.F.; Peng, H.; Gao, J.J.; Pan, M.M. Methods for calculation of water environment capacity of small and medium river channels. *Adv. Mater. Res.* **2013**, *2115*, 2745–2750. [CrossRef]

8. Gang, Z.; Kun, L.; Guo, F.U.; Guang-Jun, M. Calculation method of river water environmental capacity. *J. Hydraul. Eng. ACSE* **2014**, *45*, 227–234.
9. Li, Y.; Zhao, L.; Li, Y. Index system of water environment capacity promotion in Jilin province in China through ways of structure, engineering and supervision. *Adv. Mater. Res.* **2015**, *3848*, 1155–1159. [CrossRef]
10. Yang, J.F.; Lei, K.; Khu, S.; Meng, W.; Qiao, F. Assessment of water environmental carrying capacity for sustainable development using a coupled system dynamics approach applied to the Tieling of the Liao River Basin, China. *Environ. Earth Sci.* **2015**, *73*, 5173–5183. [CrossRef]
11. Li, K.; Zhang, L.; Li, Y.; Zhang, L.J.; Wang, X.L. A three-dimensional water quality model to evaluate the environmental capacity of nitrogen and phosphorus in Jiaozhou Bay, China. *Mar. Pollut. Bull.* **2015**, *91*, 306–316. [CrossRef] [PubMed]
12. Tao, Y.; Chen, Y.X.; Zhao, X.L.; Yang, L.B.; Lei, K.; Zhou, G. Analysis of seasonality difference on water environmental capacity of Ashi River based on EFDC. *Environ. Eng.* **2017**, *35*, 66–68. (In Chinese)
13. Zhou, X.Y.; Lei, K.; Meng, W.; Khu, S.; Zhao, J.; Wang, M.N.; Yang, J.F. Space–time approach to water environment carrying capacity calculation. *J. Clean. Prod.* **2017**, *149*, 302–312. [CrossRef]
14. Zhang, X.; Han, L.X. A water environmental capacity calculation method of the shallow lake based on water quality response coefficient. *Sichuan Environ.* **2017**, *36*, 47–49. (In Chinese)
15. Wu, J.F.; Gao, Y.; Xie, F.; Chen, L.N. Research on water environment capacity by two-dimensional hydrodynamic and quality model. *Environ. Sci. Manag.* **2017**, *42*, 53–55. (In Chinese)
16. Gu, J.; Li, Z.Y.; Mao, X.D.; Hu, C.F.; Kuang, C.P.; Zhang, W.L. Study of cod environmental capacity in coastal waters of Beidaihe in summer. *Mar. Environ. Sci.* **2017**, *36*, 683–685. (In Chinese)
17. Wu, H.; Yang, X. Calculation of water environmental capacity based on control unit of Shuangtaizi River. *Environ. Sci. Manag.* **2017**, *42*, 61–63. (In Chinese)
18. Chinh, L.V.; Hiramatsu, K.; Harada, M.; Lan, T.T. Estimation of water environment capacity in the cau river basin, vietnam using the streeter-phelps model. *J.- Fac. Agric. Kyushu Univ.* **2017**, *62*, 163–169.
19. Huang, S.L.; Zhang, Y.; Li, Q.; Xu, D.S. Research on water environmental capacity of urban river: A case study of Tuohe River in Suzhou city, northern Anhui province. *Adv. Mater. Res.* **2012**, *1479*, 867–870. [CrossRef]
20. Fan, L.L.; Sha, H.F.; Pang, Y. Water environmental capacity of Lake Taihu. *J. Lake Sci.* **2012**, *24*, 693–697.
21. Xia, Y.; Ma, C.M.; Liu, X.J.; Liu, X.Y.; Deng, Q.L. Comprehensive carrying capacity assessment of water environment in Zhengzhou city. *Appl. Mech. Mater.* **2013**, *2574*, 390–395. [CrossRef]
22. Yan, B.Y.; Xing, J.S.; Tan, H.R.; Deng, S.P.; Tan, Y.N. Analysis on water environment capacity of the Poyang Lake. *Procedia Environ. Sci.* **2011**, *10*, 2754–2759.
23. Liu, Q.; Jiang, J.; Jing, C.; Qi, J. Spatial and seasonal dynamics of water environmental capacity in mountainous rivers of the Southeastern Coast, China. *Int. J. Environ. Res. Public Health* **2018**, *15*, 99. [CrossRef] [PubMed]
24. Missaghi, S.; Hondzo, M. Evaluation and application of a three-dimensional water quality model in a shallow lake with complex morphometry. *Ecol. Model.* **2010**, *221*, 1512–1525. [CrossRef]
25. Missaghi, S.; Hondzo, M.; Melching, C. Three-dimensional lake water quality modeling: Sensitivity and uncertainty analyses. *J. Environ. Qual.* **2013**, *42*, 1684–1698. [CrossRef] [PubMed]
26. Missaghi, S. Three Dimensional Water Quality Modeling in a Shallow Lake with Complex Morphometry; Implications for Cool Water Fish Habitat under Changing Climate. Ph.D. Thesis, University of Minnesota, Minneapolis, MN, USA, July 2014.
27. Heung, W.; Hu, B.Q. Application of interval clustering approach to water quality evaluation. *J. Hydrol.* **2013**, *491*, 1–12.
28. Heung, W.; Hu, B.Q. Application of improved extension evaluation method to water quality evaluation. *J. Hydrol.* **2014**, *509*, 539–548.
29. Peche, R.; Esther, R.E. Development of environmental quality indexes based on fuzzy logic. A case study. *Ecol. Indic.* **2012**, *23*, 555–565. [CrossRef]
30. Ju, H.C.; Yoo, S.H. Using the fuzzy set theory to developing an environmental impact assessment index for a thermal power plant. *Qual. Quant.* **2014**, *48*, 673–680. [CrossRef]
31. Chen, Q.W.; Wu, W.Q.; Blanckaert, K.; Ma, J.F.; Huang, G.X. Optimization of water quality monitoring network in a large river by combining measurements, a numerical model and matter-element analyses. *J. Environ. Manag.* **2012**, *110*, 116–124. [CrossRef] [PubMed]
32. Jordan, T.E.; Weller, D.E.; Correll, D.L. Sources of nutrient inputs to the Patuxent River estuary. *Estuaries* **2003**, *26*, 226–243. [CrossRef]

33. Shen, J.; Haas, L. Calculating age and residence time in the tidal York River using three-dimensional model experiments. *Estuarine Coast. Shelf Sci.* **2004**, *61*, 449–461. [CrossRef]
34. Li, Y.; Acharya, K.; Yu, Z. Modeling impacts of Yangtze River water transfer on water ages in Lake Taihu, China. *Ecol. Eng.* **2011**, *37*, 325–334. [CrossRef]
35. Ministry of Environmental Protection of the people's Republic of China. The surface water environmental quality standard of China. GB3838-2002.
36. Chen, L.B.; Yang, Z.F.; Liu, H.F. Numerical simulations of spread characteristics of toxic cyanide in the Danjiangkou Reservoir in China under the effects of dam cooperation. *Math. Probl. Eng.* **2014**, *2014*. [CrossRef]
37. Hzamrick, H.M. *A Three-Dimensional Environmental Fluid Dynamics Computer Code: Theoretical and Computational Aspects*; The College of William and Mary, Virginia Institute of Marine Science: Williamsburg, VA, USA, 1992.
38. Caliskan, A.; Elci, S. Effects of selective withdrawal on hydrodynamics of a stratified reservoir. *Water Resour. Manag.* **2009**, *23*, 1257–1273. [CrossRef]
39. Wang, C.; Shen, C.; Wang, P.F.; Qian, J.; Hou, J.; Liu, J.J. Modeling of sediment and heavy metal transport in Taihu Lake, China. *J. Hydrodyn. Ser. B* **2013**, *25*, 379–387. [CrossRef]
40. Zhang, C.X.; You, X.Y.; Zhao, S.M. Application of the EFDC model to waterfront planning: A case study in the Tianjin Harbor Economic Zone, China. *Eng. Appl. Comp. Fluid Mech.* **2014**, *8*, 1–13. [CrossRef]
41. Park, K.; Kuo, A.Y. *A Three-Dimensional Hydrodynamic Eutrophication Model: Description of Water Quality and Sediment Process Sub-Models*; The College of William and Mary, Virginia Institute of Marine Science: Williamsburg, VA, USA, 1995.

6

Decline of N and P Uptake in the Inner Protection Zone of a Terminal Reservoir during Inter-Basin Water Transfers

Shengtian Yang [1], Juan Bai [2,3], Changsen Zhao [1], Hezhen Lou [1,*], Zhiwei Wang [2,4], Yabing Guan [2], Yichi Zhang [2], Chunbin Zhang [2] and Xinyi Yu [2]

1 College of Water Sciences, Beijing Normal University, Beijing 100875, China; yangshengtian@bnu.edu.cn (S.Y.); hzjohnson@gmail.com (C.Z.)

2 State Key Laboratory of Remote Sensing Science, Faculty of Geographical Science, Beijing Normal University, Beijing Key Laboratory for Remote Sensing of Environment and Digital Cities, Beijing 100875, China; baijuanaction@163.com (J.B.); wangzhiweisci@gmail.com (Z.W.); guanyabing2013@163.com (Y.G.); yichizhang@mail.bnu.edu.cn (Y.Z.); zhangchunbin1102@163.com (C.Z.); yuxy0106@sina.com (X.Y.)

3 College of Geographical Science, Shanxi Normal University, Linfen 041004, China

4 Key Laboratory of Radiometric Calibration and Validation for Environmental Satellites, National Satellite Meteorological Center, China Meteorological Administration, Beijing 100081, China

* Correspondence: louhezhenbj@163.com

Abstract: Inter-basin water transfer projects are designed to relieve water scarcity around the world. However, ecological problems relating to reductions in protection zone functions can occur during inter-basin transfers. This paper uses the largest inter-basin water transfer project in the world, namely, the South-to-North Water Transfer Project (SNWTP) in China, as an example to analyze the variation of Miyun Reservoir's inner protection zone functions when water is transferred. Specifically, a riparian model (RIPAM) coupled with remote sensing data were used to calculate the nitrogen (N) and phosphorus (P) losses due to plant uptake, and these results were validated by in situ survey data. Then, correlations between water levels and N and P removal were analyzed. The results show that water table disturbances resulting from elevated water levels strongly influence the growth of plants and have obvious negative impacts on N and P removal in the inner protection zone. With the implementation of the middle route of the SNWTP, the water level of Miyun will rise to 150 m in 2020, and subsequently, the total net primary productivity (NPP) could decline by more than 40.90% from the level in 2015, while the N and P uptake could decline by more than 53.03% and 43.49%, respectively, from the levels in 2015, according to the modeling results. This will lead to declines in the inner protection zone's defense effectiveness for N and P interception and increases in risks to the security of water resources. The results of this study provide useful knowledge for managing the defense function of the terminal reservoir's inner protection zone and for ensuring that water quality is maintained during the diversion process.

Keywords: protection zone; nutrient uptake; NPP; South-to-North Water Transfer Project; Miyun Reservoir

1. Introduction

Inter-basin water transfers can relieve water scarcity and are a popular water-resource topic. To date, many famous projects have been built to alleviate uneven water resource distributions, such as the Central Arizona Project, the Colorado River Projects and California North-to-South Water Transfer Project in the United States [1,2], the Siberian Rivers Diversion in Russia [3], the Snowy Mountains Scheme in Australia [4], the National River Linking Project in India [5], and the South-to-North Water Transfer Project (SNWTP) in China [6]. However, inter-basin water transfers usually are associated with complex environmental changes [7–11], just as Davies et al. [7] proposed: "Any transfer of water within or between basins will have physical, chemical, hydrological and biological implications for both donor and recipient systems, as well as for their estuaries and local marine environments." Therefore, the environmental responses to inter-basin water transfers should be studied.

Numerous ecological changes during inter-basin water transfers have been proposed, including the introduction of nonindigenous species, secondary salinization, disturbances in water chemistry and quality, changes in hydrologic regimes, and alterations of habitats [8,11–18]. The major concerns of these studies were about the protection of water source regions and transfer canals so as to ensure water safety. However, more attention should be devoted to the problems faced in receiving areas of inter-basin water projects because artificial projects can raise the water level of a receiving area and flood its drinking water protection zones, which may result in a decline of surface water quality.

Drinking water protection zones have been established by law to maintain the quality of surface waters and aquatic ecosystems [19,20]. They consist of the following three parts: inner, middle, and outer protection zones [20]. The inner protection zone is located near the drinking water source, and its boundaries are closely related to the riparian buffer boundaries. The middle protection zone runs along the inner protection zone and covers certain buffer areas. The outer protection zone covers the remaining catchment area not covered by the inner protection zone and middle protection zone [19,20]. Among them, the inner protection zone is known for its ability to control non-point source pollution (NPS) by intercepting and retaining excess particulates and dissolved nutrients originating from the surrounding uplands. Plants in the inner protection zone play a significant role in nitrogen (N) and phosphorus (P) cycling, and they can reduce N and P concentrations in soils [21,22].

The absorption of nutrients by plants largely depends upon vegetation production and other environmental factors [23]. Even small changes in the water level may induce detectable changes in vegetation production and associated N and P removal efficiencies [24,25]. When an inter-basin water transfer project is implemented, significant amounts of water will be diverted to the terminal reservoir. The elevated water level will flood the current riparian buffer, and a new one will form with a new vegetation circumstance. Such variations have the potential to render the riparian buffer ineffective at intercepting NPS and may change the inner protection zone from a nutrient sink into a nutrient source [26]. However, though studies on the responses of riparian vegetation to flooding have been done in natural lakes and rivers with some success [27,28], the impacts of inter-basin water transfers on the N and P uptake functions of a terminal reservoir's protection zone have not yet been investigated.

In this study, we use the SNWTP in China, which is the largest inter-basin transfer project in the world, as an example to assess the impacts of water transfers on N and P uptake in a terminal reservoir's inner protection zone. The total N and P losses due to plant uptake were calculated by the combined use of an N and P model and remote sensing data, and then, the relationship between the water level and N and P removal was examined. Finally, the potential risks posed by water level fluctuations and associated impacts on N and P uptake in the inner protection zone were evaluated.

2. Materials and Methods

2.1. Study Area

The SNWTP in China consists of three routes (i.e., the eastern, middle, and western routes) (Figure 1); each route covers a distance of more than 1000 km, and in total they deliver 44.8 billion m^3 of fresh water per year to northern China. It is the largest inter-basin transfer scheme in the world; therefore, it is often taken as an example of inter-basin transfers and has been intensively studied in terms of the ecological and environmental consequences since it was proposed [12,13,29–34].

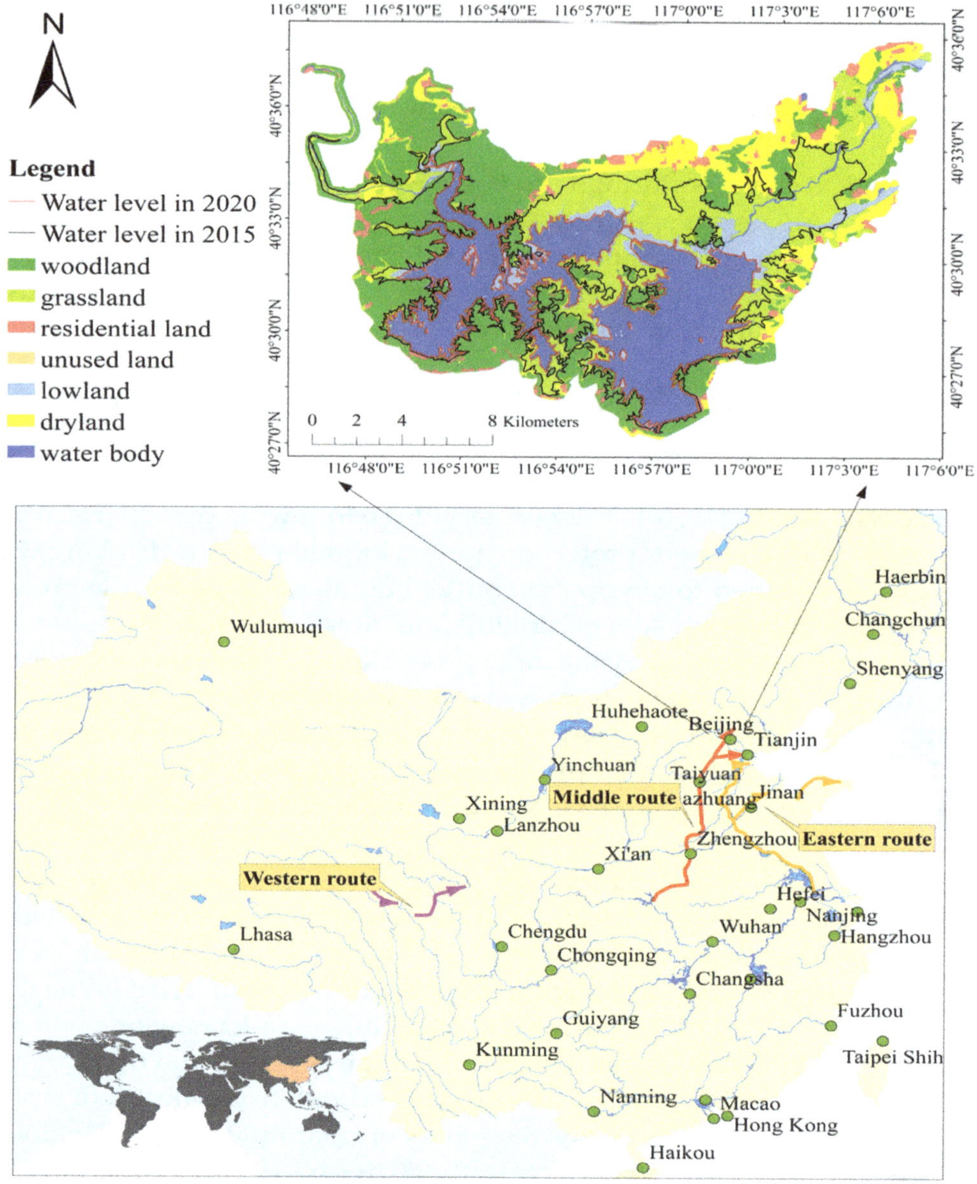

Figure 1. Schema of the South-to-North Water Transfer Project in China and land use information for the terminal Miyun Reservoir of the middle route.

Along the middle route, the Miyun Reservoir serves as the terminal point and stores surplus water transferred from the Dangjiangkou Reservoir. The middle route of the SNWTP was implemented in 2014. Surplus water is lifted 133 m in elevation so that it can enter the Miyun Reservoir through the Jingmi diversion canal with the help of nine pumping stations. The Miyun Reservoir is situated in the mountainous area of Miyun County, and the reservoir extends from 116°47′ E to 117°05′ E and from 40°26′ N to 40°35′ N (Figure 1). It has a large water surface area of 188 km^2 and a watershed area of 15,788 km^2; the total storage capacity is 4.375 billion m^3. The study area features a temperate continental monsoon climate, with a mean annual precipitation amount of 628 mm and mean annual air temperature of 11.3 °C. There is an uneven seasonal distribution of precipitation, and more than 70% of the annual precipitation occurs between June and September. The main land use types are dry land, woodland, grassland, water bodies, flood land, and residential land (Figure 1). The inner protection zone along the Miyun Reservoir mainly consists of the area encompassed by the highway surrounding the reservoir and other nearby water zones as designated by the government.

The Miyun Reservoir is the main surface drinking water source for Beijing, and it has played an important role in the social and economic development of the capital [35]. Issues related to water quality in the reservoir have been receiving increasing attention. The upper catchment of the Miyun Reservoir has exhibited eutrophication trends in recent years [36]. Thus, there is an urgent need to strengthen the management of water protection zones. The water level of the Miyun Reservoir was 136 m in 2016. With the implementation of the middle route of the SNWTP, the capacity of the terminal reservoir will be increased from 1.2 billion m^3 to 2 billion m^3 in 2020, which will raise the water level to 150 m. The transferred water will inundate part of the original inner protection zone. Vegetation in the remaining part will likely be affected by the rising water level, and these changes could pose water quality risks for the terminal reservoir.

2.2. Simulation of N and P Uptake

The effects of water level variation on the inner protection zone's defense functions were evaluated through modeling the N and P removal of vegetation during five years (1999, 2000, 2009, 2010, and 2015). The Miyun Reservoir began to receive transferred water in 2015, so we chose 2015 as the current year. In the following years, its water level will increase to 150 m, which is almost equal to the water level in 1999; hence, we chose 1999 as the objective year for reference. Besides, the difference of water levels between 1999 and 2000 was quite significant (149 m in 1999 and 142 m in 2000), as well as the difference between 2009 and 2010 (137 m in 2009 and 134 m in 2010), so these years were chosen as typical years. N and P losses were calculated by coupling remote sensing data with a RIPArian Model (RIPAM, EcoHAT, Beijing, China). RIPAM couples simple process-based modules and remote sensing data to estimate soil denitrification rates, soil nitrogen emissions, phosphorus removed by soil erosion, and nitrogen and phosphorus removal by vegetation uptake [37–40]. The framework of RIPAM that relates to N and P uptake is shown in Figure 2, and its main functions and parameters are listed in Table 1.

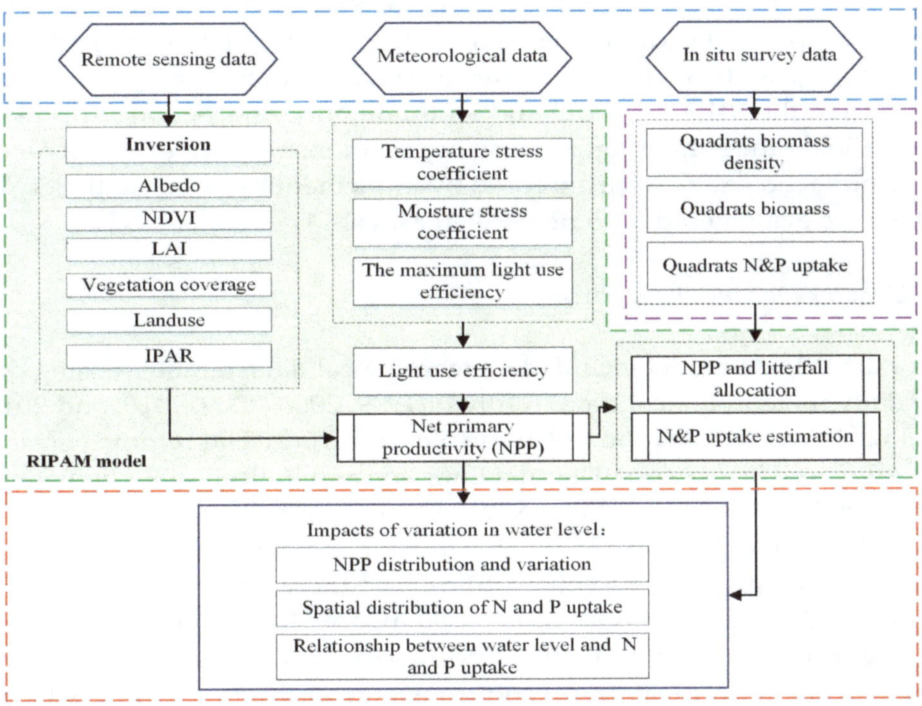

Figure 2. Framework of the N and P uptake simulation. IPAR: the amount of photosynthetically active radiation intercepted by plants; LAI: leaf area index; NPP: net primary productivity; NDVI: normalized difference vegetation index.

Table 1. Functions and parameters related to N and P uptake in the RIPAM model.

No.	Module Name	Equations	Reference
1	NPP simulation	$NPP = GPP - R_a$ $GPP = \varepsilon \times IPAR \times f_1(T) \times f_2(\beta)$ $R_a = GPP \times (7.825 + 1.145T_a)/100$	GLO-PEM model [41]
2	NPP allocation	$F_B = dB_L/d_t + dF_{lit}/d_t$ $B_L = LAI/SLA$ $R_B = K_{ra}(NPP - F_B)$ $W_B = NPP - F_B - R_B)$	ForNBM model [42]
3	Nutrient absorption	$X_{uptake} = \min(X_{avail}, X_{dem})$ $X_{dem} = (1 - K_{retra}) \times folX \times F_B + X_W \times W_B + X_r \times R_B$ $X_{dem} = NPP \times X_{cont}$	ForNBM model [42,43]
4	Litterfall decomposition	$F_{lit} = \begin{cases} a_{fh}F_B, & T_{air} > T_{fall}, \text{ deciduous species} \\ F_B, & T_{air} < T_{fall}, \text{ deciduous species} \\ a_{fs}F_B, & \text{evergreen species} \end{cases}$	ForNBM model [42]

Annotation: NPP: net primary productivity (g C·m^{-2}); GPP: gross primary productivity (g C·m^{-2}); R_a: primary production for autotrophic respiration (g C·m^{-2}); IPAR: the amount of photosynthetically active radiation intercepted by plants (MJ·m^{-2}); ε: the real efficiency of radiation utilization (g C·MJ^{-1}); $f_1(T)$ is a function of temperature stress; $f_2(\beta)$ is a function of water stress; F_B: the NPP foliage obtains (g C·m^{-2}); B_L: the foliage biomass (g·m^{-2}); F_{lit}: the leaf litter (g C·m^{-2}); t: the time step (month); SLA: specific leaf area; R_B: the NPP roots obtain (g C·m^{-2}); K_{ra}: return coefficient of leaf nutrients, constant; W_B: the NPP wood obtains (g C·m^{-2}); X_{uptake}: the nutrient uptake rate (g·m^{-2}); X_{avail}: the amount of available nutrients (X = N or P) (g·m^{-2}); X_{dem}: the nutrients demanded by biomass growth (g·m^{-2}); K_{retra}: the dimensionless foliage nutrient re-translocation proportion (Kretra = 0.35 for this study); folX: the nutrient (X = N or P) concentration of the foliage (g·g^{-1}); X_W: the nutrient concentration of the wood (g·g^{-1}); Xr: the nutrient concentration of the roots (g·g^{-1}); a_{fh}: the coefficient for deciduous species; afs: the coefficient for evergreen species; T_{air}: air temperature (°C); and T_{fall}: the threshold temperature for deciduous species litter (°C).

In other published papers [37–40], the RIPAM model was tested in the Guanting Reservoir, and the simulations of net primary productivity (NPP), NPP allocation, and nutrient uptake all had considerable accuracy during the growing season (take N uptake rates for example, its R^2 value is 0.95). Since the Miyun Reservoir and Guanting Reservoir are located within a distance of less than 100 km and their geographical conditions and climate are similar [36,39], we adopted the model parameters of the Guanting Reservoir to apply to the Miyun Reservoir [37,39,44]. The database required included remote sensing data, meteorological data, and in situ survey data. Remote sensing data and meteorological data were used to drive the RIPAM model and simulate NPP and N and P uptake. In situ survey data were used to validate the simulation results. The relationship between water levels and N and P uptake in the inner protection zone were analyzed by implementing zonal statistics. Then the impacts of N and P uptake were evaluated due to rising water levels.

2.3. Multi-Source Data for Driving the RIPAM

Input data for the RIPAM model included meteorological data, remote sensing data, and soil data (Table 2). (1) The daily meteorological data in July of 1999, 2000, 2009, 2010, and 2015 were collected from the national basic weather station at Miyun Reservoir; (2) The remote sensing data included Landsat data and Advanced Spaceborne Thermal Emission and Reflection Radiometer (ASTER)–Global Digital Elevation Model (GDEM) data. Five time series of Landsat images in July or August of 1999, 2000, 2009, 2010, and 2015 were collected to infer land surface parameters. ASTER–GDEM was used to calculate the contour map with intervals of 10 m, and the whole study area was zoned into the following four belts: 130–140 m, 140–150 m, 150–160 m, and 160–170 m; (3) Soil data were derived from the Second National Soil Survey results and field investigations in the study area. All data were reproduced at 30 m × 30 m spatial resolutions and projected to Albers by using the World Geodetic System-84 (WGS 84) datum. Some data were resolved by using ENVI4.8 (Harris Corporation, Melbourne, FL, USA), ArcGIS10.1 (Environmental Systems Research Institute, Redlands, CA, USA), or programs written in IDL (Interactive Data Language).

Table 2. Information for the multi-source data.

Data Type	Data Name	Resolution	Data Source
Meteorological data	Precipitation, wind speed, air pressure, air temperature, daily max/min temperature, relative humidity, sunshine duration, sun radiation, surface temperature	-	National basic weather station (http://data.cma.cn/)
Remote sensing data	Digital elevation model (DEM)	30 m	ASTER-GDEM (http://www.gscloud.cn/)
	Albedo, NDVI, LAI, land use, vegetation coverage, land surface temperature (LST)	30 m	Landsat (http://www.gscloud.cn/)
Soil data	Bulk density, soil texture, soil nutrient content (TN, NO_3-N, NH_4-N), pH	-	Second National Soil Survey field investigations

2.4. In Situ Survey Data for Validation

In practice, biomass is seldom estimated over a large scale. In the present paper, biomass data in situ were collected at the times of Landsat image capture (from 5 to 20 July 2015) within the inner protection zone. The whole zone can be classified into the following three phases: reservoir, terrestrial, and water phase. The water phase is located between the low-water mark and the reservoir bank, the riparian phase contains the zone between the low-water mark and high-water mark, and the remainder is the terrestrial phase. Several 90 m × 90 m quadrants were selected in each phase, and Figure 3 shows some examples of the in situ quadrants. The location of each quadrant was recorded with the Global Positioning System (GPS) to integrate field and Landsat data. The harvesting method [45] was used to measure the average biomass density of grasses and shrubs. In each quadrant, we selected 36 sample areas of 1 m × 1 m for grass or nine sample areas of 5 m × 5 m for shrubs, and the average biomass density was calculated. Forest biomass was estimated using diameter–height models [46] based on the sample trees. In this study, 39 quadrants were selected for the analysis.

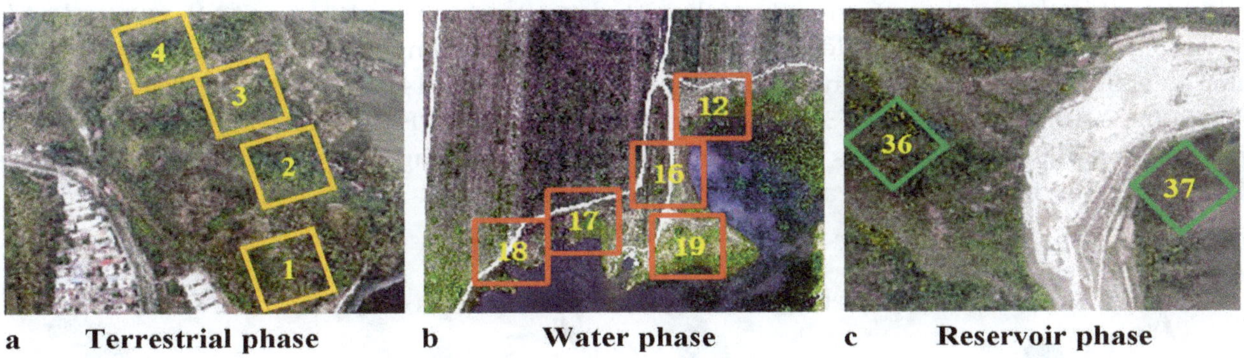

Figure 3. Examples of the in situ quadrants in different phases. (**a**) Quadrants selected in the terrestrial phase; (**b**) quadrants selected in the water phase; (**c**) quadrants selected in the reservoir phase.

The average biomass density of each quadrant was calculated, as shown in Table 3. Additionally, the data were converted into N and P uptake by using a simple ratio coefficient to validate the accuracy of the RIPAM model (see Section 3.4 for more details).

Table 3. Information for the biomass density of each quadrant. Quadrants 1–11 were selected in the terrestrial phase; quadrants 12–20 were selected in the water phase; quadrants 21–39 were selected in the reservoir phase.

Quadrant ID	Biomass Density ($g \cdot m^{-2}$)	Quadrant ID	Biomass Density ($g \cdot m^{-2}$)	Quadrant ID	Biomass Density ($g \cdot m^{-2}$)
1	1222.13	14	172.35	27	4699.23
2	1114.83	15	188.35	28	3139.40
3	773.61	16	171.57	29	4275.75
4	1321.03	17	192.58	30	13,825.23
5	1375.40	18	184.78	31	10,893.78
6	1185.17	19	177.98	32	7510.89
7	1332.65	20	142.32	33	12,186.66
8	1453.43	21	5291.21	34	10,159.57
9	859.52	22	3275.53	35	11,741.42
10	1001.63	23	1796.98	36	12,123.08
11	1488.09	24	2043.25	37	5930.64
12	172.73	25	1528.09	38	10,361.92
13	152.52	26	2679.09	39	7192.59

3. Results

3.1. NPP Distribution and Variation

The NPP of Miyun's inner protection zone during the growing season of 1999, 2000, 2009, 2010, and 2015 was simulated by the RIPAM model. According to previous research in this study area [47], it is changes in water level but not climate or land use that influence the condition of vegetation in the inner protection zone. The total NPP of the whole zone amounted to 8454.96 t in 1999, 11,922.37 t in 2000, 17,271.43 t in 2009, 13,653.78 t in 2010, and 14,306.82 t in 2015. The spatial distribution and average NPP value during these five years are shown in Figure 4a–f. Higher values were mainly located in the northern part of the study area, where the values were more than 100 g $C \cdot m^{-2}$, while the NPP of the vegetation surrounding the reservoir was relatively low. The numerical growth of the NPP was significant in the northwestern and southwestern parts of the study area. As the water level rose, the average NPP value showed a decreasing trend, with values ranging from 98.66 g $C \cdot m^{-2}$ in 2015 at a water level of 133 m to 75.10 g $C \cdot m^{-2}$ in 1999 at a water level of 149 m (Figure 4f). The average water level in 2009, 2010, and 2015 was 135 m, and the average total NPP was 102.02 g $C \cdot m^{-2}$. The average water level in 1999 and 2000 was 145 m, and the corresponding average total NPP decreased to 83.40 g $C \cdot m^{-2}$, which amounts to a decline rate of 18.26%.

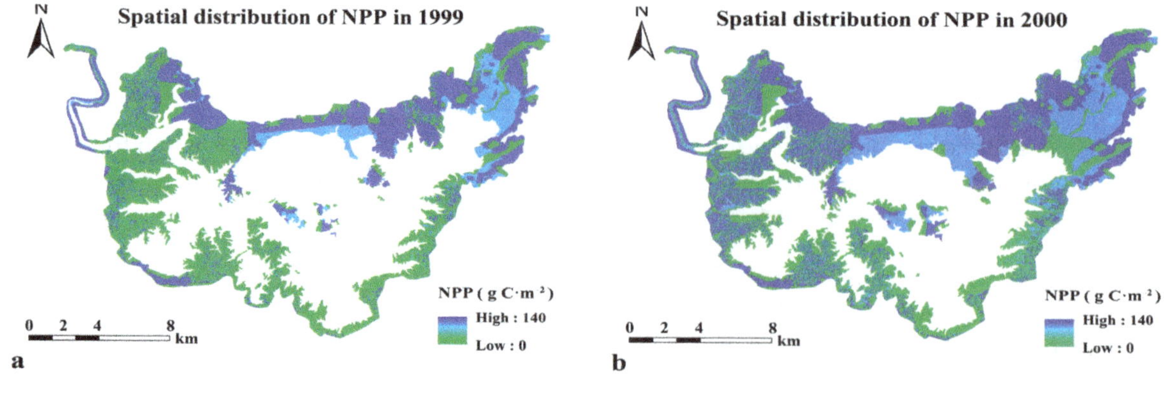

Figure 4. *Cont.*

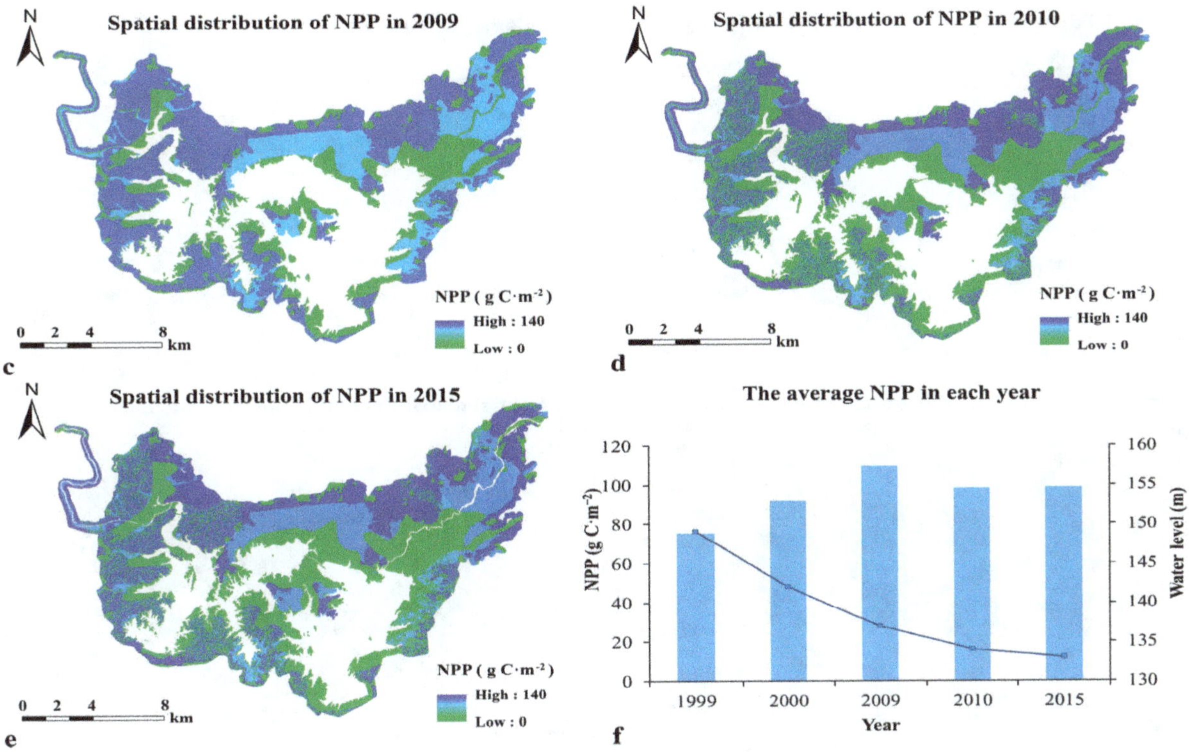

Figure 4. Spatial distribution of NPP in Miyun's inner protection zone during different years.

3.2. Spatiotemporal Variation of N and P Uptake

The RIPAM model was used to simulate the spatial distribution of N and P uptake in the inner protection zone during the growing season of 1999, 2000, 2009, 2010, and 2015. Total N uptake of the whole zone amounted to 0.57 t in 1999, 0.74 t in 2000, 0.86 t in 2009, 0.78 t in 2010, and 1.22 t in 2015, while total P uptake of the whole zone amounted to 3.79 t in 1999, 5.93 t in 2000, 6.96 t in 2009, 6.33 t in 2010, and 6.71 t in 2015. The spatial distribution of N uptake for each year is shown in Figure 5a–e. The modeling results show that plants located in the western and northern parts of the study area had a stronger ability to absorb N; average monthly N uptake rates in the inner protection zone were around 2.62 $g{\cdot}m^{-2}$. The western part is mountainous, and the major land cover is forest. For the northern part, which has a vast area of grassland, N consumption by plants was relatively weaker, except for dry land, which had higher values. In addition, areas adjacent to the water body displayed a consistently low ability to consume N. The statistics in Figure 5f show that the average N removal due to plant uptake increased from 1999 to 2015. This result is consistent with the NPP data. The value of N consumed by plants increased from 1.80 $g{\cdot}m^{-2}$ in 1999 to 3.83 $g{\cdot}m^{-2}$ in 2015.

The spatial distribution of P uptake for each year is shown in Figure 6a–e. The results show that plants located in the northwestern, northeastern, and southwestern parts of the study area had a stronger ability to absorb P; average monthly P uptake rates in the inner protection zone were around 0.041 $g{\cdot}m^{-2}$. These three areas are all mountainous, and the major land cover is forest. For the northern part, which has a vast area of grassland, the ability of plants to consume P was relatively weaker. Also, areas adjacent to the water body displayed a consistently low ability to consume P. The statistics in Figure 6f show that average P removal due to plant uptake increased from 1999 to 2015. This result is consistent with the NPP and N uptake data. The value of P increased from 0.026 $g{\cdot}m^{-2}$ in 1999 to 0.046 $g{\cdot}m^{-2}$ in 2015.

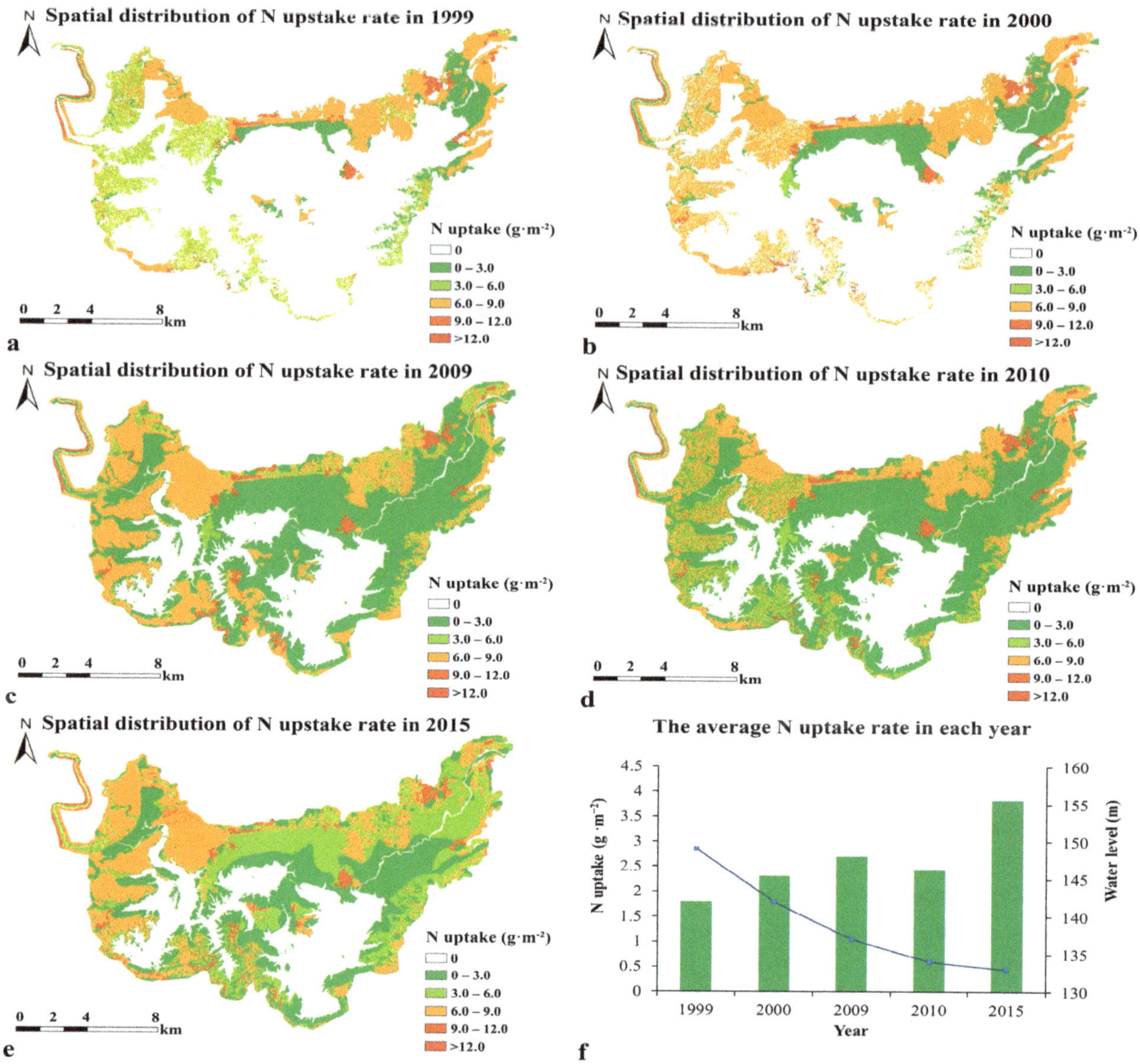

Figure 5. Spatial distribution of the N uptake rate in Miyun's inner protection zone during different years.

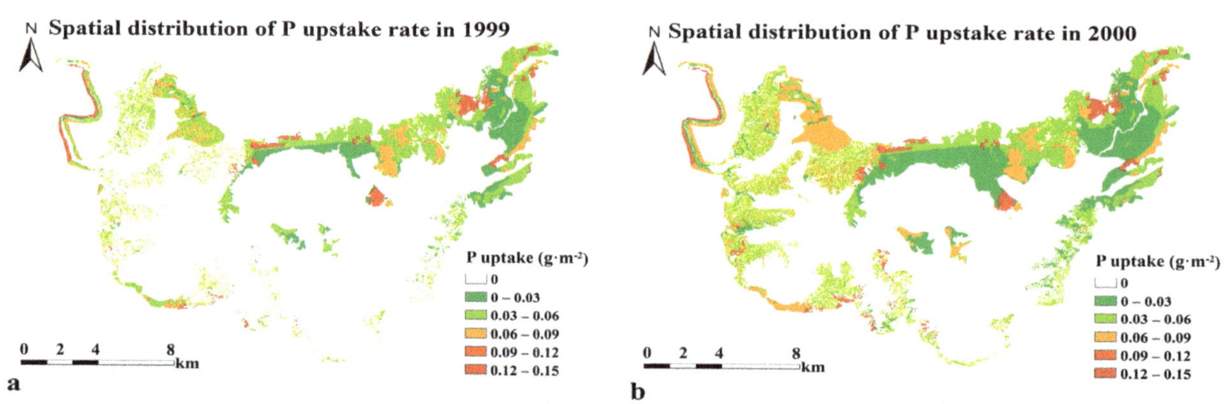

Figure 6. *Cont.*

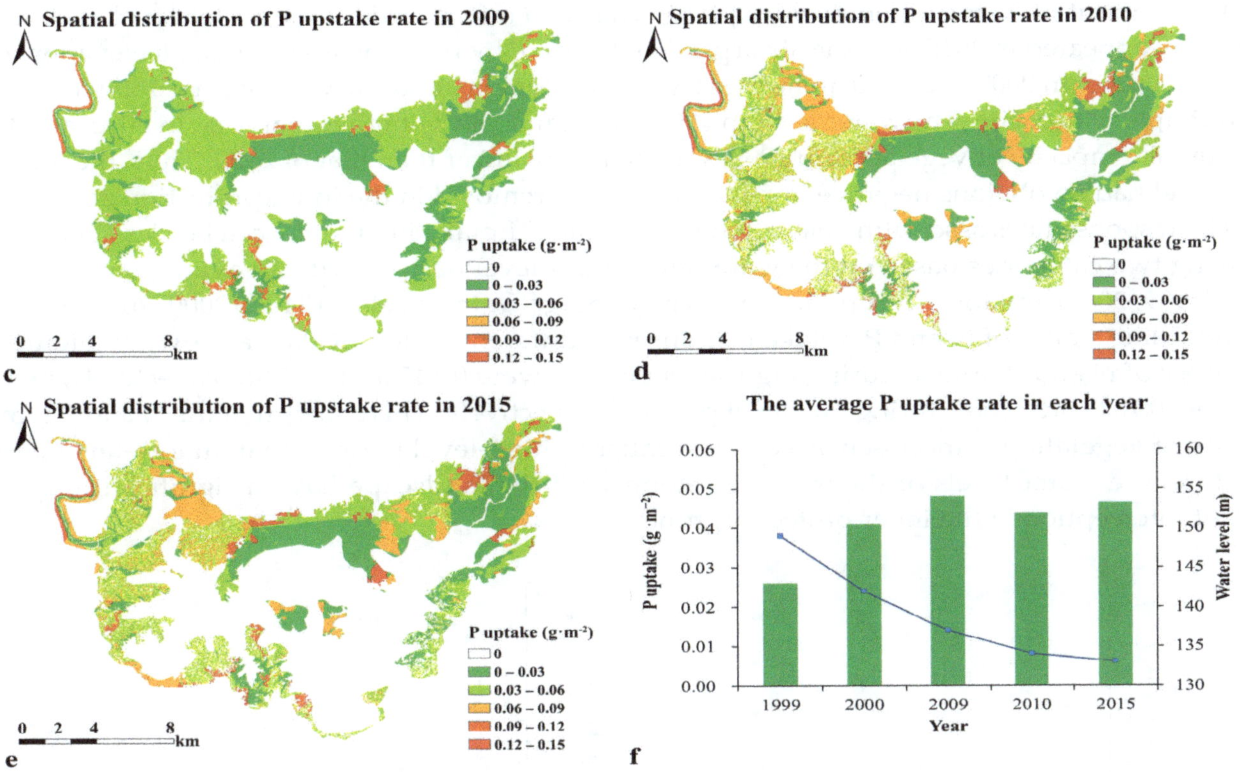

Figure 6. Spatial distribution of the P uptake rate in Miyun's inner protection zone during different years.

3.3. *Negative Relationship between the Water Level and N and P Uptake*

Water table dynamics strongly influence the N and P removal rates by plants [25] in the inner protection zone and are impacted by both landscape and stream water levels directly [48]. By overlaying the contour map on the spatial distribution of N and P removal by plants and implementing zonal statistics (130–140 m, 140–150 m, 150–160 m, and 160–170 m), we found that the N and P uptake increased synchronously with the elevation gradient (Figure 7a,b). The mean N and P uptake were generally low between elevations of 130–160 m, where values for N varied from 16.44 t to 124.74 t and values for P varied from 16.44 t to 124.74 t. Meanwhile, the values were higher at 160–170 m, where the mean consumed N increased from 124.74 t to 427.20 t and the mean consumed P increased from 124.74 t to 427.20 t. Thus, the uptake of N and P by vegetation presented a positive correlation with elevation.

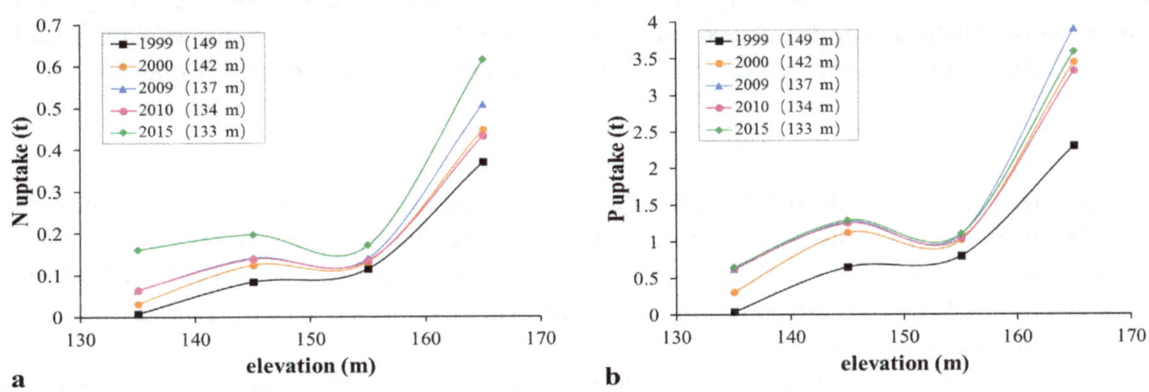

Figure 7. Zonal statistics for N (**a**) and P (**b**) consumed by plants during different years. The water level of each year is shown in parentheses following the year.

The amount of N and P removal by plants was also influenced by the water level. The lowest water level appeared in 2015, and the absorption of N and P by plants was strong in all elevation zones during this time. In 2009 and 2010, the water levels were similar, and the vegetation exhibited a similar N and P uptake capacity. The water level in 1999 was significantly higher than that in 2000, and the N and P uptake capacity of vegetation in 1999 was obviously lower than that of 2000. Thus, the increased water level had an obvious negative impact on N and P removal in the inner protection zone.

The response lines associating elevation with N and P uptake in Figure 7 can be classified into the following two categories based on the water level: high-level periods with an average water level of 145 m (1999, 2000) and low-level periods with an average water level of 134 m (2009, 2010, and 2015). The trend-line slopes of N and P uptake responses to different water levels are shown in Figure 8a,b. The slopes of N and P uptake during high-level periods were 0.145 and 0.0014, respectively, whereas they were 0.102 and 0.0009 during low-level periods, respectively. This shows that the N and P removal capacity of vegetation is more sensitive and weaker in high-level periods than in low-level periods. Therefore, the water levels of the reservoir along with the landscape have a significant impact on nutrient interception in the inner protection zone.

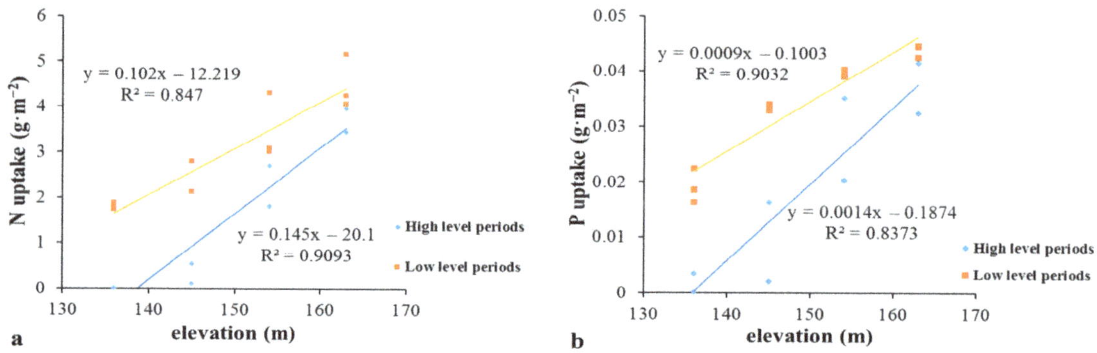

Figure 8. The distribution of N (**a**) and P (**b**) uptake rate responses to different water levels.

3.4. Validation of N and P Uptake Based on In Situ Survey Data

Generally, there are correlations between vegetation biomass density and N and P uptake and biomass can be used as a surrogate for nutrient removal potential [49,50]. To validate the accuracy of the RIPAM model, we converted the biomass density to N and P uptake by using a simple ratio coefficient. Xiao [51] did research on the ratios between biomass and N and P uptake for woodland areas in the Miyun Reservoir watershed, and the average ratio between biomass and N uptake was 0.00487, while that between biomass and P was 0.000579 [51]. For grasslands, Song [52] found that the ratios between biomass and N and P uptake were 0.0152 and 0.0012, respectively [52]. In this study, 39 quadrants were selected (shown in Section 2.4), and the N and P uptake rate of each quadrant was obtained by multiplying the biomass density with the corresponding correlation as follows:

$$X_{uptake} = \text{Biomass} \times X_{rate}, \tag{1}$$

where X_{uptake} refers to the N or P uptake rate of each quadrant (g·m^{-2}), Biomass is the biomass density of each quadrant (g·m^{-2}) (Table 3), and X_{rate} is the ratio between the biomass and N and P uptake according to Xiao [50] and Song [51]. It should be noted that the X_{rate} is varied with elevation and water level. Because of the limitation of the measured data, we used a fixed N or P uptake rate here and this would influence the results.

As each quadrant had a size of 90 m × 90 m, which corresponds to 3 pixels × 3 pixels in the RIPAM model, the average value of each 3 pixel × 3 pixel region in the simulated N and P uptake data was calculated. Then, scatter diagrams of N and P uptake between the RIPAM model and in situ quadrants were obtained, and linear regression was applied to the simulated data and in situ survey data (Figure 9a,b). We found that the R^2 and relation coefficient for N uptake were 0.88 and 0.78, respectively; the R^2 and relation coefficient for P uptake were 0.82 and 0.6, respectively. Both were a good fit. This shows that the two simulations both had considerable accuracy.

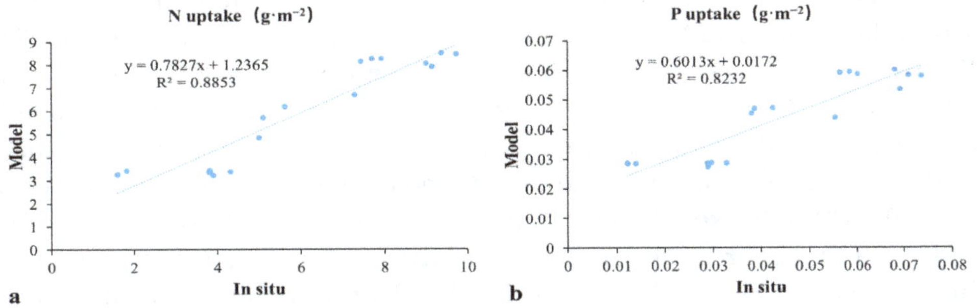

Figure 9. The N (**a**) and P (**b**) uptake rate of quadrant scatter diagrams between the RIPAM model and in situ survey.

4. Discussion

4.1. Reason for the Decline of N and P Uptake

As mentioned in Section 3.3, water level dynamics can have obvious negative impacts on N and P uptake. The growth conditions are a primary influencing factor for nutrient uptake by plants. For vegetation in the inner protection zone, soil moisture is no longer a limiting factor to its growth. In high-water-level periods, prolonged inundation may cause mechanical injury to plants and adversely affect the growth of vegetation. Additionally, high soil moisture will have more inhibitory effects than stimulatory effects on vegetation development. These both will lead to low NPP [24,28].

According to the theory described above, NPP in the inner protection zone was evaluated zonally. It was found that the variation trend of NPP was similar with that of the N and P uptake by plants during the five years that were analyzed (Figure 10a,b). Generally, there was a rising trend in NPP with increasing altitude. The NPP increased slowly from 130 m to 150 m, with mean values varying from 149.43 t to 2780.30 t. At an altitude of 160 m, there was a decline in the NPP. Then, it rose exponentially from 160 m to 170 m, whereby the mean NPP increased from 2301.45 t to 6604.80 t. The NPP was also influenced by the water level. The mean NPP in each elevation zone was higher during the low water level years (corresponding to 2009, 2010, and 2015) than that during the high water level years (corresponding to 1999 and 2000). This can be explained by the differences in the vegetation distribution, which were driven largely by the tolerances of species to specific geomorphic processes. As shown in Figure 1, grasslands and dry lands are the dominant land use types at elevations of 130–160 m; they both are home to herbaceous species and are easily influenced by elevated water tables or flooding. Meanwhile, forests, which are the dominant land use type at 160–170 m, are less affected.

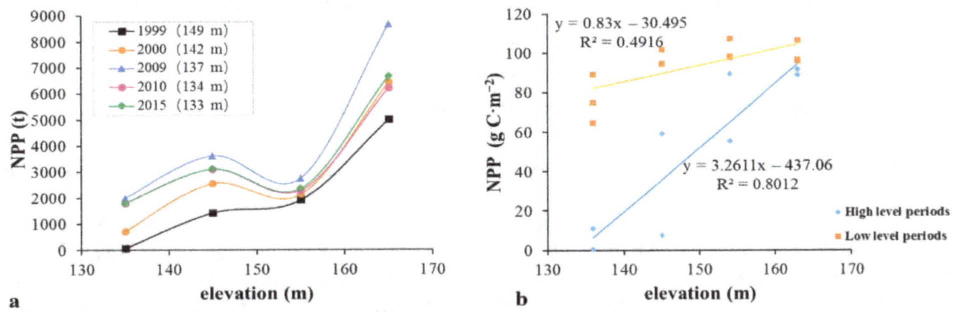

Figure 10. (**a**) Zonal statistics for NPP in different years and (**b**) NPP responses to different water levels. The numbers in parentheses following the years are the water level of each year.

4.2. Prediction of the Inner Protection Zone's Defense for N and P Interception

With the implementation of the middle route of the SNWTP, the storage capacity of the Miyun Reservoir is expected to increase from 1 billion m^3 to 2 billion m^3, and the water level will rise to 150 m in 2020, which is higher than the level in 1999. According to the contour map, the extent of the inner protection zone will decrease to 122.23 km^2, which is less than 56% of the area in 2015, and changes will be especially prominent in the northern floodplain, which represents one of the most important sections that curtails N and P losses to the water source from the upstream watershed.

The rest of the zone will also be influenced by the elevated water tables. According to the formula in Figure 10b, NPP will decrease, especially for vegetation at elevations of 130–150 m, which is sensitive to flooding. The total NPP will likely be less than the amount of 8454.96 t detected in 1999 when the water level was 149 m, and it is predicted to decline by more than 40.90% compared to the level in 2015. According to the formulas in Figure 8a,b, N and P uptake will be less than the amounts of 0.57 t N and 3.79 t P detected in 1999, and each will decline by more than 53.03% and 43.49%, respectively, compared to the levels in 2015. Thus, the ecological functions of the original inner protection zone will be impaired and this will lead to a decline of the inner protection zone's defense effectiveness in terms of N and P interception.

Although some studies indicate that transferred water is safe and has little impact on the trophic state of the receiving water [14], its ecological effectiveness as a nutrient sink providing pollution control and sediment retention will be impacted if the extant inner protection zone is damaged. The inner protection zone may even be turned from an NPS sink to an NPS source [26]. This could pose risks to the security of the water resource. Thus, vegetation cover in the extant inner protection zone of the terminal reservoir will need to be restored if the protection zone is to maintain its usual effectiveness.

5. Conclusions

As the end point of the middle route of the South-to-North Water Transfer Project (SNWTP), the Miyun Reservoir is the most important water source for Beijing. Importantly, water quality in the reservoir can have a direct impact on the wellbeing of people in Beijing and therefore warrants protection. In particular, steps will need to be taken to avoid eutrophication problems. This study used the RIPAM model to simulate N and P uptake by plants in Miyun Reservoir's inner protection zone during five years (1999, 2000, 2009, 2010, and 2015), and the results were validated by in situ survey data (R^2 of N uptake was 0.88, and R^2 of P uptake was 0.82). The correlations between the water level and the N and P uptake capacities showed that water table dynamics strongly influence the N and P removal by plants; thus, variations in the water level can have obvious negative impacts on N and P removal in the inner protection zone.

With the implementation of the middle route of the SNWTP, the water level of the Miyun Reservoir will rise to 150 m in 2020, and the total NPP is predicted to decline by more than 40.90% compared to the level in 2015; additionally, the N and P uptake is predicted to decline by more than 53.03% and 43.49%, respectively, compared to the levels in 2015. This will lead to a decline in the inner protection zone's defense effectiveness in terms of N and P interception and pose risks to the security of the water resource unless steps are taken to avoid such negative consequences. In summary, this study provides useful knowledge for assessing the risks of nutrient removal reductions in a terminal reservoir from a large inter-basin water transfer project before its implementation.

Acknowledgments: We acknowledge all of the editors and reviewers for their valuable advice. The authors are thankful for A-Xing Zhu and Guiming Zhang in the Department of Geography at UW-Madison for their valuable advice. The authors are also thankful for the financial support from the National Natural Science Foundation of China (Grant Nos. U1603241, 51779099), the National Key Project for R&D (Nos. 2016YFC0402403, 2016YFC0402409), the project titled Impact of Rising Water Levels on the Riparian Ecological Environment in Miyun Reservoir (230200019), and the Central non-profit research institutes fundamental research of Yellow River Institute of Hydraulic Research (HKY-JBYW-2017-10).

Author Contributions: Shengtian Yang, Juan Bai and Zhiwei Wang conceived and designed the experiments; Juan Bai, Zhiwei Wang, Hezhen Lou, Chunbin Zhang, Yabing Guan and Xinyi Yu performed the experiments;

Shengtian Yang, Zhiwei Wang, and Hezhen Lou analyzed the data; Chunbin Zhang contributed analysis tools; Yichi Zhang prepared the figures; Shengtian Yang, Juan Bai, and Zhiwei Wang wrote the main manuscript text. All authors discussed the structure and commented on the manuscript at all stages.

References

1. Cantor, L.M. The California state water project-a reassessment. *J. Geogr.* **1980**, *79*, 133–140.
2. Gastelum, J.R.; Cullom, C. Application of the Colorado River simulation system model to evaluate water shortage conditions in the central Arizona project. *Water Resour. Manag.* **2013**, *27*, 2369–2389. [CrossRef]
3. Voropaev, G.; Velikanov, A. Partial southward diversion of Northern and Siberian Rivers. *Int. J. Water Resour. Dev.* **1984**, *2*, 67–83. [CrossRef]
4. Hudson, J.W. *The Role of Water in the Biology of the Antelope Ground Squirrel Citellus Leucurus*; University of California Press: Berkeley, CA, USA, 1962.
5. Misra, A.K.; Saxena, A.; Yaduvanshi, M.; Mishra, A.; Bhadauriya, Y.; Thakur, A. Proposed river-linking project of India: A boon or bane to nature. *Environ. Geol.* **2007**, *51*, 1361–1376. [CrossRef]
6. Liu, C.; Zheng, H. South-to-North water transfer schemes for China. *Int. J. Water Resour. Dev.* **2002**, *18*, 453–471. [CrossRef]
7. Davies, B.R.; Thoms, M.; Meador, M. An assessment of the ecological impacts of inter-basin water transfers, and their threats to river basin integrity and conservation. *Aquat. Conserv.-Mar. Freshw. Ecosyst.* **1992**, *2*, 325–349. [CrossRef]
8. Yang, J.; Zhang, S.; Chang, L.; Li, F.; Li, T.; Gao, Y. Gully erosion regionalization of black soil area in Northeastern China. *Chin. Geogr. Sci.* **2017**, *27*, 78–87. [CrossRef]
9. Liu, K.; Ding, H.; Tang, G.; Na, J.; Huang, X.; Xue, Z.; Yang, X.; Li, F. Detection of catchment-scale gully-affected areas using unmanned aerial vehicle (uav) on the Chinese loess plateau. *ISPRS Int. J. Geo-Inf.* **2016**, *5*, 238. [CrossRef]
10. Zhuang, W. Eco-environmental impact of inter-basin water transfer projects: A review. *Environ. Sci. Pollut. Res. Int.* **2016**, *23*, 12867–12879. [CrossRef] [PubMed]
11. Li, Z.; Zhang, Y.; Zhu, Q.; Yang, S.; Li, H.; Ma, H. A gully erosion assessment model for the Chinese loess plateau based on changes in gully length and area. *Catena* **2017**, *148*, 195–203. [CrossRef]
12. Tang, C.; Yi, Y.; Yang, Z.; Cheng, X. Water pollution risk simulation and prediction in the main canal of the South-to-North water transfer project. *J. Hydrol.* **2014**, *519*, 2111–2120. [CrossRef]
13. Yang, W.; Zhang, L.; Shan, L.; Chen, X.; Chen, S. Response of extreme hydrological events to climate change in the water source area for the middle route of south-to-north water diversion project. *Adv. Meteorol.* **2016**, *2016*, 1–15. [CrossRef]
14. Zeng, Q.; Qin, L.; Li, X. The potential impact of an inter-basin water transfer project on nutrients (nitrogen and phosphorous) and chlorophyll a of the receiving water system. *Sci. Total Environ.* **2015**, *536*, 675–686. [CrossRef] [PubMed]
15. Zhang, Q. The South-to-North water transfer project of China: Environmental implications and monitoring strategy1. *J. Am. Water Resour. Assoc.* **2009**, *45*, 1238–1247. [CrossRef]
16. Zhang, Y.; Li, G.-M. Influence of south-to-north water diversion on major cones of depression in north China plain. *Environ. Earth Sci.* **2013**, *71*, 3845–3853. [CrossRef]
17. Li, Z.; Fang, H. Impacts of climate change on water erosion: A review. *Earth-Sci. Rev.* **2016**, *163*, 94–117. [CrossRef]
18. Pandey, A.; Himanshu, S.K.; Mishra, S.K.; Singh, V.P. Physically based soil erosion and sediment yield models revisited. *Catena* **2016**, *147*, 595–620. [CrossRef]
19. Walcher, M.; Bormann, H. On the transferability of the concept of drinking water protection zones from EU to Latin American Countries. *Water Resour. Manag.* **2015**, *29*, 1803–1822. [CrossRef]
20. Brenčič, M.; Prestor, J.; Kompare, B.; Matoz, H.; Kranjc, S. Integrated approach to delineation of drinking water protection zones. *Geologija* **2009**, *52*, 175–182. [CrossRef]
21. Sabater, S.; Butturini, A.; Clement, J.-C.; Burt, T.; Dowrick, D.; Hefting, M.; Matre, V.; Pinay, G.; Postolache, C.; Rzepecki, M. Nitrogen removal by riparian buffers along a European climatic gradient: Patterns and factors of variation. *Ecosystems* **2003**, *6*, 0020–0030. [CrossRef]
22. Correll, D. Buffer zones and water quality protection: General principles, Buffer zones: Their processes and potential in water protection. In Proceedings of the International Conference on Buffer Zones, Harpenden, Hertfordshire, UK, September 1996; pp. 7–20.

23. Klemas, V. Remote sensing of riparian and wetland buffers: An overview. *J. Coast. Res.* **2014**, *297*, 869–880. [CrossRef]
24. Nilsson, C.; Svedmark, M. Basic principles and ecological consequences of changing water regimes: Riparian plant communities. *Environ. Manag.* **2002**, *30*, 468–480. [CrossRef]
25. Liu, X.; Vidon, P.; Jacinthe, P.A.; Fisher, K.; Baker, M. Seasonal and geomorphic controls on n and P removal in riparian zones of the US Midwest. *Biogeochemistry* **2014**, *119*, 245–257. [CrossRef]
26. Schilling, K.E.; Li, Z.; Zhang, Y.-K. Groundwater–surface water interaction in the riparian zone of an incised channel, Walnut Creek, Iowa. *J. Hydrol.* **2006**, *327*, 140–150. [CrossRef]
27. Džubáková, K.; Molnar, P.; Schindler, K.; Trizna, M. Monitoring of riparian vegetation response to flood disturbances using terrestrial photography. *Hydrol. Earth Syst. Sci.* **2015**, *19*, 195–208. [CrossRef]
28. Kozlowski, T.T. Physiological-ecological impacts of flooding on riparian forest ecosystems. *Wetlands* **2002**. [CrossRef]
29. Stone, R.; Jia, H. Going against the flow. *Science* **2006**, *313*, 1034–1037. [CrossRef] [PubMed]
30. Zhang, J.; Meng, F.; Lu, Y.; Jing, Y.; Zhang, H.; Zhang, B.; Zhang, C. Ecological assessment of lakeshore wetland rehabilitation on eastern route of south-to-north water transfer project. *Front. Environ. Sci. Eng. China* **2008**, *2*, 306–310. [CrossRef]
31. Cheng, S.; Song, H. Conservation buffer systems for water quality security in south to north water transfer project in China: An approach review. *Front. For. China* **2009**, *4*, 394–401. [CrossRef]
32. Gu, W.; Shao, D.; Jiang, Y. Risk evaluation of water shortage in source area of middle route project for south-to-north water transfer in China. *Water Resour. Manag.* **2012**, *26*, 3479–3493. [CrossRef]
33. Liang, Y.-S.; Wang, W.; Li, H.-J.; Shen, X.-H.; Xu, Y.-L.; Dai, J.-R. The South-to-North water diversion project: Effect of the water diversion pattern on transmission of oncomelania hupensis, the intermediate host of *Schistosoma japonicum* in China. *Parasites Vectors* **2012**, *5*, 52. [CrossRef] [PubMed]
34. Barnett, J.; Rogers, S.; Webber, M.; Finlayson, B.; Wang, M. Sustainability: Transfer project cannot meet China's water needs. *Nature* **2015**, *527*, 295–297. [CrossRef] [PubMed]
35. Li, X.-S.; Wu, B.-F.; Zhang, L. Dynamic monitoring of soil erosion for upper stream of Miyun reservoir in the last 30 years. *J. Mt. Sci.* **2013**, *10*, 801–811. [CrossRef]
36. Jiao, J.; Du, P.; Lang, C. Nutrient concentrations and fluxes in the upper catchment of the Miyun reservoir, China, and potential nutrient reduction strategies. *Environ. Monit. Assess.* **2015**, *187*, 110. [CrossRef] [PubMed]
37. Dong, G.; Yang, S.; Gao, Y.; Bai, J.; Wang, X.; Zheng, D. Spatial evaluation of phosphorus retention in riparian zones using remote sensing data. *Environ. Earth Sci.* **2014**, *72*, 1643–1657. [CrossRef]
38. Wang, X.; Mannaerts, C.M.; Yang, S.; Gao, Y.; Zheng, D. Evaluation of soil nitrogen emissions from riparian zones coupling simple process-oriented models with remote sensing data. *Sci. Total Environ.* **2010**, *408*, 3310–3318. [CrossRef] [PubMed]
39. Wang, X.; Wang, Q.; Yang, S.; Zheng, D.; Wu, C.; Mannaerts, C.M. Evaluating nitrogen removal by vegetation uptake using satellite image time series in riparian catchments. *Sci. Total Environ.* **2011**, *409*, 2567–2576. [CrossRef] [PubMed]
40. Wang, X.; Yang, S.; Mannaerts, C.M.; Gao, Y.; Guo, J. Spatially explicit estimation of soil denitrification rates and land use effects in the riparian buffer zone of the large Guanting reservoir. *Geoderma* **2009**, *150*, 240–252. [CrossRef]
41. Prince, S.D.; Goward, S.N. Global primary production: A remote sensing approach. *J. Biogeogr.* **1995**, *22*, 815–835. [CrossRef]
42. Zhu, Z.X.; Arp, P.A.; Meng, F.R.; Bourque, C.P.A.; Foster, N.W. A forest nutrient cycling and biomass model (FORNBM) based on year-round, monthly weather conditions, part 1: Assumption, structure and processing. *Ecol. Model.* **2003**, *169*, 347–360. [CrossRef]
43. Arp, P.A.; Oja, T. A forest soil vegetation atmosphere model (ForSVA) 1. Concepts. *Ecol. Model.* **1997**, *95*, 211–224. [CrossRef]
44. Zhang, L.-T.; Li, Z.-B.; Wang, S.-S. Spatial scale effect on sediment dynamics in basin-wide floods within a typical agro-watershed: A case study in the hilly loess region of the Chinese loess plateau. *Sci. Total Environ.* **2016**, *572*, 476–486. [CrossRef] [PubMed]
45. Husson, E.; Lindgren, F.; Ecke, F. Assessing biomass and metal contents in riparian vegetation along a pollution gradient using an unmanned aircraft system. *Water Air Soil Pollut.* **2014**, *225*. [CrossRef]
46. Kachamba, D.J.; Orka, H.O.; Gobakken, T.; Eid, T.; Mwase, W. Biomass estimation using 3D data from unmanned aerial vehicle imagery in a tropical woodland. *Remote Sens.* **2016**, *8*, 968. [CrossRef]

47. Yang, S.; Bai, J.; Zhao, C.; Lou, H.; Zhang, C.; Guan, Y.; Zhang, Y.; Wang, Z.; Yu, X. The assessment of the changes of biomass and riparian buffer width in the terminal reservoir under the impact of the South-to-North water diversion project in China. *Ecol. Indic.* **2018**, *85*, 932–943. [CrossRef]
48. Jung, M.; Burt, T.; Bates, P. Toward a conceptual model of floodplain water table response. *Water Resour. Res.* **2004**, *40*. [CrossRef]
49. Rheinhardt, R.; Brinson, M.; Meyer, G.; Miller, K. Integrating forest biomass and distance from channel to develop an indicator of riparian condition. *Ecol. Indic.* **2012**, *23*, 46–55. [CrossRef]
50. Jiang, F.Y.; Chen, X.; Luo, A.C. A comparative study on the growth and nitrogen and phosphorus uptake characteristics of 15 wetland species. *Chem. Ecol.* **2011**, *27*, 263–272. [CrossRef]
51. Xiao, Y. Study on the Adjustment and Control Mechanism of Forest to Nonpoint Source Pollution in Beijing Mountain Area. Ph.D. Thesis, Beijing Forestry University, Beijing, China, 29 June 2008.
52. Song, X. Study on the Variation of N, P, K Nutrients Pool of Different Type of Alpine Meadow in Eastern Qilian Mountains. Master's Thesis, Gansu Agricultural University, Lanzhou, China, 10 June 2008.

7

Optimal Operation of Cascade Reservoirs for Flood Control of Multiple Areas Downstream

Chao Zhou [1], Na Sun [2], Lu Chen [2,*], Yi Ding [1], Jianzhong Zhou [2], Gang Zha [2], Guanglei Luo [2], Ling Dai [2] and Xin Yang [2]

1 Changjiang Institute of Survey, Planning, Design and Research, Wuhan 430010, China; zhouchao2@cjwsjy.com.cn (C.Z.); dingyi@cjwsjy.com.cn (Y.D.)

2 School of Hydropower and Information Engineering, Huazhong University of Science and Technology, Wuhan 430074, China; sunna@hust.edu.cn (N.S.); jz.zhou@hust.edu.cn (J.Z.); zhagang@hust.edu.cn (G.Z.); m201773774@hust.edu.cn (G.L.); dailing2016@hust.edu.cn (L.D.); d201880948@hust.edu.cn (X.Y.)

* Correspondence: chen_lu@hust.edu.cn

Abstract: The purpose of a flood control reservoir operation is to prevent flood damage downstream of the reservoir and the safety of the reservoir itself. When a single reservoir cannot provide enough storage capacity for certain flood control points downstream, cascade reservoirs should be operated together to protect these areas from flooding. In this study, for efficient use of the reservoir storage, an optimal flood control operation model of cascade reservoirs for certain flood control points downstream was proposed. In the proposed model, the upstream reservoirs with the optimal operation strategy were considered to reduce the inflow of the reservoir downstream. For a large river basin, the flood routing and time-lag cannot be neglected. So, dynamic programming (DP) combined with the progressive optimality algorithm (POA) method, DP-POA, was proposed. Thus, the innovation of this study is to propose a two-stage optimal reservoir operation model with a DP-POA algorithm to solve the problem of optimal co-operation of cascade reservoirs for multiple flood control points downstream during the flood season. The upper Yangtze River was selected as a case study. Three reservoirs from upstream to downstream, Xiluodu, Xiangjiaba and the Three Gorges reservoirs (TGR) in the upper Yangtze River, were taken into account. Results demonstrate that the two-stage optimization algorithm has a good performance in solving the cascade reservoirs optimization problem, because the inflow of reservoir downstream and the division volumes were largely reduced. After the optimal operation of Xiluodu and Xiangjiaba reservoirs, the average reduction of flood peak for all these 13 typical flood hydrographs (TFHs) is 13.6%. Meanwhile, the cascade reservoirs can also store much more storm water during a flood event, and the maximum volumes stored in those two reservoirs upstream in this study can reach 25.2 billion m^3 during a flood event. Comprising the proposed method with the current operation method, results demonstrate that the flood diversion volumes at the flood control points along the river decrease significantly.

Keywords: optimal flood control operation; cascade reservoirs; dynamic programming with progressive optimality algorithm (DP-POA); the upper Yangtze River Basin

---

1. Introduction

For safety of lives and property from risks related to floods, proper planning of relevant applications is important for development [1–13]. To date, dams have become numerous and have had a profound effect on social and economic development. As of 2007, China was the world's leader in the construction of large dams [14]. Reservoirs are often operated with a number of purposes related to environmental, economic and public services. Generally, these purposes include flood control, hydropower generation, navigation, sediment control, water supply, recreation, and fisheries, among which flood control is the most significant function for many of the reservoirs [15].

The main purpose of flood control is to prevent flood damage downstream of the reservoir and the safety of the reservoir itself. The key variables governing the operation of flood control reservoirs are the available storage capacity and the expected inflow magnitude from an incoming flood [15]. Some of the research on multi-reservoir flood control operations is for design of the reservoirs [16], in which the objective functions are the minimum of the initial investment cost and potential flood damage costs. Until now, most of the research on multi-reservoir flood control operation concerns minimizing the maximal water level for the dam itself or the releases at the downstream flood control points, such as the research carried out by Lee et al. [17], Qi et al. [18] and Chen et al. [11]. The other aspect for multi-reservoir flood control operation is to address the conflicts between flood control and conservation. A great number of research works on dynamic control flood limiting water level have been carried out in recent years to balance the conflict between flood control and conservation, such as the work done by Li et al. [19], Chen et al. [20], and Zhou et al. [9,21].

Accordingly, releases are restricted by the maximum allowable non-damaging channel capacity at the downstream flood control points [15]. Sometimes the available storage capacity of one reservoir is not enough to protect the flood control points downstream. In this situation, more reservoirs are needed to work together to protect these flood control points from flooding, especially the downstream control points with great importance. Thus, a new optimal reservoir operation model, which can make efficient use of reservoir storage to protect downstream flood control points from flooding during the flood season is needed. He et al. [22] studied the flood control operation of Jinsha River cascade reservoirs combined with the Three Gorges for flood control of the Chenglingji areas downstream based on the equal water storage method, in which the optimal operation technique was not considered. The storage capacities of these reservoirs are underutilized. Zhou et al. [9] proposed a virtual reservoir approach that aggregates multi-reservoir systems into a virtual reservoir for flood control decision making. However, in their method, two reservoirs were taken as a virtual reservoir, which cannot reflect the hydraulic connection between reservoirs. Therefore, in this paper, a two-stage optimal operation of cascade reservoirs for flood control of multiple areas downstream, which considers the optimization and hydraulic connections among cascade reservoirs, was proposed.

Take three reservoirs, as shown in Figure 1, for example. A, B, and C are three reservoirs from upstream to downstream, and points 1, 2, and 3 are three flood control points downstream of the reservoirs, whose locations can also be seen in Figure 1. Points 1 and 2 are protected by Reservoirs A and B. Point 3 is protected by Reservoirs A, B and C, since Reservoir C cannot have enough capacity for flood control of Point 3. Reservoirs A and B can help Reservoir C realize the flood control by minimizing the inflow of Reservoir C. Therefore, two objective functions are needed. For Reservoirs A and B, the objective function is to minimize the inflow of Reservoir C; for Reservoir C, the objective function is to minimize the excess flood volumes diverted to the floodplain at point C.

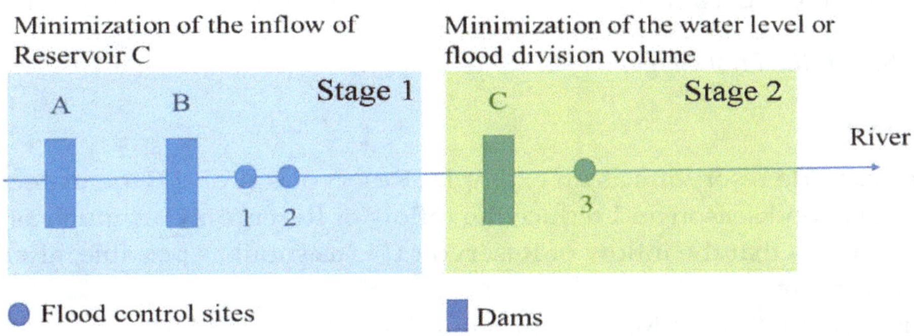

Figure 1. Diagram of dams and flood control points.

In addition, usually the optimal flood control reservoir operation model for a large river basin does not consider the flood routing during the processes for obtaining the optimal solution. When the river basin is large, these factors cannot be ignored. Since the Muskingum method was used for flood routing [23–25], the delayed time is usually considered, which causes the time-lags issue during the optimal processes [10]. In this case, the number of decision variables and constraints is increased exponentially, and it is difficult for a single traditional optimization method or modern intelligence algorithm to tackle the time-lags problem [10]. In the broad area of reservoir operation, development and use of prescriptive models based on optimization techniques have been extensively used [26,27]. An adaptive optimal algorithm also should be considered for the large river basins with flood routing and time-delayed issues.

There are various studies on optimal operations of single reservoir and multi-reservoir systems for flood control in the literature [9,28–30]. The dynamic programming (DP) method is the most widely used optimization technique for reservoir operation. It also has been used for reservoir operation by [31–34]. DP can be used for solving problems containing discrete variables, non-convex, non-continuous, and non-differentiable objective functions. However, the "curse of dimensionality", the phenomenon that when the number of reservoirs rises, the computation scale of DP increases exponentially, limits its application of DP. To deal with the dimensionality problems, modified DP algorithms, such as discrete differential dynamic programming (DDDP) [35–37], dynamic programming successive approximation (DPSA) [38,39] and the progressive optimality algorithm (POA) have been widely used for finding the optimal or near-optimal reservoir release hydrographs for flood control [40].

The POA, proposed by Howson and Sancho [41], was used to resolve the *n*th single reservoir operation problems, which divides the multi-stage problem into multiple two-stage sub-problems. Turgeon [42] used the principle of progressive optimality to minimize the total production cost of a multi-dimensional and multistage optimization problem. Guo et al. [43] applied the POA algorithm to solve the joint operation of the multi-reservoir systems of the Three Gorges and the Qingjiang cascade reservoirs. However, the disadvantages of the POA method are that the initial solutions have great influence on the final optimal results. In this study, a combined algorithm of DP and POA was proposed to solve the problem of flood routing and the influence of initial solutions.

The objective of this study was, therefore, to establish an optimal flood control operation model of cascade reservoirs for multiple flood control points downstream. The objective of the whole model is to minimize the flood diversion volumes at the flood control points. An adaptive combined algorithm is also proposed to solve the problem of flood routing. The upper Yangtze River, China, was selected as a case study. Three reservoirs were considered in this study.

2. Optimal Flood Control Reservoir Operation Model

A two-stage optimal reservoir operation model was proposed. Take the three reservoirs for an example, as shown in Figure 1. In the first stage, the optimal operation models of Reservoirs A and B were established, in which the objective function is to reduce the inflow of the downstream reservoir. In the second stage, the optimal operation model of Reservoir C is established, in which the objective function is to minimize the highest water level and the flood diversion volumes at flood control point 3.

2.1. Optimal Operation in the First Stage

2.1.1. Objective Function

Take the three cascade reservoirs as an example. Reservoirs A and B are expected to store much more water during the flood season and reduce the inflow of Reservoir C as much as possible. In other words, the best situation is that the inflow of Reservoir C is as small as possible, after the utilization of the two reservoirs upstream.

Thus, the objective function of Reservoirs A and B can be defined as:

$$\min F = \min\left\{\sum_{t=1}^{T}\left[Q'(t) + q(t)\right]^2\right\} \tag{1}$$

where $Q'(t)$ is the discharge of Reservoir B considering the flood routing along the river; and $q(t)$ is the discharge of the interval basin, which also includes the flow from other tributaries of the river.

2.1.2. Constraints for Model in the First Stage

Four constraints were considered in this study. The first one is water balance equation, which must obey during the reservoir operation process. The second one is the constraint concerning water level, which gives the allowable maximum and minimum water levels during the flood season. The third one is the discharge limit, which gives the allowable maximum and minimum discharges during the flood season. The fourth is the limit of the discharge for downstream flood control point, which protects downstream areas from flooding.

(a) Water balance equation:

$$V_{i,t} = V_{i,t-1} + (I_{i,t} - Q_{i,t})\Delta t \tag{2}$$

where $V_{i,t}$ is the reservoir storage at time t for the ith reservoir; and $I_{i,t}$ and $Q_{i,t}$ are the inflow and outflow at time t for the ith reservoir, respectively.

(b) Constraint of water level:

$$Z_{i,t}^{\min} \le Z_{i,t} \le Z_{i,t}^{\max} \tag{3}$$

where $Z_{i,t}$ represents the water level of the ith reservoir at time t; $Z_{i,t}^{\min}$ and $Z_{i,t}^{\max}$ represent the minimum and maximum water level of the ith reservoir at time t, respectively.

(c) Constraint of reservoir discharge:

$$Q_i^{\min} \le Q_{i,t} \le Q_i^{\max}(Z_t) \tag{4}$$

where $Q_{i,t}$ is the discharge of the ith reservoir at time t; $Q_i^{\max}(Z_t)$ means the maximum discharge ability corresponding to the water lever Z_t at time t for the ith reservoir; and $Q_i^{\min}$ means the minimum discharge for the ith reservoir. The maximum discharge ability is related to the water level, and the reservoir discharge $Q_{i,t}$ cannot exceed the maximum discharge ability.

Besides Equation (4), the discharge variation between adjacent time periods should be also considered by:

$$|Q_{i,t} - Q_{i,t-1}| \le \Delta Q_i \tag{5}$$

where ΔQ_i is the allowed discharge variation between adjacent time periods for the ith reservoir.

(d) Constraint of discharge for downstream flood control points

$$q_{j,t} = Q'_{i,t} + \Delta q_t \le q_j^{\max}, j = 1,2,3 \tag{6}$$

where $q_{j,t}$ is the discharge at time t for the jth flood control point downstream; $Q'_{i,t}$ is the discharge of the ith reservoir considering the flood routing at the jth flood control point; Δq_t is the inflow in the interval zone between the ith reservoir and the flood control point downstream; and $q_j^{\max}$ is the allowed maximum discharge for the jth flood control point downstream.

2.2. *Optimal Operation in the Second Stage*

2.2.1. Objective Function

Flood control reservoir operation is generally a complex problem, which needs to consider the safety of a reservoir and its upstream and downstream areas. This paper established a flood control operation model aiming to control the highest water level of Reservoir C and reduce the flood diversion magnitudes of the downstream areas. Two objective functions were considered and given as follows. When the release does not exceed the safety water level downstream, the highest water level of Reservoir C is taken as an objective function. Although those reservoirs operate together, it is possible that the water level at flood control points still exceeds the safety water level for a high return period flood (e.g., 1000-year). In this situation, the objective function is to minimize the flood diversion magnitude in the downstream areas. These two objective functions are given as follows.

(1) Minimize the highest water level to ensure the safety of the reservoir itself and its upstream areas

$$\min F'_1 = \min\{\max Z_t\}, t = 1, 2, \cdots, T_d \tag{7}$$

where Z_t is the water level at time t; and T_d is the time duration for the flood control operation.

(2) Minimize the flood diversion magnitudes for downstream areas:

$$\min F'_2 = \min\left\{\sum_{t=1}^{T_s} \Delta q(t) \cdot \Delta t\right\} \tag{8}$$

where $\Delta q(t)$ is the excess water volume into the flood diversion areas around the downstream control site at time t; Δt is the interval time; and T_s is the time duration for flood diversion, when the water level exceeds the safety water level downstream.

2.2.2. Constraints for the Model in the Second Stage

The constraints include the water balance, water level limits, reservoir discharge limits and allowable discharges for downstream flood control points. Equations for the constraints are the same as those in stage one (Section 2.1).

2.3. Optimization Algorithm

Considering the "curse of dimensionality", the time-lag issue for multi-reservoir operations, and the influence of the initial solutions, the dynamic programming combined with the progressive optimality algorithm (DP-POA) method was proposed to find the optimal operation polices for multi-reservoir systems, in which the DP method was used first, and the output of the DP method was the input of the POA method. The skeleton of this algorithm is shown in Figure 2. The equations for DP and POA methods are given below.

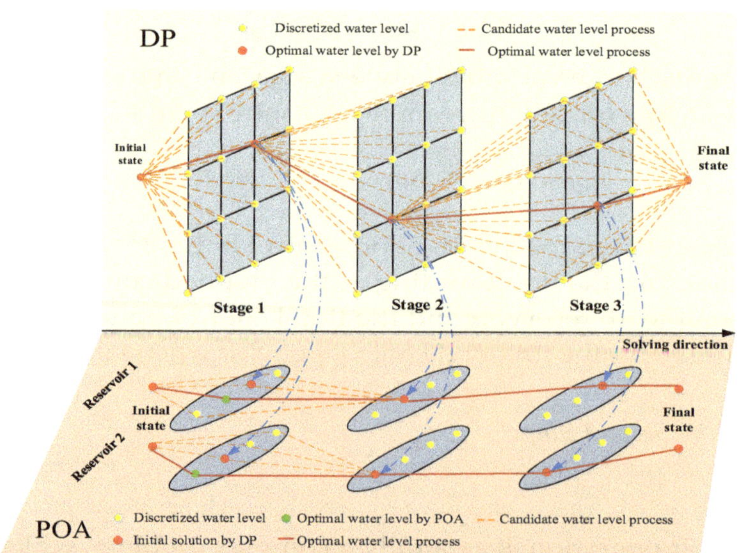

Figure 2. The skeleton of the dynamic programming combined with progressive optimality (DP-POA) algorithm.

2.3.1. Dynamic Programming (DP)

A multiple-period reservoir operation is formulated as a recursive function in DP:

$$\begin{aligned} \max \quad & G_t(s_t) = g_t(s_t, r_t) + G_{t+1}(s_{t+1}) \\ s.t. \quad & s_t + q_t = r_t + s_{t+1} \end{aligned} \tag{9}$$

where s_t, r_t, and q_t are storage, release, and inflow at period t, respectively; $g_t(s_t, r_t)$ is single-period utility function; and $G_t(s_t)$ is the maximum cumulative utility from period t (current period) to period T (end of the operation period). Equation (9) is a generalized formula that characterizes the reservoir operations of hydropower, water supply, flood control, and so on.

2.3.2. Progressive Optimality Algorithm (POA)

POA often starts with an initial solution obtained by engineering judgment, heuristic search algorithms and other methods. In this method, the initial solution was given based on the results of the DP method. Then, POA solves all the two-stage optimization sub-problems in a serial way. For the jth sub-problem, use Equation (10) to find an improved decision vector between stages $j-1$ and $j+1$.

$$F^*(S_{j-1}, S_j, S_{j+1}) = \min_{S_j \in C_j} \{F(S_{j-1}, S_j) + F(S_j, S_{j+1})\} \tag{10}$$

where S_j is the possible water level at the jth period, C_j is the set of water level at the jth period, $F(S_{j-1}, S_j)$ is the objective water level between stages $j-1$ and j, $F^*(S_{j-1}, S_j, S_{j+1})$ denotes the optimal objective water level between stages $j-1$ and $j+1$.

3. Study Area

3.1. Introduction of Study Area

The upper Yangtze River is selected as a case study. The largest reservoir in China is created by the Three Gorges Dam (TGD), which stores 39.3 billion m^3 of water and has a surface area of 1045 km^2. The TGD is located on the Yangtze River, which is about 6300 km long and occupies 18.8% of China's land area. The function of the TGD includes flood control, power generation and navigation and so on. Since the Yangtze River includes many devastating floods over the centuries killing thousands of people and causing millions of dollars in damage, among these functions, flood control is the most important one.

Floods in the middle and lower reaches of the Yangtze River mainly stem from the upper region of the Yichang gauging station, which is also the control site of TGD [44]. Usually the flood volume of the upper Yichang site is about 50% of the total flow volume of the Yangtze River, about 90% of the Jingjiang River reach that is part of the Yangtze River from the Zhicheng to Chenglinji gauging station and regarded as the most key area for flood prevention [44–46]. The Three Gorges Reservoir (TGR), which uses the available empty space of the reservoir to absorb the flood, can reduce the downstream flooding around the Jingjiang River and Dongting Lake areas and guarantee the safety of the Jingjiang River downstream when a 100-year flood occurs in TGD.

However, the flood control capacity of TGR with a capacity of 22.2 billion m^3 is still not enough. If a flood event throughout the Yangtze River Basin is to occur, the TGD cannot hold back floods downstream of the dam (such as the 1954 flood that killed 30,000 people outright and more through starvation and disease). Thus, more reservoirs located upstream are needed to help TGD reduce the flood. The Xiluodu and Xiajiaba are two dams built in the Jinsha River, which is part of the Yangtze River upstream as shown in Figure 3.

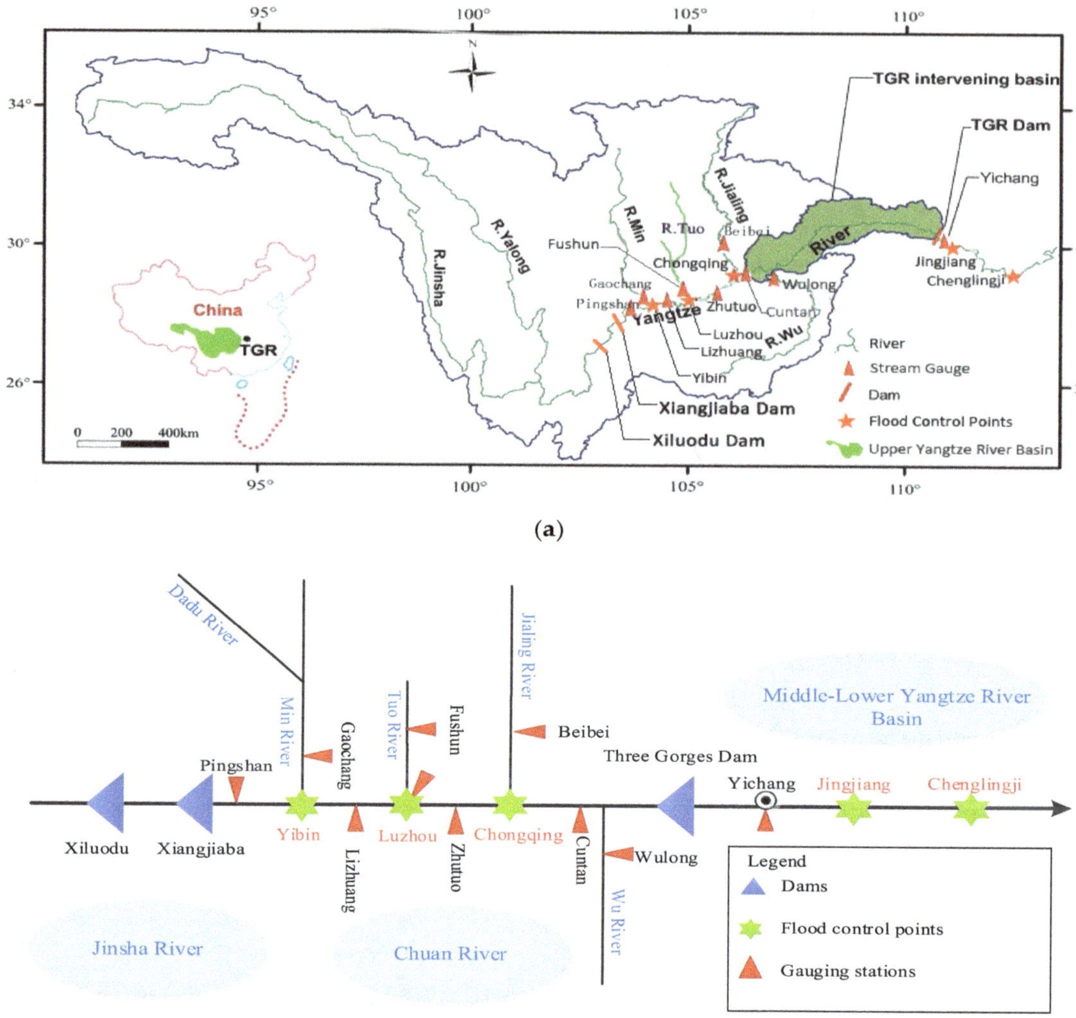

Figure 3. Locations of the dams, flood control points and gauging stations in the Yangtze River Basin. (**a**) The upper Yangtze River Basin; (**b**) skeleton of the upper Yangtze River Basin.

The Xiluodu dam, located on the upper reach of the Yangtze River between Yunnan and Sichuan Provinces, is the third tallest arch dam in the world and a key component of the Jinsha River Project. It is also the second largest dam in China, next only to the TGD in terms of size, construction cost and generating capacity. The overall elevation of the dam crest is 600 m, with the concrete double-curvature arch being 285.5 m high and 700 m long. The reservoir capacity is about 12.7 billion m^3 with 4.7 billion m^3 for flood control. The Xiangjiaba dam is located downstream of the Xiluodu dam, which is only 157 km far from it. It is a large gravity dam on the Jinsha River. The flood control capacity is 0.9 billion m^3. The total flood control capacity of both of the two dams are nearly 5.6 billion m^3. The characteristics of the three dams are summarized in Table 1.

Five flood control areas considered in this study from upstream to downstream are Yinbin, Luzhou, Chongqing, Jingjiang and Chenglingji. The Xiluodu and Xiangjiaba reservoirs need to protect the cities Yibin and Luzhou from flooding, when the flood with a 20-year return period occurs, and protect the city Chongqing from flooding, when the 50-year return period flood occurs. The Xiluodu, Xiangjiaba and TGR operate together to protect the Jingjiang area from flooding, when a 100-year flood event occurs. The design standard for the TGR is 1000 years corresponding to the water level of 175 m. Therefore, the flood events with 50-year, 100-year and 1000-year return periods were considered in this study. Three major cities from upstream to downstream (Yibin, Luzhou, and Chongqing) are located along the Yangtze River. Among these cities, Chongqing is the most populous Chinese municipality, and also the largest directly controlled municipality in China, and comprises 26 districts, eight counties, and four autonomous counties. Yibin and Luzhou are two prefecture-level cities in the south-eastern part of Sichuan province, China. Yibin is located at the junctions of the Min River and Yangtze Rivers. Luzhou, situated at the confluence of the Tuo River and the Yangtze River, and is an important port on the Yangtze River. According to the 2015 census, these three major cities have a total population of about 41 million.

In recent years, the economy of the region around the Jingjiang River reach has continued to develop. Concretely, the current agricultural acreage area and aquatic area are 37,000 and 6500 square hectometers, respectively, while fixed assets in the district are valued at 9 billion yuan, and gross domestic product is 8.7 billion yuan. This area has become an important grain, cotton, and aquaculture base in China. Therefore, once a catastrophic flood occurred, the affected population is very large, and the economic losses are extremely high. Another flood control points is Chenglingji station which is the outlet of the Dongting Lake. There are many retention basins around the Chenglingji station, which is home to a lot of people. According to the Chinese flood control regulations, as the backbone of flood management measures for the mid-downstream Yangtze River Basin, the Three Gorges Reservoir provides enormous benefits to flood control for the Yangtze River. The operation of TGR has improved the flood control capacity of the Jingjiang River Reach downstream from 20-year to 100-year, and prevents devastating damage through combining the use of retention basins for the case of floods of 100–1000 years' frequency. Additionally, it also reduces the possibility of using retention basins near the Chenglingji area to ensure the safety of life and property.

Table 1. Characteristics of the Xiluodu, Xiangjiaba and Three Gorges Dam (TGD).

Dam	Drainage Area (10^4 km^2)	Normal Water Level (m)	Flood Control Water Level (m)	Dead Water Level (m)	Regulating Storage Capacity (Billion m^3)	Flood Control Reservoir Capacity (Billion m^3)
Xiluodu	45.4	600	560	540	6.5	4.7
Xiangjiaba	45.9	380	370	370	0.9	0.9
TGD	100	175	145	145	16.5	2.2

3.2. *Data of Study Area*

The Xiluodu and Xiajiaba reservoirs can provide certain flood capacity to store the flood occurring in the upper Yangtze River, especially the flood in the Jinsha River. However, under the condition of preventing the downstream cities—Yibin, Luzhou and Chongqing—from flooding as shown in Figure 3, how to use the two reservoirs rationally and effectively to reduce the inflow of TGR is a challenging task. In addition, besides those two reservoirs, how to use the flood capacity of TGR to realize the flood control of the Jingjiang River downstream effectively is another important task. Thus, since the flood control problems are much more complicated for cascade reservoirs with multiple flood control points, judicious operation methods during a flood event are needed.

The gauging station used in this study is shown in Figure 3 as well. From Xiangjiaba to Yichang, there are four main tributaries, namely Min, Tuo, Jialing, and Wu Rivers. The control sites for Min, Tuo, Jiangling, and Wu Rivers are Gaochang, Fushun, Beibei, and Wulong, respectively. The Yangtze River upper than Yibin is also named the Jinsha River, and the Yangtze River from Yibin to Yichang is also named the Chuan River because these areas belonged to Sichuang Province before. There are three flood control points located on the Chuang River reach, namely Yibin City, Luzhou City and Chongqing City; and there are two flood control points located downstream of the TGD, namely Jingjiang and Chenglingji areas. The Xiluodu and Xiangjiaba dams were employed to protect the cities of Yibin, Luzhou, and Chongqing from flooding. The TGD combined with the two dams operated together to guarantee the safety of the Jingjiang and Chenglingji areas.

Since the Xiluodu and Xiangjiaba dams are on the Jinsha River, the characteristics of the flood in the Jinsha River were discussed. The runoff in the Jinsha River Basin mainly concentrate in the period of June to October, which can account for 80% of the total annual runoff in the upper Jinsha River and 75% in the lower Jinisha River. The mean runoff and its percentage for each month at the Pingshan gauging station in the Jinsha River are given in Table 2. It is indicated from Table 2 that the river flow in July, August and September accounting for 54.0% of the total amount of the annual flow, and in June to November accounting for 81.2%.

Table 2. Runoff and its percentage for each month at the Pingshan gauging station.

Months	Jan.	Feb.	Mar.	Apr.	May	Jun.	Jul.	Aug.	Sept.	Oct.	Nov.	Dec.
Runoff (m^3/s)	1660	1420	1350	1540	2320	5080	9580	10,200	10,000	6540	3450	2180
Volume (10^8 m^3)	44.5	34.7	36.2	39.8	62.1	132	257	274	260	175	89.5	58.3
Percentage (%)	3.04	2.38	2.47	2.72	4.25	9.00	17.5	18.7	17.8	12.0	6.12	3.99

4. Application

4.1. Design Flood Hydrographs

During the flood control reservoir operation, usually the extreme or large flood events were considered. Therefore, the input of the model is the T-year design flood hydrograph. One of the methods to derive the design flood hydrographs (DFH) is the typical flood hydrograph (TFH) method which has been widely used by practitioners [47,48]. The DFH can be obtained according to the following steps.

Step 1: The annual maximum (AM) sampling method was used in this study, and the AM peak or n-day volumes data were derived.

Step 2: The Pearson type three distribution was used to fit the AM data sets. The design flood peak or volumes for certain return periods, such as 50 years, 100 years and 1000 years, were obtained.

Step 3: The typical flood hydrographs (TFH) were determined. Usually, the observed flood hydrograph with highest peak or largest volume was selected as a TFH [49].

Step 4: The DFH was constructed by multiplying each discharge ordinate of the TFH by an amplifier [12,13]. Certain amplitude method was used to make the flood peak or volumes of TFH equaling to the design values. In this study, only the volume amplitude method was considered, in accordance with the practical situation. Assume that the T-year design and observed n-day flood volumes are W_d^T and W_O respectively, and the amplifier with return period T is defined as k_T, $k_T = W_d^T/W_O$. The TFH was transformed to DFH by multiplying the parameter k_T. However, some other attributes, such as duration, were also considered for derivation of DFHs in some research works (e.g., Serinaldi and Grimaldi [50]; Vandenberghe et al. [51]; Gräler et al. [52]; Brunner et al. [53]), which were not considered in this study. The design flood hydrographs (DHFs) for the whole season can be obtained finally.

The floods in the years 1954, 1968, 1969, 1980, 1981, 1982, 1983, 1988, 1996, 1998, 1999, 2002 and 2010 were selected as TFHs in this study. Since the summer rainfall in the 1980s and 1990s is much higher than in the previous decades, eight flood hydrographs were selected during these periods [54]. The 1954 and 1998 floods are two serious flood events that happened in the past 100 years, which should be selected for flood control analysis. In the years of 1968 and 2002, the Wu River, which is a main tributary of the upper Yangtze river, experienced a severe flood [55]. Thus, the flood hydrographs of these two years were selected. In the year of 1969, extreme flood events happen simultaneously on the three main tributaries: the Jinsha, Min, and Tuo Rivers. In the year of 2010, the entire Yangtze River basin suffered from tremendous flooding, which led to the number of affected people as about

1.4 billion, and the direct economic loss of 197.6 billion Chinese yuan. According to the shape of these flood hydrographs, they can be divided into 3 types, namely the long duration type, single flood peak type and multiple flood peak type, which contain all the potential characteristics of flood events occurring in the upper Yangtze River Basin. The floods of 1954, 1996 and 1998 show long duration, which lasted for more than one month. For example, the 1954 and 1998 China floods lasted from middle of June to the beginning of September. There is only one main flood peak in the years of 1980, 1981, 1982, 1999 and 2002. In addition, there are multiple flood peaks in the years of 1968, 1969, 1983, 1988 and 2010. All the selected THFs are shown in Figure 4.

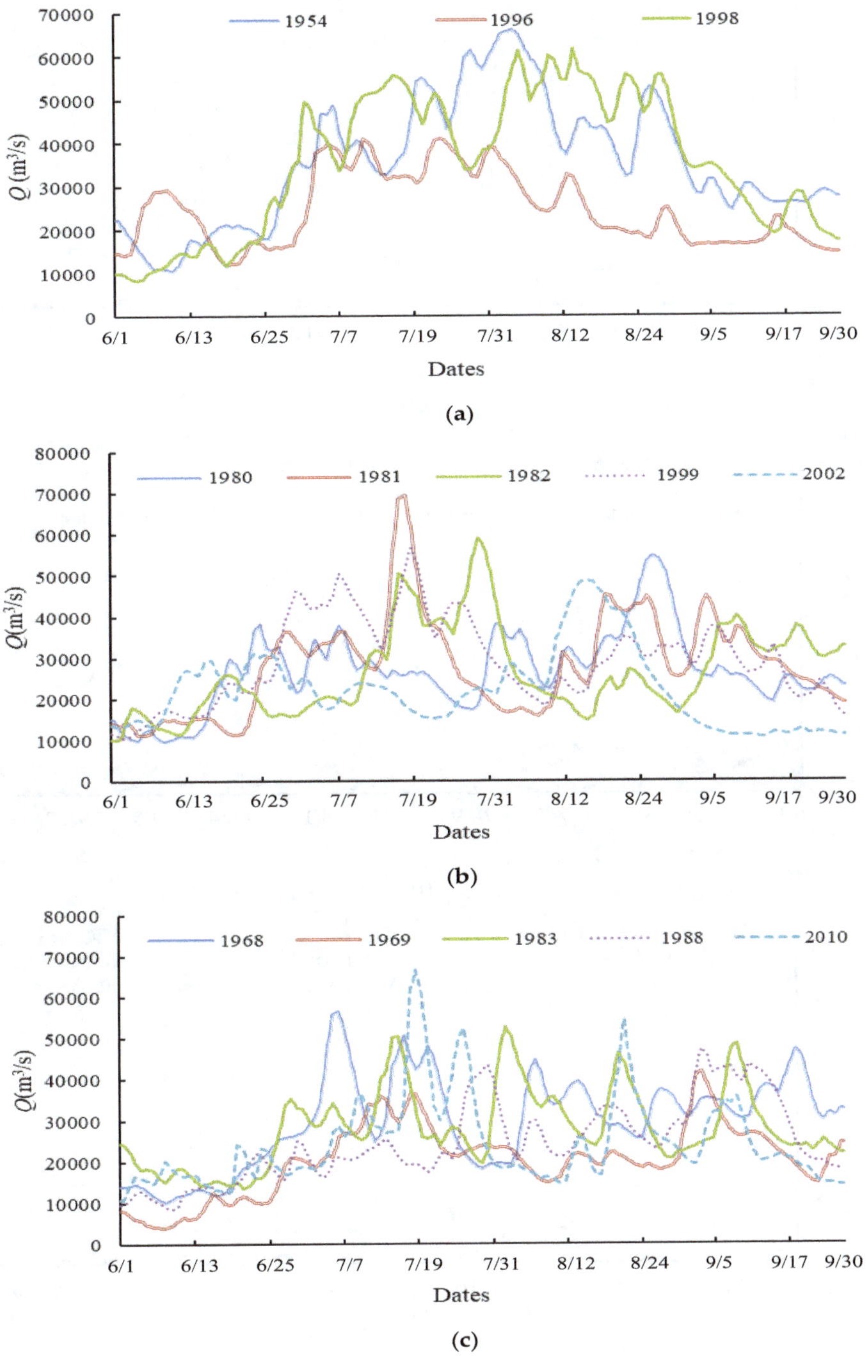

Figure 4. All the typical flood hydrographs (THFs) used in this study. (**a**) Observed flood hydrographs of the years 1954, 1996, and 1998; (**b**) observed flood hydrographs of the years 1980, 1981, 1982, 1999 and 2002; (**c**) observed flood hydrographs of the years 1968, 1969, 1983, 1988 and 2010.

The duration of the flood hydrograph plays a key role in determination of the DFH. In this study, the hydrograph lasts for the whole flood season (1 June to 30 September) [56–58]. All typical flood hydrographs mentioned above were used to derive the DFHs. The volume amplitude method, namely the 30-day volume amplifier, was selected in this study to obtain the DFHs, in accordance with practical use. In other words, the 30-day volume amplifier was employed for amplifying the hydrograph of the whole flood season. Considering the length of the paper, only the design flood hydrographs based on the TFH of 1954 were shown in Figure 5.

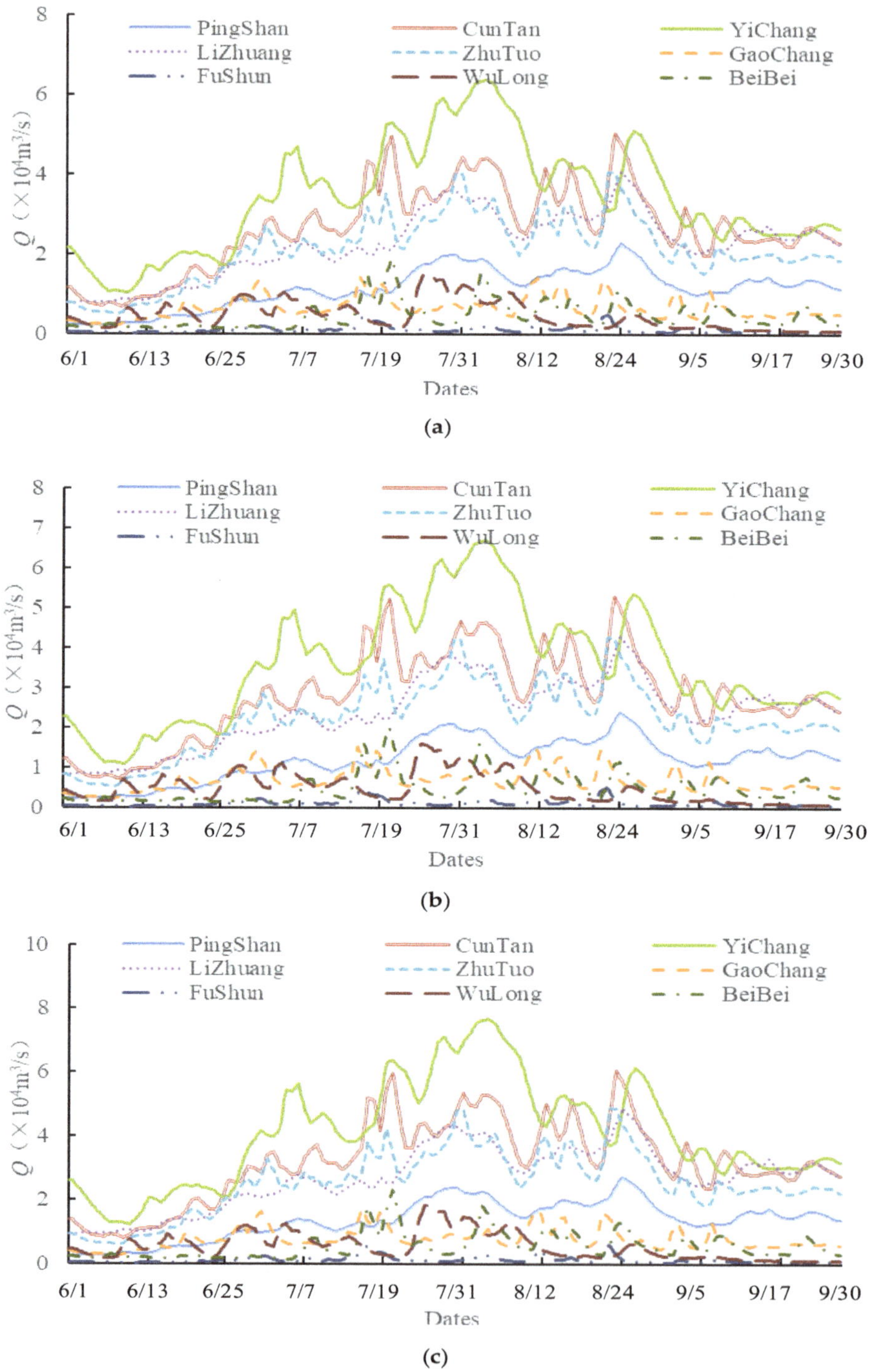

Figure 5. The 50-year, 100-year and 1000-year design flood hydrographs based on the typical flood hydrograph of 1954. (**a**) 50-year; (**b**) 100-year; (**c**) 1000-year.

4.2. *Optimal Reservoir Operation Results*

4.2.1. Operation Results of the Xiluodu and Xiangjiaba Reservoirs

The Xiluodu and Xiangjiaba Dams were used to reduce the inflow of TGR, which is also an effective method to alleviate the flood control burden of the Jingjiang and Chenglingji areas. A two-stage optimization algorithm was employed to solve the optimization model of the Xiluodu and Xiangjiaba cascade reservoirs, because of the "curse of dimensionality" and time-lag issue for multi-reservoir systems. First, the DP method was used to obtain the optimal solution set. Second, the output of the DP method was the input of the POA method, which avoids the influence of initial solutions on the optimal solutions. The years of 1954, 1968, 1969, 1980, 1981, 1982, 1983, 1988, 1996, 1998, 1999, 2002, and 2010 were selected as TFHs. The real-time water level and discharge at each time period were calculated by the flood hydrograph routing method.

Operation results of each DFH are given in Table 3, in which the inflow values of TGR and the flood stored in the Xiluodu and Xiangjiaba Dams are shown. The flood reductions are also given in Table 3.

As shown in Table 3, the Xiluodu and Xiangjiaba cascade reservoirs can obviously reduce the the inflow of TGR. For the 1000-year design flood, amplified based on the TFH of 1998, the maximum inflow of the Three Gorges decreased from 109,705.2 m^3/s to 92,570.6 m^3/s. The flood peak was reduced by 17,134.6 m^3/s and its corresponding reducing rate was 15.6%. The average reduction of flood peak for all these 13 TFHs can reach 13.6%. Therefore, the decrease of the peak inflow can relieve the flood control burdens of TGD. Additionally, the cascade reservoirs can also store much more water during a flood event, and the maximum volumes stored in those two reservoirs upstream in this study can reach 252.1 $\times$ 10^8 m^3. It also can be seen from Table 3 that the two-stage optimization algorithm has a good performance in solving the cascade reservoirs optimization model. Furthermore, when the POA method was used in the second stage, the model shows higher efficiency. More water was stored in the two reservoirs, and the inflow of TGR was further reduced. For instance, the maximum inflow of the 100-year design flood based on 1998 TFH reduced to 59,905.7 m^3/s in the first stage, and to 57,890.4 m^3/s in the second stage.

Figure 6 presents the 1000-year design flood hydrographs of 1954, and the inflow of TGR after the reservoir operation. The inflows regulated by the cascade reservoirs are much flatter with smaller peak values, which means that the inflow is easier to control.

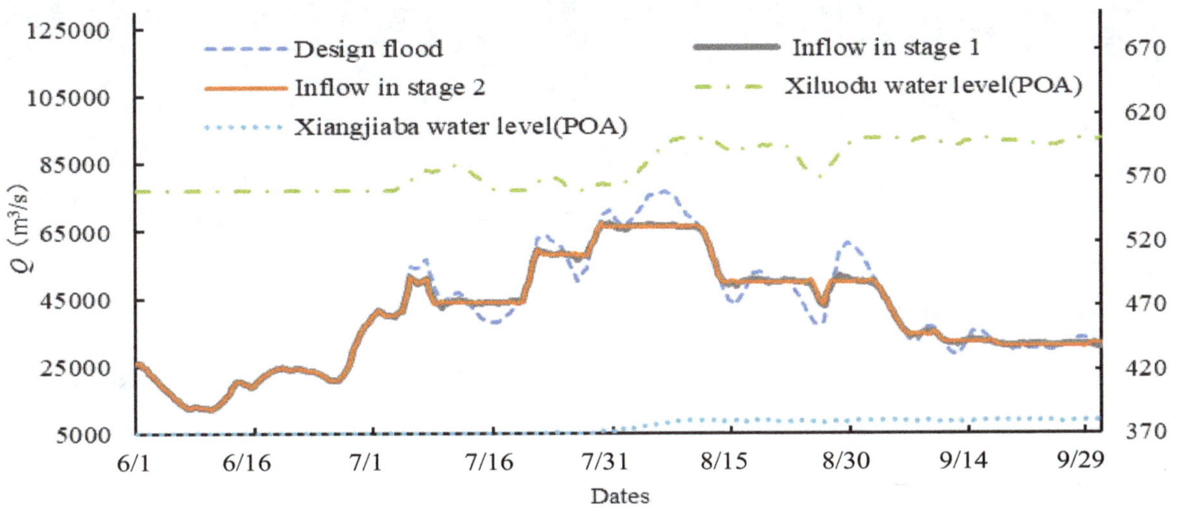

Figure 6. Design flood hydrograph of the year 1954, inflow of TGR and water levels of the Xiluodu and Xiangjiaba Dams.

Table 3. Operation results of the Xiluodu–Xiangjiaba cascade reservoirs.

TFH [1]	Return Periods (Year)	Inflow of the Three Gorges (m^3/s)					Flood Stored in the Xiluodu and Xiangjiaba Reservoirs (10^8 m^3)		
		Natural Runoff	Stage 1	Flood Reduction	Stage 2	Flood Reduction	Stage 1	Stage 2	Comparisons
1954	50	76,676	67,500.34	9175.66	67,284.74	215.60	144.28	159.37	15.08
	100	67,025.4	58,388.07	8637.32	58,147.99	240.08	138.16	152.22	14.06
	1000	63,720.4	55,267.93	8452.47	54,993.17	274.75	133.9	148.0	14.1
1968	50	96,390	84,091.62	12,298.37	84,091.0	0.57	240.2	252.1	11.9
	100	83,916	74,102.2	9813.77	74,101.3	0.92	234.2	243.5	9.3
	1000	79,833.6	70,840.2	8993.4	70,839.8	0.35	231.6	239.9	8.3
1969	50	92,599	77,086.3	15,512.6	76,456.3	630.1	160.0	174.3	14.3
	100	80,908.9	64,984.7	15,924.1	64,588.8	395.9	150.9	162.7	11.8
	1000	76,928.4	61,000.5	15,927.9	60,846.7	153.7	147.6	158.6	11.0
1980	50	91,182	79,087.6	12,094.38	78,731.4	356.17	163.8	175.7	11.9
	100	79,661.4	66,305.01	13,356.39	66,116.25	188.75	167.32	179.13	11.8
	1000	75,730.2	63,189.1	12,541.04	63,160.1	29.03	165.79	176.63	10.8
1981	50	111,200	96,768.65	14,431.35	96,689.93	78.71	177.64	188.79	11.1
	100	97,091.5	88,907.63	8183.8	88,408.22	499.4	151.5	161.02	9.4
	1000	92,921.5	81,542.78	11,378.72	81,504.39	38.38	156.96	165.87	8.9
1982	50	92,040	79,071.08	12,968.92	79,030.08	41	162.40	174.88	12.48
	100	80,417	69,906.41	10,510.5	69,889.25	17.16	149.2	163.11	13.9
	1000	76,523	66,847.61	9675.39	66,832.96	14.64	143.3	157.4	14.06
1983	50	95,732	87,083.95	8648.05	87,083.9	0.05	182.1	195.1	13.0
	100	82,582	74,997.29	7584.7	74,997.29	0	167.09	186.42	19.3
	1000	78,479.2	70,371.11	8108.08	70,371.1	0	188.91	206.8	17.89
1988	50	80,106	69,897.8	10,208.17	68,171.5	1726.338	155.74	169.56	13.8
	100	69,772.8	59,948.75	9824.05	59,915.72	33.03	144.4	156.49	12.04
	1000	66,360	57,210.63	9149.37	57,210.63	0	142.0	153.2	11.2
1996	50	78,501	69,472.82	9028.18	69,058.74	414.1	153.67	181.93	28.25
	100	68,554.8	60,624.35	7930.4	60,622.61	1.74	140.67	174.7	34.02
	1000	65,184.6	57,982.89	7201.7	57,982.76	0.13	136.29	170.24	33.9
1998	50	109,705.2	92,571.2	17,133.9	92,570.6	0.6	192.3	201.2	8.8
	100	67,438.1	59,905.7	7532.3	57,890.4	2015.3	159.4	176.6	17.2
	1000	64,168	56,802.3	7365.6	54,815.2	1987.0	157.7	174.3	16.6
1999	50	90,153	73,305.4	16,847.5	73,306.6	-1.1	180.9	197.6	16.7
	100	78,642.9	64,300.6	14,342.6	64,143.8	156.4	168.5	185.0	16.4
	1000	74,787.3	61,446.0	13,341.2	61,318.9	127.1	164.1	181.5	17.4
2002	50	81,648	70,413.2	11,234.7	69,874.5	538.7	127.5	136.5	9.0
	100	71,344.8	60,551.9	10,792.8	60,001.7	550.2	117.2	124.5	7.3
	1000	67,845.6	57,273.2	10,572.3	56,824.3	448.8	113.5	120.7	7.2
2010	50	109,705.2	99,823.4	9881.8	99,823.0	0.39	190.3	201.8	11.4
	100	96,407.6	88,551.2	7856.3	88,550.7	0.52	176.4	189.3	12.9
	1000	91,088.56	84,048.3	7040.2	84,047.9	0.5	167.5	179.7	12.2

[1] TFH: typical flood hydrograph.

4.2.2. Operation Results of the Three Gorges Reservoir

Considering the flood control in the middle and lower reaches of the Yangtze River, this paper compared two operation methods, namely single reservoir operation model (TGR) and the cascade operation model (Xiluodu, Xiangjiaba and TGR).

Tables 4 and 5 present the flood diversion volumes of each DFH at the Jingjiang and Chenglingji flood control points, respectively, when different operation models were used. As shown in Table 4, the flood in the Jingjiang area does not need to be diverged for all 50-year DFHs. That means both the single and cascade models can ensure the water level of the Jingjiang area not exceeding the safety water level, when a 50-year flood event occurs. For 100-year DHFs, the single model can protect the Jingjiang areas from flooding for most of DFHs, except for the year of 1968, which means that when a single reservoir is used, the 100-year flood control standard for the Jingjiang area cannot always be guaranteed. When the cascade model used, there is no need to diverse the flood volumes and the safety of this areas can be ensured for the flood events with the 100-year return period. Generally, the differences between single reservoir operation model and the cascade models are not so obvious for the 50-year and 100-year DFHs. But for the 1000-year DFHs, it is clear from Table 4 that the flood diversion volumes obtained by the cascade operation model are less than those by the single reservoir model. For example, for the 1000-year design flood amplified based on the TFH of 1988, the flood diversion at the Jingjiang flood control point is 20.1 billion m^3 calculated by the single operation model, but reduced to zero when the Xiluodu and Xiangjiaba reservoirs were taken into accounts. It can be seen from Table 5 that the volumes decrease significantly when the cascade reservoir models are used. Tables 4 and 5 show that the cascade optimal operation model works better than single operation model in reducing the flood diversion volumes in the Jingjiang and Chenglingji areas.

Table 4. Flood diversion volumes of each design flood hydrograph (DFH) at the Jingjiang flood control point (billion m^3).

TFH	Return Periods (Year)	Without Operations	For the Jingjiang Area		For the Chenglingji Area	
			Single	Cascade	Single	Cascade
1954	50	4.8	0	0	0	0
	100	8.4	0	0	0	0
	1000	23.7	0	0	0	0
1968	50	17.0	0	0	0	0
	100	24.0	4.2	0	4.2	0
	1000	56.5	33.9	37.9	56.4	37.9
1969	50	12.6	0	0	0	0
	100	17.7	0	0	0	0
	1000	37.5	17.3	12.2	17.2	12.2
1980	50	8.5	0	0	0	0
	100	11.1	0	0	0	0
	1000	24.8	0	0	0	0
1981	50	14.0	0	0	0	0
	100	18.5	0	0	0	0
	1000	38.1	0	0	0	0
1982	50	8.3	0	0	0	0
	100	11.3	0	0	0	0
	1000	26.0	17.8	0	16.9	0
1983	50	19.0	0	0	0	0
	100	25.5	0	0	0	0
	1000	55.2	55.2	39.6	55.1	39.7
1988	50	5.9	0	0	0	0
	100	10.5	0	0	0	0
	1000	26.7	20.1	0	20.1	0

Table 4. *Cont.*

TFH	Return Periods (Year)	Without Operations	For the Jingjiang Area		For the Chenglingji Area	
			Single	Cascade	Single	Cascade
1996	50	9.7	0	0	0	0
	100	15.9	0	0	0	0
	1000	42.1	18.9	13.4	18.9	13.4
1998	50	6.0	0	0	0	0
	100	12.9	0	0	0	0
	1000	29.4	0	0	0	0
1999	50	9.4	0	0	0	0
	100	15.2	0	0	0	0
	1000	37.5	14.3	8.9	14.3	8.9
2002	50	6.8	0	0	0	0
	100	9.8	0	0	0	0
	1000	20.0	0	0	0	0
2010	50	13.4	0	0	0	0
	100	17.2	0	0	0	0
	1000	29.4	0	0	0	0

Table 5. Flood diversion volumes of each design flood hydrograph (DFH) at the Chenglingji flood control point (billion m^3).

TFH	Return Periods (Year)	Without Operations	For the Jingjiang Area		For the Chenglingji Area	
			Single	Cascade	Single	Cascade
1954	50	31.2	35.4	26.4	9.6	4.0
	100	37.1	45.3	36.6	19.3	13.7
	1000	54.3	74.2	70.1	59.8	54.2
1968	50	59.9	73.5	63.8	64.5	59.0
	100	69.8	88.4	80.8	79.7	78.4
	1000	93.9	111.1	101.7	93.9	102.1
1969	50	37.7	50.0	38.8	33.2	28.3
	100	45.2	61.7	50.5	50.5	45.6
	1000	70.8	88.8	87.3	88.9	87.4
1980	50	28.7	35.7	26.0	12.1	6.6
	100	36.2	46.1	35.9	23.5	18.0
	1000	61.6	83.9	74.2	77.2	71.6
1981	50	39.8	49.4	38.8	20.5	14.4
	100	46.3	59.1	48.2	35.8	30.3
	1000	68.3	98.0	91.2	89.9	84.4
1982	50	28.9	35.1	25.6	14.8	3.7
	100	35.5	43.2	34.3	24.6	13.6
	1000	51.7	67.9	61.9	56.4	52.5
1983	50	47.6	64.4	52.5	52.4	46.9
	100	55.5	78.9	69.9	72.5	66.9
	1000	83.0	83.0	92.2	82.8	91.7
1988	50	30.5	34.4	25.4	17.8	12.4
	100	34.4	41.1	33.6	30.3	24.8
	1000	48.2	54.1	62.8	52.4	62.6
1996	50	47.6	57.1	51.5	32.5	26.9
	100	54.0	68.0	63.9	43.7	38.1
	1000	68.7	89.4	88.8	63.1	63.0
1998	50	55.5	61.2	54.1	37.5	32.0
	100	62.0	73.7	68.1	51.9	46.4
	1000	41.5	61.5	53.0	39.3	33.7

Table 5. *Cont.*

TFH	Return Periods (Year)	Without Operations	For the Jingjiang Area		For the Chenglingji Area	
			Single	Cascade	Single	Cascade
1999	50	43.5	52.2	41.1	24.7	19.1
	100	50.6	62.3	52.9	42.6	37.1
	1000	71.7	89.6	88.9	82.6	82.5
2002	50	18.2	22.9	19.4	1.6	0.0
	100	20.8	27.1	24.0	5.6	0.1
	1000	32.0	43.1	40.5	19.8	14.6
2010	50	23.2	32.1	23.1	3.0	0.0
	100	27.9	39.3	29.3	9.0	3.3
	1000	41.5	61.5	52.7	39.3	33.7

Table 6 gives the highest water level of TGR of each TFH with different return periods, for the flood control of the Jingjiang area, and Table 7 for the flood control of both the Jiangjiang and Chenglingji areas. Tables 6 and 7 illustrate that the highest water level of TGR generally is lower, when the cascade optimal operation model was considered instead of the single operation model. For the cascade optimal operation model considering the Jingjiang flood control point only, the highest water level nearly reached 170 m for the 1000-year DFH amplified by three TFHs. For a 100-year design flood, the maximum highest water level was 164.1 m, and the minimum was only 150.8 m. These results indicate that when the cascade optimal operation model is in use, the safety of TGR can be ensured, but the reservoir's flood control capacity is underutilized only considering the Jingjiang flood control area. Furthermore, as shown in Table 7, when both the Jingjiang and Chenglingji flood control points were taken into account, the highest water level of TGR rises obviously, which proves that the flood control capacity of TGR is further utilized. Thus, both the Jingjiang and Chenglingji flood control areas can be protected in a practical flood control operation.

Table 6. The highest water level of Three Gorges Reservoir considering the flood control for the Jingjiang area (m).

TFH	1954			1981			1998		
Return Periods (Year)	1000	100	50	1000	100	50	1000	100	50
Single mode	175.0	159.3	153.9	174.4	166.1	163.9	174.7	160.7	153.4
Cascade mode	169.3	150.8	145.0	169.4	164.1	156.4	169.8	152.4	145.3
Comparison	5.7	8.5	8.9	5.0	2.0	7.5	4.9	8.3	8.1

Table 7. The highest water levelof Three Gorges Reservoir considering the flood control for the Jingjiang and Chenglingji areas (m).

TFH	1954			1981			1998		
Return Periods (Year)	1000	100	50	1000	100	50	1000	100	50
Single mode	175	175	175	175	175	175	175	175	175
Cascade mode	171.9	175	175	170.8	172.0	173.0	172.3	175	175
Comparison	3.1	0	0	4.2	3.0	2.0	2.7	0	0

Figure 7 presents the discharge and water level hydrographs of TGR calculated by single and cascade optimal operation models, considering the flood control in the Chenglingji area, for the 1000-year design flood of the year 1954. This figure demonstrates that results obtained by the cascade optimal operation model is generally better than that obtained by single operation, since the peak value is much smaller and the discharge curve is much flatter. Especially, the highest water level of TGR was also lower.

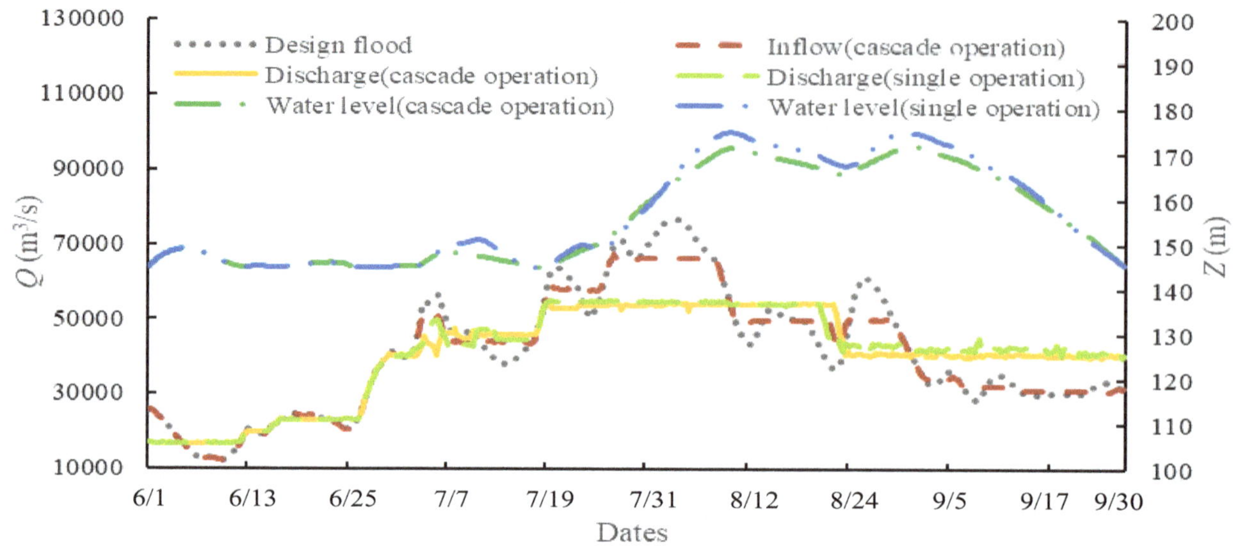

Figure 7. Discharge and water level hydrographs of TGR obtained by single and cascade optimal operation models for the 1954 DFH.

4.3. Comparisons with the Current Operation Rules of Xiluodu, Xiangjiaba and TGD in the Flood Season

To testify the feasibility of the cascade optimal operation model, this paper also compares the optimal operation model with current operation method of these three dams, where the TFHs in use are the same as those mentioned in Section 4.1 in this paper.

The flood diversion volumes at the Jingjiang and Chenglingji flood control points were calculated by the optimal and current operation models, since the objective of the reservoir operation is to minimize the flood diversion volumes at the control points. As shown in Tables 8 and 9, compared with the current operation model, the flood diversion volumes calculated by the optimal operation rules decrease significantly. From this point of view, the optimal model, which has the smaller flood diversion volumes, performs better than the current operation model. For the 50-year and 100-year design flood, when both of the two methods are used, the safety of the Jingjiang area can be ensured. But for the 1000-year DFHs the cascade optimal operation method has a better performance than the current used model, obviously. For instance, for the 1000-year design flood based on the TFH of 1999, the flood diversion volumes in the Jingjiang area obtained by the current operation method are 13.8 billion m^3 (considering the Jingjiang flood control point only) and 18.2 billion m^3 (considering the Jingjiang and Chenglingji flood control points); while they are the optimal operation models used, these diversion values can be reduced to 8.9 billion m^3 for both of the flood control points. This improvement is much more significant for the Chenglingji area. Generally, the cascade optimal operation model can be with less flood diversion volumes than the current operation model for all DFHs with different return periods. Hence, it can be concluded that the cascade optimal operation model can improve the flood control capacity in the middle and lower reaches of the Yangtze River, compared with the current operation model. All the results obtained by the two models do not consider the uncertainties in flood control reservoir operation, which can be discussed in future research.

Table 8. Flood diversion volumes of some design flood hydrographs at the Jingjiang flood control point (billion m^3).

TFH	Return Periods (Year)	Without Operations	For the Jingjiang Area		For both the Jingjiang and Chenglingji Areas	
			Current	Optimal	Current	Optimal
1954	50	4.8	0	0	0	0
	100	8.4	0	0	0	0
	1000	23.7	0	0	5.7	0
1980	50	8.5	0	0	0	0
	100	11.1	0	0	0	0
	1000	24.8	0	0	4.1	0
1981	50	14.0	0	0	0	0
	100	18.5	0	0	0	0
	1000	38.1	0	0	5.2	0
1996	50	9.7	0	0	0	0
	100	15.9	0	0	0	0
	1000	42.1	18.7	13.4	22.8	133.8
1999	50	9.4	0	0	0	0
	100	15.2	0	0	0	0
	1000	37.5	13.8	8.9	18.2	8.9
2002	50	6.8	0	0	0	0
	100	9.8	0	0	0	0
	1000	20.0	0	0	2.8	0

Table 9. Flood diversion volumes of some design flood hydrographs at the Chenglingji flood control point (billion m^3).

TFH	Return Periods (Year)	Without Operations	For the Jingjiang Area		For both the Jingjiang and Chenglingji Areas	
			Current	Optimal	Current	Optimal
1954	50	31.2	35.5	26.4	26.8	4.0
	100	37.1	45.2	36.6	36.2	13.7
	1000	54.3	75.1	70.1	65.3	54.2
1980	50	28.7	35.7	26.0	21.1	6.6
	100	36.2	46.2	35.9	32.7	18.0
	1000	61.6	84.2	74.2	69.5	71.6
1981	50	39.8	50.7	38.8	38.2	14.4
	100	46.3	60.4	48.2	48.2	30.3
	1000	68.3	99.2	91.2	86.4	84.4
1996	50	47.6	57.1	51.5	49.8	26.9
	100	54.0	68.6	63.9	59.9	38.1
	1000	68.7	89.8	88.8	74.0	63.0
1999	50	43.5	51.9	41.1	41.5	19.1
	100	50.6	62.6	52.9	52.4	37.1
	1000	71.7	90.4	88.9	79.4	82.5
2002	50	18.2	23.7	19.4	17.8	0
	100	20.8	28.3	24.0	20.7	0.1
	1000	32.0	44.8	40.5	32.3	14.6

Figure 8 displays the 1000-year design inflow flood hydrograph of 1954, in which one hydrograph is operated by the current operation model and the other by the Xiluodu and Xiangjiaba cascade optimal operation model. Figure 9 shows the discharge and water level curves of TGR obtained respectively by current and cascade optimal operation models considering the flood control of the Chenglingji area. It is obvious that when using the cascade optimal operation model, both the inflow and discharge hydrographs of TGR are flatter and have smaller peak values than those obtained by the current model. For example, the maximum discharge value of 1980 DHF with the return period 1000 years reduces from 76,000 m^3/s to 54,995 m^3/s, which can ensure the safety of the Jingjiang River reach. These results also indicate that the capacities of the Xiluodu and Xiangjiaba reservoirs were further utilized when the cascade optimal operation model was used, which alleviates the flood control burdens of TGR.

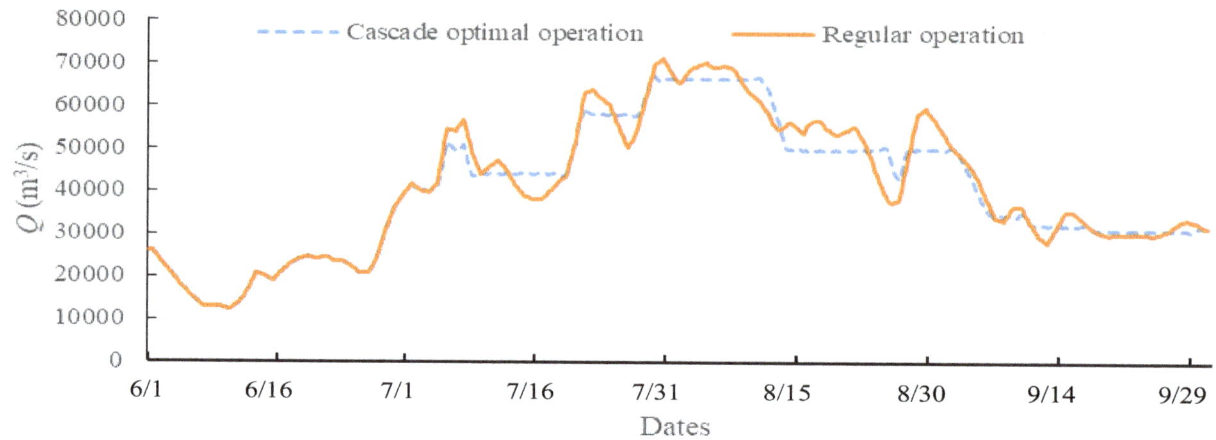

Figure 8. Inflow of TGR obtained by the two operation models for the 1000-year DHF amplified by the 1954 TFH.

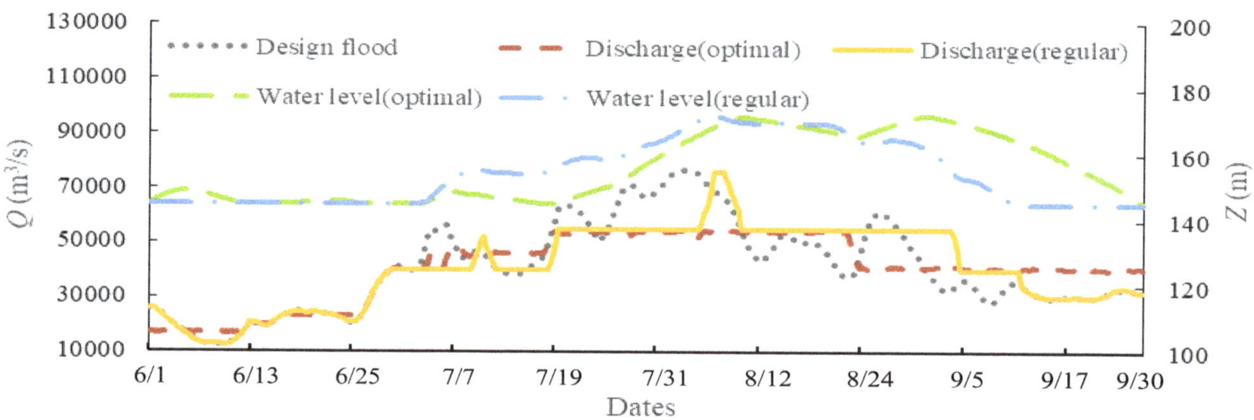

Figure 9. Discharge and water level hydrographs of Three Gorges Reservoir (TGR) obtained by regular and optimal operation models.

In order to further testify the reliability of the model, the historical observations, including the flood hydrographs of the years 1954, 1968, and 1998, were considered in this section and taken as the inputs of the current and optimal models. The highest water level of TGR routed by the proposed and current models are shown in Table 10, which indicates that the highest water level of TGR generally is lower based on the optimal model.

Table 10. The highest water level of TGR by two models (m).

TFH	1954	1968	1998
Cascade regular operation model	157.3	145.3	147.9
Cascade optimal operation model	148.8	145	145.2
Comparisons	8.5	0.3	2.7

According to the comparisons given in Table 10, it can be seen that the cascade optimal operation model can produce superior results to the current operation model. Therefore, the flood control capacity of the middle and lower reaches of the Yangtze river can be enhanced. The cascade optimal operation model built in this paper worked well in a multi-reservoir flood control operation system, and can be considered in practical application.

5. Conclusions and Discussions

This study established optimal flood control operation models of cascade reservoirs for flood control of multiple points downstream. The current multi-reservoir flood control co-operation models mainly concern minimizing the maximal water level for the dam itself or the releases at the downstream flood control points, or balancing the conflict between the flood control and conservation during the flood season. In this study, a two-stage optimal operation model of cascade reservoirs was proposed to ensure the safety of multiple downstream areas, in which the objective functions are to minimize the inflow of the reservoir C downstream in the first stage and the water level or flood diversion volume in the second stage. For a large river basin, the river reach is very long, the time-lag issue and Muskingum flood routing along the river cannot be neglected. A DP-POA optimal algorithm was proposed in this study for the establishment of the optimal model to avoid the curse of dimensionality and influence of initial solutions. The upper Yangtze River was selected as a case study. The flood capacity of TGR cannot hold back floods downstream, when extreme flooding occurred throughout the whole upper Yangtze River Basin. Therefore, more reservoirs are needed to help TGR realize the purpose of flood control. The two reservoirs, Xiluodu and Xiangjiaba located in the Jinsha River Basin, were considered in this study to reduce the inflow of TGR. The main conclusions of this paper are summarized as follows.

(1) The first innovation of this paper is the proposed two-stage optimal operation model with different objective functions. In the first stage, the optimal operation model of Reservoirs A and B was established to reduce the inflow of the downstream reservoir. In the second stage, the optimal operation model of the reservoir downstream was established to minimize the highest water level and the flood diversion volumes at flood control points. The second innovation of this paper is the proposed DP-POA algorithm, which can solve the curse of dimensionality and influence initial solutions effectively. The results of Figures 6 and 7 indicate that optimal reservoir operation processes can be obtained. The Muskingum flood routing method combined with the time delay makes the operation's results much more objective. The two-stage optimization algorithm has a good performance in solving the cascade reservoirs optimization model. When the POA method was used in the second stage, the model shows higher efficiency as more water is stored in the two reservoirs and the inflow of the reservoir downstream is further reduced.

(2) Based on the proposed model, the Xiluodu and Xiangjiaba Reservoirs can effectively reduce the peak inflow and volumes of TGR. After the optimal reservoir operation of Xiluodu and Xiangjiaba, the average reduction of flood peaks for all these 13 TFHs are 13.6%. Meanwhile, the cascade reservoirs can also store much more water during a flood event, and the maximum volumes stored in those two reservoirs upstream in this study can reach 25.2 billion m^3 during a flood event.

(3) The cascade optimal operation model works better than the single operation model. The highest water level of TGR is generally lower when the cascade optimal operation model is considered instead of the single operation model. For the 1000-year DFHs, it is obvious that the flood diversion volumes obtained by the cascade operation model are less than those by the single reservoir model. After the optimal operation of cascade reservoirs, the peak value is much smaller and the discharge curve is much flatter, which is beneficial for the flood control of the middle and lower reaches of the Yangtze River.

(4) Compared with the current operation rules, the cascade optimal operation model can produce a superior result to the current operation model. For the 50-year and 100-year design floods, when both of the two methods were used, the safety of the Jingjiang area can be ensured. But for the 1000-year

DFI Is, the cascade optimal operation method has a better performance than the current one obviously. For the 1000-year design flood amplified based on the TFH of 1999, the diversion values can decline to 8.9 billion m^3 for both of the Jingjiang and Chengliangji flood control points, when the cascade optimal operation model was used.

This paper only discussed the downstream flood control points that are protected by cascade reservoirs upstream. The proposed work can give some guidance for real-time flood management. Flood control reservoir operation rules can be developed based on the optimal results. In future, more reservoirs, including cascade reservoirs and parallel reservoirs, will be considered, and a more efficient algorithm will be proposed. Furthermore, the main purpose of this paper is to discuss the protection of the flood control areas downstream based on the multi-reservoir systems. The overtopping risk of the dam is not discussed in the paper.

In addition, many uncertainties also present in this complex problem have an influence on the flood control reservoir operation. Shi et al. [59], Zhang et al. [60] and Chen et al. [61] indicated that the most important uncertainty factor that affects the flood control reservoir operation was the flood forecast error. The predicted flow is the input of the flood control reservoir operation model, which has a great influence on the operation results. Besides flood forecast errors, Li et al. [19] and Huang et al. [62] demonstrated that the uncertainties in flood hydrograph shape also had a significant influence on the highest water level and maximum discharge. Besides these, Chen et al. [63] indicated that the uncertainties in the discharge capacity curve and water level-storage curve also influenced reservoir operation greatly. Onyutha and Willems [4] assessed the uncertainties in estimation of flood quantiles and/or hydrographs, especially of very high return periods which is useful for various types of risk-based water management applications related to floods. These uncertainties were not considered in this deterministic modelling, because of the burden of the optimal calculation. Stochastic frameworks, which can work better for these uncertainty problems, will be studied in the future research.

Author Contributions: C.Z. proposed the ideas of this paper and guided the writing and calculation work; L.C. designed the experiments and wrote most parts of the paper; N.S. prepared the manuscript and made some corrections; Y.D. and J.Z. made some corrections; G.Z., L.D., G.L. and X.Y. together did the calculation work and draw the figures.

Acknowledgments: This research was founded by the National Key R&D Program of China, grant number [2017YFC0405900, 2016YFC0402210; 2016YFC0402205]; the National Natural Science Foundation of China, grant number [51679094]; and the project with the title of "impact of flood control multi-reservoir co-operation in the upper Yangtze River Basin on the Three Gorges Reservoir and its corresponding risk analysis".

References

1. Webster, T.; Mcguigan, K.; Collins, K.; Macdonald, C. Integrated river and coastal hydrodynamic flood risk mapping of the lahave river estuary and town of Bridgewater, Nova Scotia, Canada. *Water* **2014**, *6*, 517–546. [CrossRef]
2. Schnebele, E.; Cervone, G.; Kumar, S.; Waters, N. Real time estimation of the calgary floods using limited remote sensing data. *Water* **2014**, *6*, 381–398. [CrossRef]
3. Morita, M. Flood risk impact factor for comparatively evaluating the main causes that contribute to flood risk in urban drainage areas. *Water* **2015**, *6*, 253–270. [CrossRef]
4. Onyutha, C.; Willems, P. Uncertainties in flow-duration-frequency relationships of high and low flow extremes in lake victoria basin. *Water* **2013**, *5*, 1561–1579. [CrossRef]
5. Assani, A.; Landry, R.; Labrèche, M.; Frenette, J.J.; Gratton, D. Temporal variability of monthly daily extreme water levels in the st. Lawrence river at the sorel station from 1912 to 2010. *Water* **2014**, *6*, 196–212. [CrossRef]

6. Joo, J.; Lee, J.; Kim, J.H.; Jun, H.; Jo, D. Inter-event time definition setting procedure for urban drainage systems. *Water* **2013**, *6*, 45–58. [CrossRef]
7. Keast, D.; Ellison, J. Magnitude frequency analysis of small floods using the annual and partial series. *Water* **2013**, *5*, 1816–1829. [CrossRef]
8. Box, P.; Thomalla, F.; van den Honert, R. Flood risk in australia: Whose responsibility is it, anyway? *Water* **2013**, *5*, 1580–1597. [CrossRef]
9. Zhou, Y.L.; Guo, S.L.; Chang, F.J.; Liu, P.; Chen, A.B. Methodology that improves water utilization and hydropower generation without increasing flood risk in mega cascade reservoirs. *Energy* **2018**, *143*, 785–796. [CrossRef]
10. Zhang, W.; Liu, P.; Chen, X.; Wang, L.; Ai, X.; Feng, M.; Liu, D.; Liu, Y. Optimal operation of multi-reservoir systems considering time-lags of flood routing. *Water Resour. Manag.* **2016**, *30*, 523–540. [CrossRef]
11. Chen, C.; Yuan, Y.; Yuan, X. An improved nsga-iii algorithm for reservoir flood control operation. *Water Resour. Manag.* **2017**, *31*, 4469–4483. [CrossRef]
12. Chen, L.; Singh, V.; Huang, K. Bayesian technique for the selection of probability distributions for frequency analyses of hydrometeorological extremes. *Entropy* **2018**, *20*, 117. [CrossRef]
13. Chen, L.; Singh, V.P. Entropy-based derivation of generalized distributions for hydrometeorological frequency analysis. *J. Hydrol.* **2018**, *557*, 699–712. [CrossRef]
14. Narasaiah, M.L. *Microcredit and Agricultural Development*; Discovery Publish House: New Delhi, India, 2007.
15. Liu, X.; Chen, L.; Zhu, Y.; Singh, V.P.; Qu, G.; Guo, X. Multi-objective reservoir operation during flood season considering spillway optimization. *J. Hydrol.* **2017**, *552*, 554–563. [CrossRef]
16. Yazdi, J.; Neyshabouri, S.A.A.S. Optimal design of flood-control multi-reservoir system on a watershed scale. *Nat. Hazards* **2012**, *63*, 629–646. [CrossRef]
17. Lee, S.Y.; Hamlet, A.F.; Fitzgerald, C.J.; Burges, S.J. Optimized flood control in the columbia river basin for a global warming scenario. *J. Water Resour. Plan. Manag. ASCE* **2009**, *135*, 440–450. [CrossRef]
18. Qi, Y.; Yu, J.; Li, X.; Wei, Y.; Miao, Q. Reservoir flood control operation using multi-objective evolutionary algorithm with decomposition and preferences. *Appl. Soft Comput.* **2017**, *50*, 21–33. [CrossRef]
19. Li, X.; Guo, S.; Liu, P.; Chen, G. Dynamic control of flood limited water level for reservoir operation by considering inflow uncertainty. *J. Hydrol.* **2010**, *391*, 126–134. [CrossRef]
20. Chen, J.; Guo, S.; Li, Y.; Liu, P.; Zhou, Y. Joint operation and dynamic control of flood limiting water levels for cascade reservoirs. *Water Resour. Manag.* **2013**, *27*, 749–763. [CrossRef]
21. Zhou, Y.; Guo, S.; Liu, P.; Xu, C. Joint operation and dynamic control of flood limiting water levels for mixed cascade reservoir systems. *J. Hydrol.* **2014**, *519*, 248–257. [CrossRef]
22. He, X.; Zhang, L.; Liu, L. Study on flood control operation of jinsha river cascade reservoirs combined with the three gorges along the yangtze river. *Mater. Sci. Eng.* **2017**, *199*, 012032.
23. Sahoo, B. Field application of the multilinear muskingum discharge routing method. *Water Resour. Manag.* **2013**, *27*, 1193–1205. [CrossRef]
24. Khan, M.H. Muskingum flood routing model for multiple tributaries. *Water Resour. Res.* **1993**, *29*, 1057–1062. [CrossRef]
25. Gill, M.A. Flood routing by the muskingum method. *J. Hydrol.* **1978**, *36*, 353–363. [CrossRef]
26. Yeh, W.W.G. Reservoir management and operations models: A state-of-the-art review. *Water Resour. Res.* **1985**, *21*, 1797–1818. [CrossRef]
27. Choudhury, P. Reservoir flood control operation model incorporating multiple uncontrolled water flows. *Lakes Reserv. Res. Manag.* **2010**, *15*, 153–163. [CrossRef]
28. Hall, W.A.; Butcher, W.S.; Esogbue, A. Optimization of the operation of a multiple-purpose reservoir by dynamic programming. *Water Resour. Res.* **1968**, *4*, 471–477. [CrossRef]
29. Loucks, D.P.; Sigvaldason, O.T. *Multiple-Reservoir Operation in North America*; American Society of Civil Engineers (ASCE): Reston, VA, USA, 1982; pp. 711–728.
30. Labadie, J.W. Optimal operation of multireservoir systems: State-of-the-art review. *J. Water Resour. Plan. Manag.* **2004**, *130*, 93–111. [CrossRef]
31. Butcher, W.S. Stochastic dynamic programming for optimum reservoir operation. *Water Resour. Bull.* **1971**, *7*, 115–123. [CrossRef]
32. Yeh, W.W.G.; Becker, L.; Chu, W.-S. Real-time hourly reservoir operation. *J. Water Resour. Plan. Manag. Div. Am. Soc. Civ. Eng.* **1979**, *105*, 187–203.

33. Yakowitz, S. Dynamic programming applications in water resources. *Water Resour. Res.* **1982**, *18*, 673–696. [CrossRef]
34. Kelman, J.; Stedinger, J.R.; Cooper, L.A.; Hsu, E.; Sun-Quan, Y. Sampling stochastic dynamic programming applied to reservoir operation. *Water Resour. Res.* **1990**, *26*, 447–454. [CrossRef]
35. Heidari, M.; Chow, V.T.; Kokotović, P.V.; Meredith, D.D. Discrete differential dynamic programing approach to water resources systems optimization. *Water Resour. Res.* **1971**, *7*, 273–282. [CrossRef]
36. Chow, V.T.; Cortes-Rivera, G. *Application of DDDP in Water Resources Planning*; University of Illinois at Urbana-Champaign. Water Resources Center: Saint Paul, MN, USA, 1974.
37. Cheng, C.T.; Wang, S.; Chau, K.W.; Wu, X.Y. Parallel discrete differential dynamic programming for multireservoir operation. *Environ. Model. Softw.* **2014**, *57*, 152–164. [CrossRef]
38. Larson, R.E.; Korsak, A.J. *A Dynamic Programming Successive Approximations Technique with Convergence Proofs*; Pergamon Press, Inc.: Oxford, UK, 1970; pp. 245–252.
39. Erkmen, I.; Karatas, B. Short-term hydrothermal coordination by using multi-pass dynamic programming with successive approximation. In Proceedings of the Mediterranean Electrotechnical Conference (MELECON '94), Antalya, Turkey, 12–14 April 1994; Volume 923, pp. 925–928.
40. Liu, P.; Cai, X.M.; Guo, S.L. Deriving multiple near-optimal solutions to deterministic reservoir operation problems. *Water Resour. Res.* **2011**, *47*, 20. [CrossRef]
41. Howson, H.R.; Sancho, N.G.F. A new algorithm for the solution of multi-state dynamic programming problems. *Math. Program.* **1975**, *8*, 114–116. [CrossRef]
42. Turgeon, A. Optimal short-term hydro scheduling from the principle of progressive optimality. *Water Resour. Res.* **1981**, *17*, 481–486. [CrossRef]
43. Guo, S.L.; Chen, J.H.; Li, Y.; Liu, P.; Li, T.Y. Joint operation of the multi-reservoir system of the three gorges and the qingjiang cascade reservoirs. *Energies* **2011**, *4*, 1036–1050. [CrossRef]
44. Chen, L.; Singh, V.P.; Guo, S.L.; Hao, Z.C.; Li, T.Y. Flood coincidence risk analysis using multivariate copula functions. *J. Hydrol. Eng.* **2012**, *17*, 742–755. [CrossRef]
45. Chen, L.; Ye, L.; Singh, V.; Asce, F.; Zhou, J.; Guo, S. Determination of input for artificial neural networks for flood forecasting using the copula entropy method. *J. Hydrol. Eng.* **2014**, *19*, 217–226. [CrossRef]
46. Chen, L.; Singh, V.P.; Guo, S.; Zhou, J.; Zhang, J. Copula-based method for multisite monthly and daily streamflow simulation. *J. Hydrol.* **2015**, *528*, 369–384. [CrossRef]
47. Nezhikhovsky, R.A. *Channel Network of the Basin and Runoff Formation*; Hydrometeorological Publishing: Leningrad, Russia, 1971.
48. Yue, S.; Ouarda, T.B.M.J.; Bobée, B.; Legendre, P.; Bruneau, P. Approach for describing statistical properties of flood hydrograph. *J. Hydrol. Eng.* **2002**, *7*, 147–153. [CrossRef]
49. Chen, L.; Guo, S.L.; Yan, B.W.; Liu, P.; Fang, B. A new seasonal design flood method based on bivariate joint distribution of flood magnitude and date of occurrence. *Hydrol. Sci. J.* **2010**, *55*, 1264–1280. [CrossRef]
50. Serinaldi, F.; Grimaldi, S. Synthetic design hydrographs based on distribution functions with finite support. *J. Hydrol. Eng.* **2011**, *16*, 434–446. [CrossRef]
51. Vandenberghe, S.; Berg, M.J.; Gräler, B.; Petroselli, A.; Grimaldi, S.; De Baets, B.; Verhoest, N.E.C. Joint return periods in hydrology: A critical and practical review focusing on synthetic design hydrograph estimation. *Hydrol. Earth Syst. Sci. Discuss.* **2012**, *9*, 6781–6828. [CrossRef]
52. Graeler, B.; van den Berg, M.J.; Vandenberghe, S.; Petroselli, A.; Grimaldi, S.; De Baets, B.; Verhoest, N.E.C. Multivariate return periods in hydrology: A critical and practical review focusing on synthetic design hydrograph estimation. *Hydrol. Earth Syst. Sci.* **2013**, *17*, 1281–1296. [CrossRef]
53. Brunner, M.I.; Viviroli, D.; Sikorska, A.E.; Vannier, O.; Favre, A.C.; Seibert, J. Flood type specific construction of synthetic design hydrographs. *Water Resour. Res.* **2017**, *53*, 1390–1406. [CrossRef]
54. GONG. Flooding 1990s along the yangtze river, has it concern of global warming? *J. Geogr. Sci.* **2001**, *11*, 43–52. [CrossRef]
55. Wang, L.N.; Chen, X.H.; Shao, Q.X.; Li, Y. Flood indicators and their clustering features in wujiang river, south china. *Ecol. Eng.* **2015**, *76*, 66–74. [CrossRef]
56. Huang, K.; Lu, C.; Zhou, J.; Zhang, J.; Singh, V.P. Flood hydrograph coincidence analysis for mainstream and its tributaries. *J. Hydrol.* **2018**, *565*, 341–353. [CrossRef]
57. Chen, L.; Singh, V.P.; Guo, S.; Zhou, J.; Zhang, J.; Liu, P. An objective method for partitioning the entire flood season into multiple sub-seasons. *J. Hydrol.* **2015**, *528*, 621–630. [CrossRef]

58. Chen, L.; Singh, V.P.; Guo, S.L.; Fang, B.; Liu, P. A new method for identification of flood seasons using directional statistics. *Hydrol. Sci. J.* **2013**, *58*, 28–40. [CrossRef]
59. Shi, H.; Li, T.; Liu, R.; Chen, J.; Li, J.; Zhang, A.; Wang, G. A service-oriented architecture for ensemble flood forecast from numerical weather prediction. *J. Hydrol.* **2015**, *527*, 933–942. [CrossRef]
60. Zhang, J.; Chen, L.; Singh, V.P.; Cao, W.; Wang, D. Erratum to: Determination of the distribution of flood forecasting error. *Nat. Hazards* **2015**, *75*, 2065. [CrossRef]
61. Chen, L.; Singh, V.P.; Lu, W.; Zhang, J.; Zhou, J.; Guo, S. Streamflow forecast uncertainty evolution and its effect on real-time reservoir operation. *J. Hydrol.* **2016**, *540*, 712–726. [CrossRef]
62. Huang, K.; Chen, L.; Wang, Q.; Dai, L.; Zhou, J.; Singh, V.P.; Huang, M.; Zhang, J. Risk analysis of flood control reservoir operation considering multiple uncertainties. *J. Hydrol.* **2018**. [CrossRef]
63. Chen, J.; Zhong, P.A.; Zhao, Y.F. Research on a layered coupling optimal operation model of the three gorges and gezhouba cascade hydropower stations. *Energy Convers. Manag.* **2014**, *86*, 756–763. [CrossRef]

8

Relationships among Animal Communities, Lentic Habitats and Channel Characteristics for Ecological Sediment Management

Mikyoung Choi [1,*], Yasuhiro Takemon [2], Kinko Ikeda [3] and Kwansue Jung [4]

[1] International Water Resources Research Institute, Chungnam National University, Daejeon 34134, Korea; choi.mk1981@gmail.com
[2] Disaster Prevention Research Institute, Kyoto University, Kyoto 611-0011, Japan; takemon.yasuhiro.5e@kyoto-u.ac.jp
[3] ASIA AIR SURVEY CO., LTD., Osaka 530-6029, Japan; kom.ikeda@ajiko.co.jp
[4] Department of Civil Engineering, Chungnam National University, Daejeon 34134, Korea; ksjung@cnu.ac.kr
* Correspondence: choi.mk1981@gmail.com

Abstract: This study used a multiscale analysis of relationships among the bitterling and mussel communities, lentic habitat structures with conditions and flooding frequency, and channel characteristics for application in ecological sediment management. From the Kizu River in Japan, 120 lentic habitats were sampled in 2007 and 2010. The floodplain vertical shape index (FVSI), which indicates the degree of convexity or concavity of the vertical shape of a floodplain, was used as channel characteristics using historical cross-section profiles obtained from 1960 to 2012. For examining the relationships between bitterlings/mussels and each habitat condition or structure, abundance values of bitterlings and mussels were transformed into habitat suitability index (HSI). Furthermore, the relationships between the number of habitat structures and FVSI were analyzed. The results indicated that bitterlings and mussels are more abundant in terrace ponds than in active ponds, especially so in terrace ponds located in the lower area of bars with a flooding frequency of 8–16 days/year (bitterlings), those located in the lower area of bars with a flooding frequency of 8 days/year, and those located in the upper area of bars with a flooding frequency of 16–22 days/year (mussels). These ponds tended to have less than 1 cm mud depth that was negatively related to abundance of mussels. These suitable habitat types tended to be located in channels with a floodplain vertical shape index between −0.35 and 0.05. We established countermeasures to prevent channel types with floodplain vertical shape index exceeding 0.05 instead of restoring the previous channel conditions.

Keywords: lentic habitats; bitterling; mussel; floodplain vertical shape index; sediment management

---

1. Introduction

The interruption of the natural flow and sediment transfer by dams causes changes in the flow, river structure, and ecosystem in the reservoirs and regions downstream of the dam. When reservoirs are filled with sediments, their storage capacity is reduced, affecting the water supply and hydroelectric power [1]. Downstream of dams, the reduced sediment supply results in degradation of the stability of channel structure, and the quality of aquatic habitats [2]. In order to remove sediments accumulated in reservoirs and to transport sediment downstream, several sediment management strategies such as replenishment, sluicing, and bypassing have been tested worldwide [3–5]. Actually, artificial sediment supplies downstream have contributed to enhancing the available spawning habitat for chinook salmon [2] and the lotic habitat quality for invertebrates and fishes [6]. In order to predict the geomorphic and environmental impacts of sediment management and for successful ecological restoration, it is essential to understand river characteristics and conditions under natural and artificial disturbances. The reach-scale channel configuration, such as braided, wandering, and straight channel, can be helpful for linkage between hydraulic conditions and aquatic ecosystems. This can be examined using hydraulic and geomorphic parameters, including discharge and slope [7], sediment load and lateral stability [8], and slope and bed materials [9]. The variations in channel configuration are in response to the flow regime and sediment transport, which further influence the geomorphic units such as ponds, bars, wetlands, backwaters, and pool-riffle sequences. Further, the assemblages of geomorphic units at the reach-scale are related to the habitat diversity and animal communities [10]. The high complexity of landforms generally equates to the high diversity of hydraulics and thus the high biodiversity [11,12]. If research related to the relationship between the channel configuration and ecosystem are efficiently conducted, it is possible to predict the impact of natural or artificial disturbances on the ecosystem [13]. Therefore, integration of ecological and geomorphological perspectives with the acknowledges the multiscale abiotic and biotic structures of a stream system is required, but is an unresolved challenge [14]. Wyrick and Rasternack [12] systematically studied the relationship between geomorphic units, hydraulic parameters, and their relevance for various habitats. Other previous studies on multiscale relationships between abiotic and biotic structures have focused on lotic habitat conditions; the relationships between lotic habitat conditions and channel configurations [15,16], landscape diversity with channel characteristics [17], or habitat diversity in specific channel types [18,19].

This study focused on multiscale relationships among animal communities, lentic habitats, and channel characteristics in the Kizu River. The Kizu River provides suitable lentic habitats such as ponds or backwaters for bitterlings and mussels on its floodplain [20]. However, the diversities of bitterlings and mussels decreased, and the protected bitterling *Acheilognathus longipinnis* disappeared. Therefore, the target animal communities of this study are considered to be bitterlings and mussels. Further, because bitterlings and mussels live only in lentic habitats, lentic habitat structures and conditions was considered as habitat scale. In the case of lentic habitat on floodplains, the environmental conditions are characterized by hydrological connectivity with the main river channels [21]. Flooding provides hydrological connectivity with the exchange of nutrients, sediments and organisms. Therefore, we considered not only structures but also the flooding frequency. First, this study investigates the relationships between the abundance of bitterling/mussel and lentic habitat structures/conditions with the flooding frequency. Second, the relationships between habitat structures and the reach-scale channel characteristics were investigated. The floodplain vertical shape index (FVSI) was used as the reach-scale channel characteristic. As the channel types are not numerical values, FVSI was developed in Reference [22] and represented the concave and convex floodplain shapes. Additionally, the historical changes of FVSI were calculated using the historic cross-section data. Finally, we discussed previous environmental conditions and the application of these characteristics to ecological sediment management.

2. Materials and Methods

2.1. Study Site

The study area was comprised of the lower reaches (0–26 km) of the Kizu River in Japan (Figure 1). The riverbed of the Kizu River degraded, and the vegetation in the floodplain expanded after dam construction over a 65-year period. A total of five dams, namely, Takayama Dam, Syourenji Dam, Murou Dam, Nunome Dam, and Hinachi Dam, which were constructed in 1969, 1970, 1974, 1992, and 1999, respectively, are located in the Kizu River basin. The approximately 6000 m^3/s peak floods occurred before the construction of dams; however, the peak of floods decreased to approximately 3000 m^3/s after construction [23]. The estimated amount of bed-load transportation downstream of the dam was approximately 183,000 m^3/y in the 1960s; further, it decreased to 23,000 m^3/y in the

2000s (Figure 2) [13]. The reduction in the peak discharge and sediment supply upon dam construction influenced the alternation of channels from braided to single or alternating channels. These river channel changes caused the degradation of lentic habitat conditions on the floodplain owing to the reduction of inundation [20].

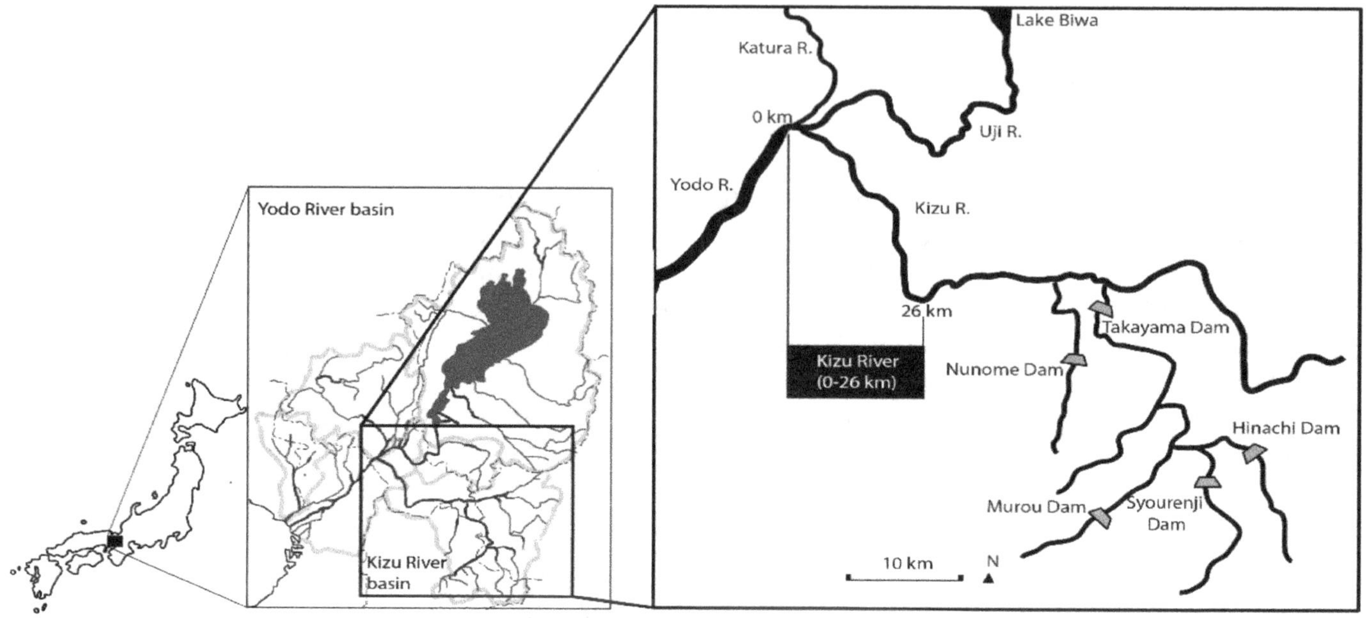

Figure 1. Study site in the Kizu River.

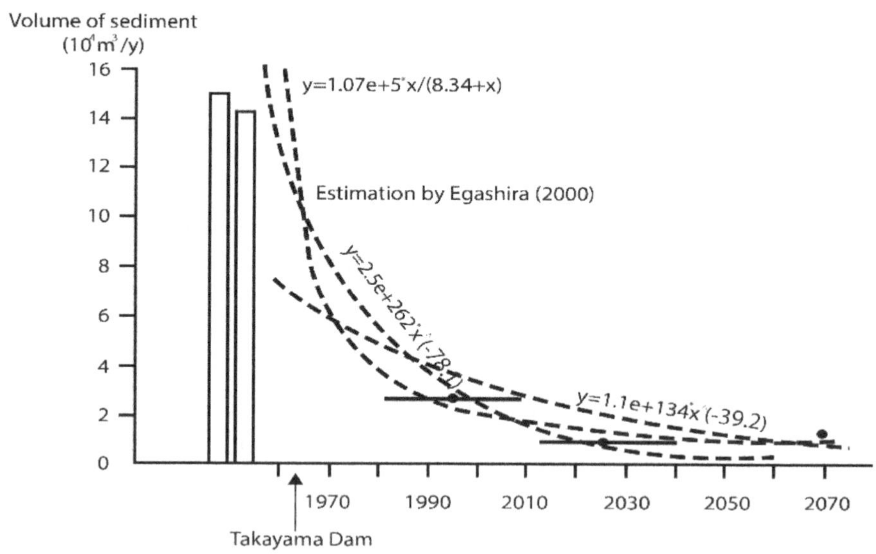

Figure 2. Historic changes in annual transport sediment volume in the Kizu River were estimated in Reference [23].

2.2. Data Acquisition of Bitterling and Mussel Communities

In the case of the Kizu River, because bitterlings and mussels live only in lentic habitats such as ponds and backwaters, the habitat types of pond and backwater were selected for sampling (Figure 3d,e). Prior to conducting the field survey, the size and location of the habitat were identified using aerial photos. In 2006, 190 ponds were detected in aerial photos, whereas 178 ponds were detected in 2010. Among the detected ponds, bitterlings and mussels were sampled in 47 ponds among the 190 ponds in 2007 and in 73 among the 178 ponds in 2010 by the Yodogawa River Bureau. The sampling was conducted in the summer season. Bitterlings were observed in 63 ponds and mussels were observed in 47 ponds among 120 ponds. There are five taxa of bitterlings (*Acheilognathus rhombeus, Acheilognathus tabira tabira, Acheilognathus cyanostigma, Rhodeus ocellatus ocellatus, Tanakia lanceolate*) and five taxa of mussels (*Anodonta calipygos Kobelt, Anodonta woodiana, Lanceolaria grayana cuspidate, Lanceolaria oxyrhyncha (Martens), Unio douglasiae nipponensis*) in sampling ponds. The total number (population) of bitterlings and mussels were divided by the surveyed time (h) and number of people who participated in the sampling (n), and this value was used as the abundance of bitterlings and mussels. The abundance value was used to calculate the best model of lentic habitat conditions for bitterlings and mussels. For examining the relationships between bitterlings/mussels and each habitat condition or structure, abundance values transformed into habitat suitability index (HSI) [19–21]. Habitat suitability index was derived by Equation (1).

$$w_i = \frac{u_i / \sum_{i=1}^{n} u_i}{a_i / \sum_{i=1}^{n} a_i} \tag{1}$$

w_i is the ratio for the ith of n habitat categories, u_i is the total abundance in category I, $\sum u_i$ is the total abundance for all habitat categories, a_i is the number of samples from categories i, and $\sum a_i$ is the total number of samples [24]. The indices range from 0 to 1 for each variable, with 0 indicating the least suitable habitat conditions and 1 indicating the optimum habitat condition [25].

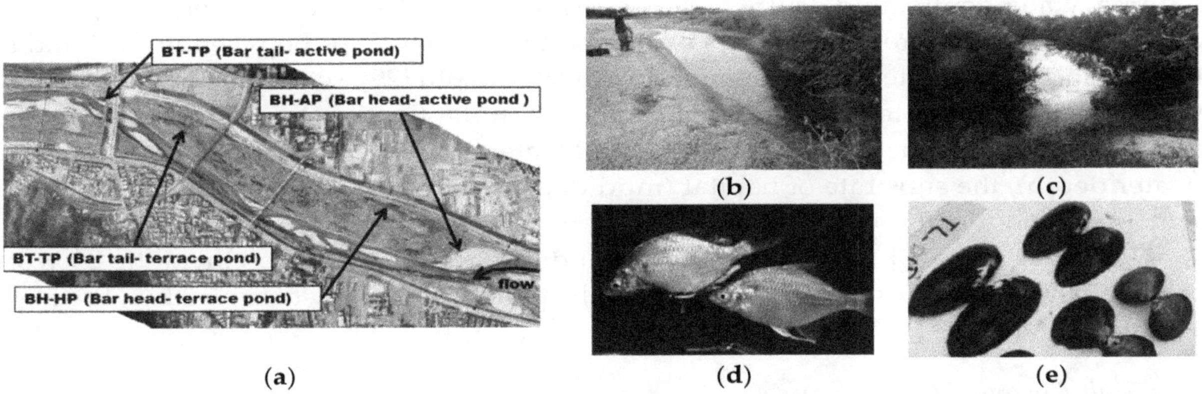

Figure 3. Lentic habitats and target species. (**a**) Image showing classification of lentic habitats; (**b**) an active pond (AP); (**c**) a terrace pond (TP); (**d**) bitterling; and (**e**) mussel.

2.3. Data Acquisition of Lentic Habitat Structure and Conditions

Lentic habitat structures were classified into two types, active pond (AP) and terrace pond (TP), depending on their locations. The aerial photos obtained in 2006 and 2010 were used for this classification. AP is defined as a pond and backwater that is located on active channels such as sand-bars or sanded islands (Figure 3b), and TP is defined as a pond located on a terrace with vegetation (Figure 3c). These ponds were further classified based on the location of upper and lower sites on a particular bar. The bar was divided into two halves based on the direction of stream flow. The ponds located in the upper area were defined as bar head-ponds, whereas those located in the lower area were defined as bar tail-ponds. This classification was based on the assumption that even if the same habitat structure type was located on the same bar, the habitat conditions or the abundance of species differed depending on whether the habitats were located on the upper or lower part of the bar. Thus, four types of lentic habitats were considered: Bar head-active pond (BH-AP), bar head-terrace pond (BH-TP), bar tail-active pond (BT-AP), and bar tail-terrace pond (BT-TP) (Figure 3). A total of 190 ponds were detected in aerial photos in 2006, whereas 178 ponds were detected in 2010. Although only a total of 120 ponds were surveyed, all the detected ponds in aerial photos

($n = 368$) had values of the geomorphic parameters of flooding frequency. The term flooding frequency refers to the inundation frequency in a pond per year. The water level data obtained during the period from 1989 to 2008 were converted into discharge using the HQ formula. Using the converted discharge, 1 day discharge (909.1 m^3, at Inooka observatory station), 8 day discharge (249.5 m^3), 16 day discharge (149.0 m^3), 22 day discharge (118.3 m^3), 45 day discharges (72.9 m^3), 71 day discharge (52.9 m^3), 185 day discharge (26.2 m^3) and 365 day discharge were calculated. The water level of the cross section (200 m intervals) according to each discharge was calculated using HEC-RAS software. Further, the values of water levels were overlapped with the digital elevation model (DEM) data using Geomedia 6.1 software. Thus, if the pond would be inundated based on the 1 day discharge, the pond would have a flooding frequency of 1 day/year. If the pond would be inundated based on the 8 day discharge, the pond would have a flooding frequency of 8 days/year. Therefore, the ponds have one of the following flooding frequencies: 1 day, 8 days, 16 days, 22 days, 45 days, 71 days, 185 days, and 365 days per year [26–28]. In this study, the flooding frequency parameter was used as the representative environmental parameter because flooding frequency is an important external condition that determines the internal habitat conditions [21]. This is especially true for lentic habitats such as wetlands or ponds. We classified the structure of the habitats by reflecting the visible geomorphic characteristics and location of the habitat to find visible targets. However, in order to compensate for the limitations of types of habitat structure identified only from aerial photos, the values of flood frequency obtained by numerical simulation were used.

In all, 7 habitat condition parameters were surveyed in 120 ponds. The parameters were as follows: Area, water depth, mud depth, mean grain size (D50), dissolved oxygen (DO), chlorophyll, and wood coverage. The previous studies examined the relationships between mussel and habitat conditions; velocity, substrate size and compaction, water depth [29]; sediment softness, velocity and sediment types [30]; water temperature and quality with the flood pulse frequency [21]. As these studies considered hydraulic habitat conditions and water quality, we determined the habitat shape (area, water depth), the substrate of habitat (mud depth and D50), and the water quality (DO and chlorophyll) to a considerable extent. The method of the survey parameters can be referenced to in the reports of References [26–28]. The wood coverage indicates the shaded shoreline ratio by wood, i.e., 100% wood coverage indicates that the shoreline of the pond was entirely occupied by vegetation such as that depicted in Figure 3c.

2.4. Reach-Scale Channel Characteristics

The reach-scale channel configuration in the Kizu River is categorized into single, semi-wandering-straight and sinuous, wandering-straight and sinuous, bifurcated-straight and sinuous, and braided sinuous using parameters of number of channels and sinuosity per 2 km [13]. Aerial photos obtained between 1948 and 2010 were used for performing this classification. The channel configuration of the Kizu River changed from braided channel (characterized by multi-channels and braided sand bars) to single or slightly wandering channel (characterized by less channels and wide vegetation area on the floodplain) between 1948 and 2010. However, because the channel types are not numerical values, we used floodplain vertical shape index (FVSI). This index indicates the degree of convexity or concavity of the vertical shape of a floodplain [22]. The FVSI value was calculated using cross-section data of 200 m intervals. The relative elevation of the riverbed from the water level was arranged in an ascending order. Further, we determined the area between the arranged shape (B) and the triangle connecting the following three points (A): The riverbed tangent to the water surface, bottom of the bank, and top elevation of the bank (Figure 4). The value is positive if B−A is greater than A, and negative if it is less than A. As one channel type of 2 km has 10 cross sections, 10 values are averaged in 2 km. A positive value means a convex floodplain vertical shape (Figure 4a), and a negative value indicates a concave vertical shape (Figure 4b).

The FVSI was calculated using cross-sectional data obtained between 1960 and 2010, as done in the classification of the reach-scale channel configuration in [13]. The classified reach-scale channel configurations were single, semi-wandering, wandering, bifurcated, braided channels in the Kizu River. Figure 5 depicts the average the FVSI values of the classified channel types by Tukey's tests. The majority of the channel configurations in the Kizu River exhibited a concave floodplain vertical shape with negative values of FVSI. Although a significant difference of FVSI was not observed between single/semi-wandering and wandering/bifurcated, and the difference of wandering/bifurcated and braided, we could identify the difference between single/semi-wandering and braided channels. The single and slightly wandering channels in the Kizu River tended to have slightly convex floodplain vertical shape ($-0.3 <$ values < 0.1), and wandering and bifurcated channels tended to have FVSI values between -0.5 and -0.1. Braided channels have a significantly concave floodplain vertical shape (values < -0.6) (Figure 5).

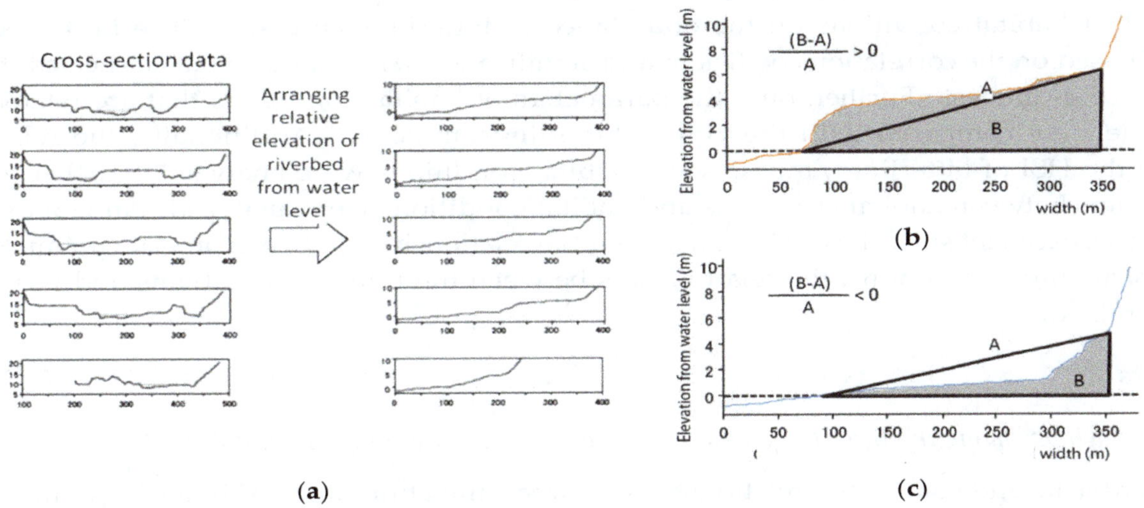

Figure 4. (**a**) Calculation of the floodplain vertical shape index (FVSI); (**b**) a positive value indicates a convex floodplain vertical shape; and (**c**) a negative value indicates a concave vertical shape.

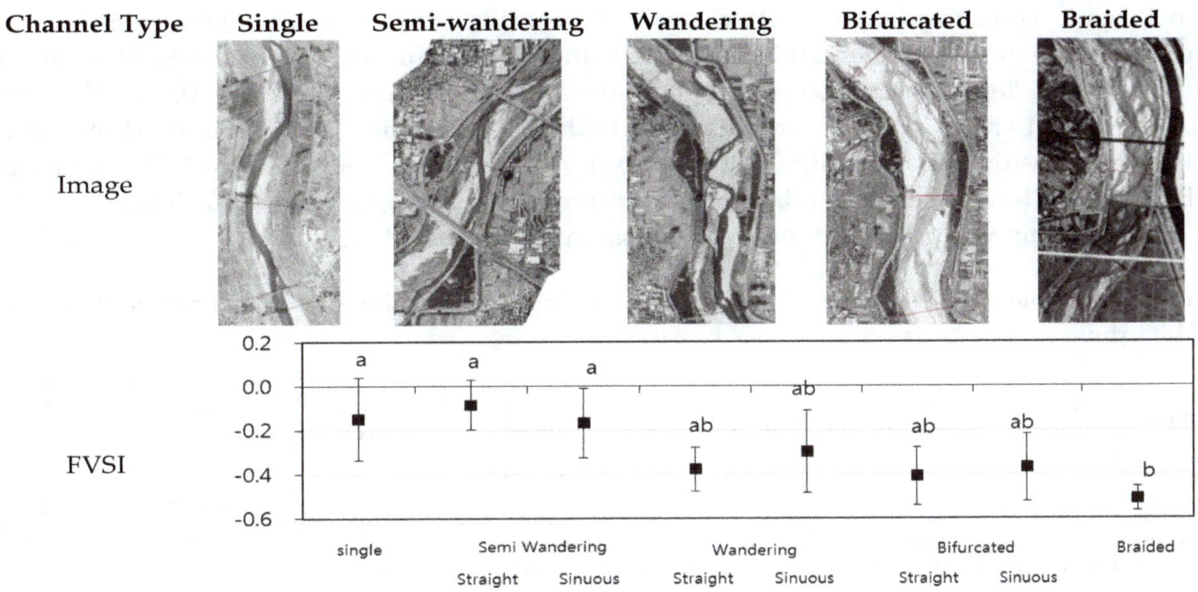

Figure 5. Classification of the channel configurations in the Kizu River [13] and their relationship with FVSI. The squares of FVSI indicate the average FVSI value per channel type (mean $\pm$ SE). Similar letters above histograms indicate significant differences based on Tukey's tests.

2.5. Statistical Analyses

In order to understand the relationships among species, habitat structures and conditions, and channel characteristics, we considered the following items in the analysis:

- A multiple regression model using the parameters of habitat conditions for the abundance of bitterlings and mussels
- Comparison of the habitat suitability index (HSI) of bitterlings/mussels and habitat conditions
- Comparison of the HSI of bitterlings/mussels and habitat structures
- Comparison of habitat structures and habitat conditions
- Comparison of habitat structures and FVSI

We generated multiple linear regression models with a stepwise selection method to examine the effect of habitat conditions on the abundance of bitterlings/mussels. The best model was selected based on the correlation coefficient and significance. An α value of 0.05 indicates statistical significance in the test. Further, only the parameters of habitat conditions that are related to the best models are compared with the HSI of bitterlings/mussels. Additionally, the relationship between the HSI of bitterlings/mussels and habitat conditions were analyzed. Furthermore, the relationships between habitat structures and habitat conditions were analyzed. An HSI exceeding 0.7 indicates excellent suitability. After analyzing the relationships between bitterlings/mussels and habitat structures/conditions, the relationships between the number of habitats and FVSI values were examined.

3. Results

3.1. Relationships among the Bitterlings/Mussels, Habitat Structures, and Habitat Conditions

In order to understand the relationships between the abundance of bitterlings/mussels and habitat conditions, the best models were selected based on surveyed ponds (n = 120) by using a multiple regression model. The multiple regression models were generated to test the influences of habitat conditions (area, water depth, D50, DO, Chlorophyll and wood coverage) on abundance of bitterlings and mussels. Only the parameters of habitat conditions related to the best models are listed in Table 1. The abundance of bitterlings was best described by a model consisting of DO and chlorophyll (correlation coefficient: 0.238, $p < 0.05$), and that of mussels was best described by a model consisting of mud depth and wood coverage (correlation coefficient: 0.272, $p < 0.05$). Abundance of bitterlings tended to have a negative relationship with DO (regression coefficient of −1.08), whereas it tended to have positive relationships with chlorophyll (regression coefficient of 2.52). Abundance of mussels tended to have a negative relationship with mud depth (regression coefficient of −1.29) and a positive relationship with wood coverage (regression coefficient of 2.67).

Table 1. Multiple regression analysis to examine the best models of habitat conditions for the abundance of bitterlings and mussels.

Abundance	Area (cm^2)	Water Depth (cm)	Mud Depth (cm)	D50 (mm)	DO (mg/L)	Chlorophyll (μg/L)	Wood Coverage (%)	Best Model	
								R	p
Bitterling					−1.08 *	2.52 *		0.238	<0.05
Mussel			−1.29				2.67 **	0.272	<0.05

The best models were selected on the basis of the correlation coefficient (R) and significance (p). Values of independent variables indicate the regression coefficient (t). Significance: * $p < 0.05$, ** $p < 0.01$, *** $p < 0.001$.

The HSI values of bitterlings and mussels with the habitat conditions that are selected in the best model are shown in Figure 6. If there are no sampling points in the class or no detection of bitterlings or mussels, the result will exhibit a zero value. Although the HSI values of bitterlings decreased in 15–20 mg/L DO, this exhibited a high habitat suitability in DO case of more than 25 mg/L DO

(Figure 6a). The HSI values of bitterlings tended to increase in chlorophyll between 0 and 500 μg/L and then significantly decreased after 500 μg/L (Figure 6b). The HSI of mussels exhibited the maximum score in mud depth of 0–1 cm (Figure 6c). The HSI values of mussels showed excellent suitability in 100% wood coverage (Figure 6d). There was no sampling pond with wood coverage of 60% and 80%, and number of mussels in ponds with wood coverage of 50, 70 and 90% was very low. Therefore, HSI of mussels showed about zero values in these groups.

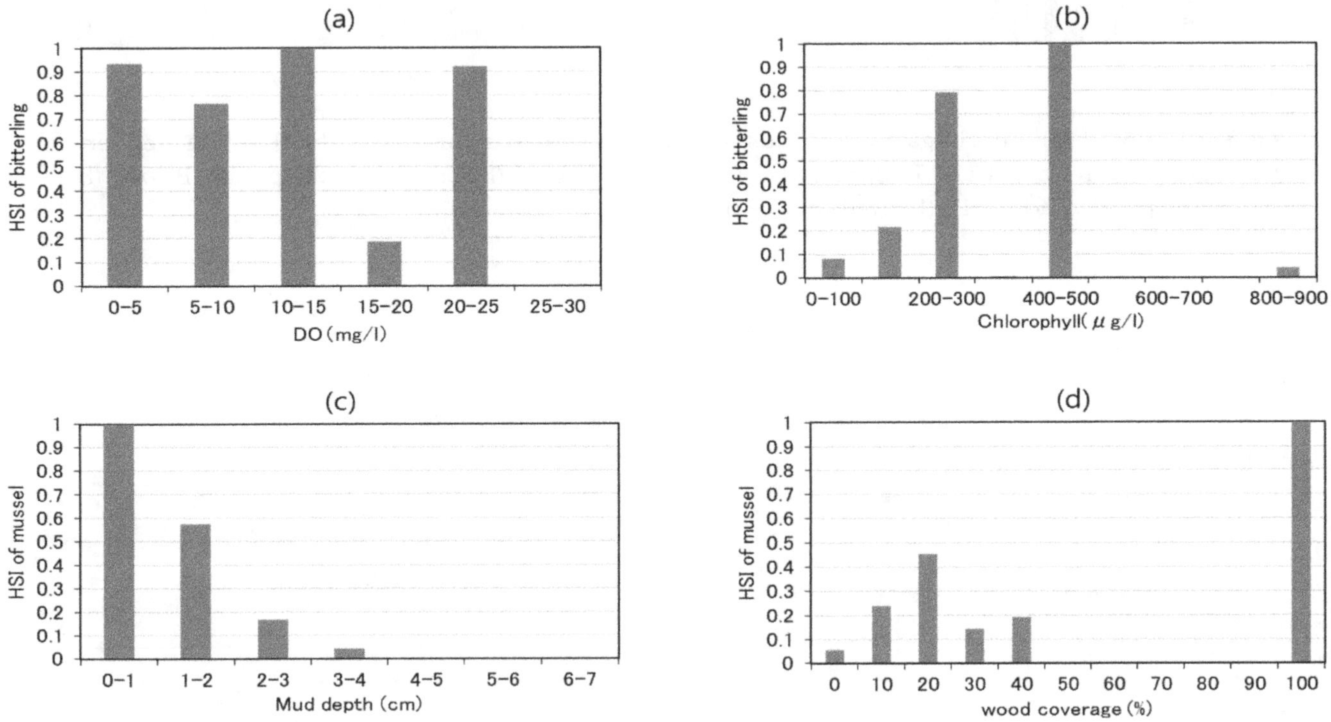

Figure 6. Habitat Suitability Index (HSI) values with habitat conditions were selected for the best models. The HSI of bitterlings and habitat conditions of (**a**) dissolved oxygen (DO) (mg/L) and (**b**) Chlorophyll (μg/L). The HSI of mussels and habitat condition of (**c**) mud depth (cm) and (**d**) wood coverage (%).

The HSI values of bitterlings and mussels are explained by considering the habitat structures with the flooding frequency observed for 120 ponds (Figure 7). Among terrace ponds, BT-TP with a flooding frequency between 8 and 16 days/year showed excellent suitability for bitterlings (Figure 7a). BT-TP and BH-TP with flooding frequencies of 8 days/year and 16–22 days/year exhibited excellent suitability for mussels, respectively (Figure 7b). That is, the terrace ponds (BH-TP and BT-TP) with a flooding frequency between 8 and 22 days/year tended to have high HSI values for bitterlings and mussels.

Figure 8 illustrates the relationships between habitat environmental conditions and habitat structures with flooding frequency. Almost all the habitat structures, except those with a flooding frequency of 45 days of BT-TP, exhibited an average of between 7 and 13 mg/L of DO (Figure 8a). Values of chlorophyll in BT-TP with flooding frequency of 1, 8, 16, and 22 days were higher than other habitat types in the same flooding frequency (Figure 8b), and the values of mud depth in BT-TP with flooding frequencies of 16, 22, and 45 days were highest as compared to other habitat structures (Figure 8c). Further, the relationship between wood coverage and habitat structures with flooding frequency could not detect obvious difference (Figure 8d).

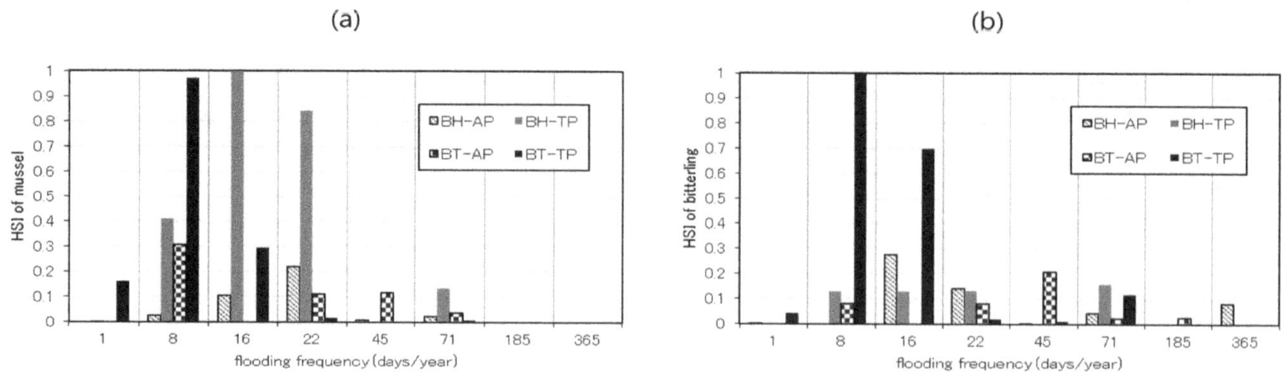

Figure 7. HSI of (**a**) bitterlings and (**b**) mussels, and habitat structures with the flooding frequency. Habitat structures are: BH-AP (bar head-active pond), BH-TP (bar head-terrace pond), BT-AP (bar tail-active pond), and BT-TP (bar tail-terrace pond).

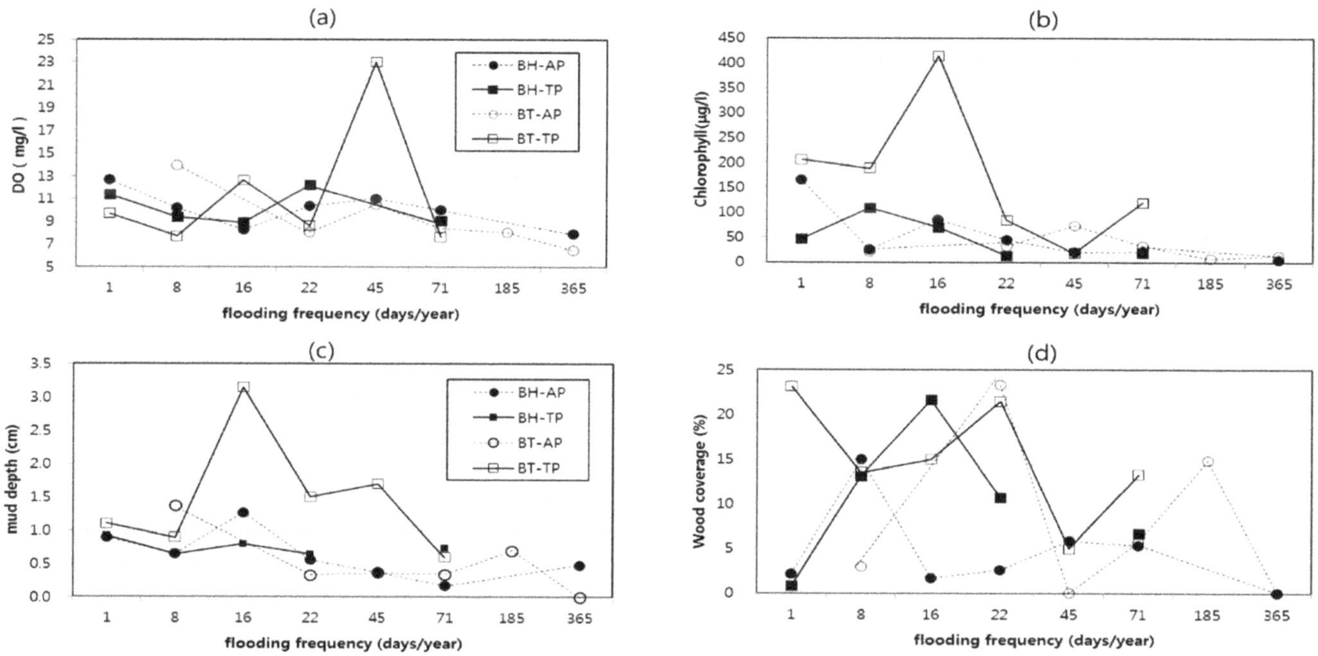

Figure 8. Relationships between the average habitat conditions: (**a**) DO; (**b**) chlorophyll; (**c**) mud depth; and (**d**) wood coverage and habitat types with flooding frequency. Habitat structures are: BH-AP (bar head-active pond), BH-TP (bar head-terrace pond), BT-AP (bar tail-active pond), and BT-TP (bar tail-terrace pond).

3.2. Historical Changes in FVSI

The values of FVSI per year were shown in Figure 9. As the study area extends to a length of 26 km, 12 FVSI values were obtained per year. Average FVSI increased from −0.44 in 1960 to −0.17 in 2010. The cross-sectional floodplain shape of the Kizu River tended to show convex shapes during the 60 years. Since 1990, only values of one or two sections had an FVSI > 0.

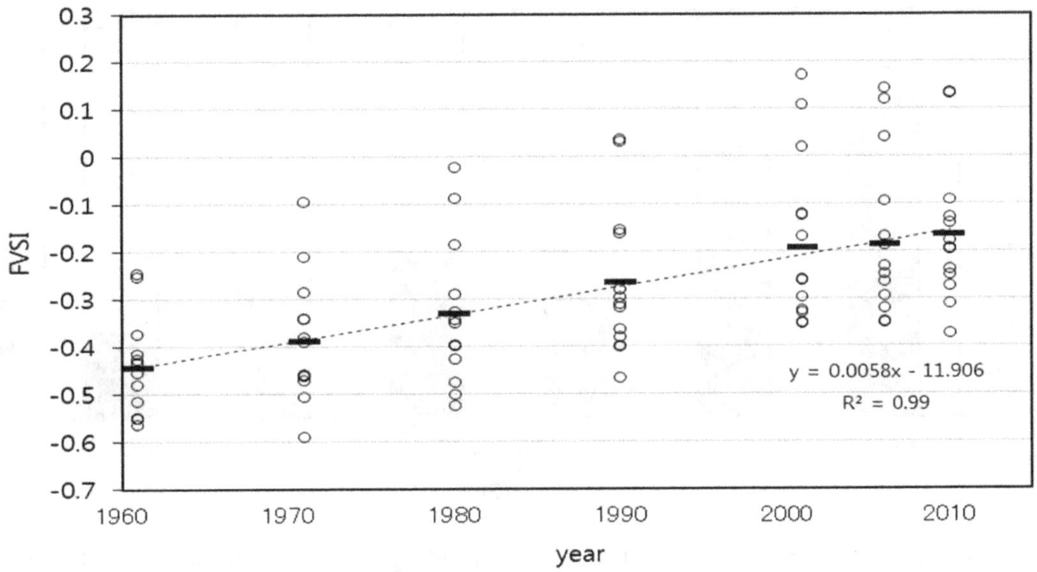

Figure 9. Historical changes in FVSI. Since the study area extends to a length of 26 km, 12 FVSI values per year obtained. The squares indicate the annual average.

3.3. Relationship between the Habitat Structures and FVSI

FVSI was classified into six groups (Figure 10). The number of ponds was counted by considering the categorized values of FVSI (Figure 9a), and ratio of ponds with different flooding frequencies was measured by FVSI (Figure 9b). All the ponds detected in aerial photos (n = 386), including 120 surveyed ponds, were used in this relationship. Reaches with FVSI values of less than −0.35 had a high number of APs (Figure 10a), adversely those of more than 0.05 had high TPs (Figure 10b). Based on the increase in FVSI values, the number of APs decreased, whereas the number of TPs increased. The ratio of ponds with flooding frequencies is depicted in Figure 11. The ratio of ponds with a flooding frequency of 1 day/year significantly increased in reaches with FVSI exceeding 0.05. However, ponds with flooding frequencies of less than 16 days/year were not observed in the case of an FVSI of less −0.35. BH-TP and BH-TP with a flooding frequency of 8 days/year and 16–22 days/year exhibited that excellent suitability tended to exist for bitterlings and mussels on a reach with an FVSI that was between −0.35 and 0.05.

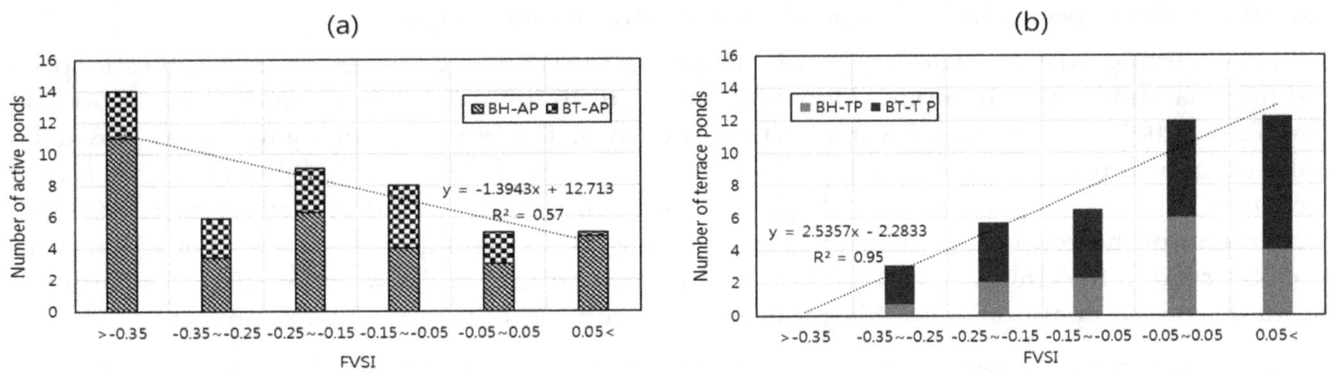

Figure 10. Relationships between the number of habitat types with (**a**) active ponds and (**b**) terrace ponds and FVSI. Habitat structures are: BH-AP (bar head-active pond), BH-TP (bar head-terrace pond), BT-AP (bar tail-active pond), and BT-TP (bar tail-terrace pond).

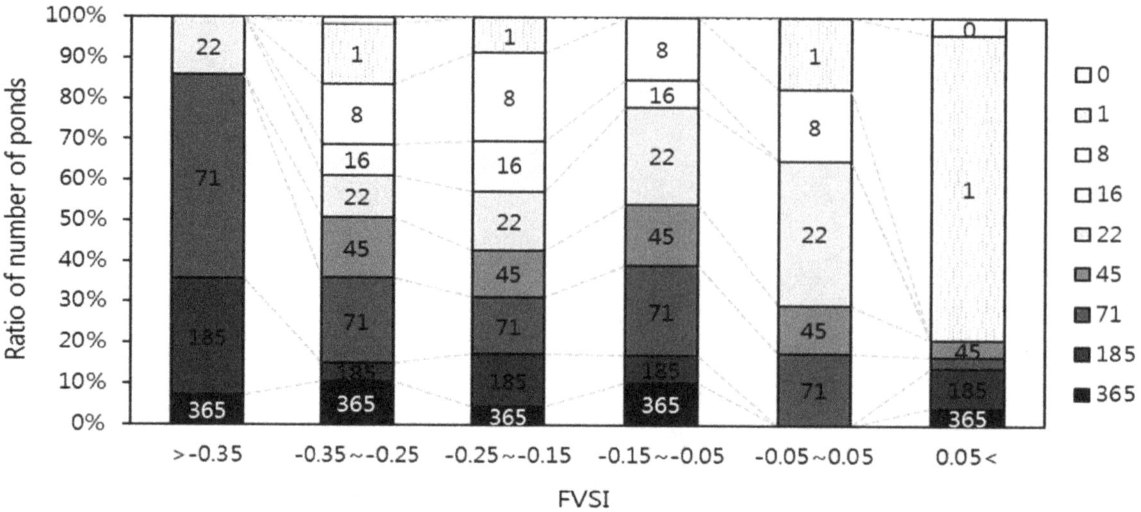

Figure 11. Relationships between the ratio of flooding frequency of habitat and FVSI.

4. Discussion

4.1. Relationships among Species Abundance and Habitat

We attempted to understand the relationships between bitterlings/mussels and habitat structures with flooding frequency. Further, to explain the effect of habitat structures and flooding frequency on the habitat conditions, and relationships between habitat structures and habitat conditions were analyzed. Relationships between bitterling and chlorophyll tended be positive, even though the suitability of the bitterling in chlorophyll over 500 was very low (Table 1; Figure 6b). According to relationship between chlorophyll and habitat structures with flooding frequency, terrace pond with flooding frequency of 1–16 days/year showed relatively high chlorophyll (Figure 8b). Among these ponds, terrace ponds with flooding frequency of 8–16 days/year had high a HSI of bitterlings (Figure 7a). That is, suitability of bitterling had the high values in terrace ponds with a flooding frequency of 8–16 days/year, because they had a high value of chlorophyll that were positively related with bitterling. As the flooding frequency increases, the values of chlorophyll decreases. Maybe this is because frequent flooding reduces the residence time of water [21]. The infrequent flooding of ponds tended to maintain concentration of chlorophyll, and thus tended to influence bitterling habitat. A previous study [31] on CCA analysis between bitterlings/mussels and habitat conditions also showed that these species and chlorophyll have positive relationships.

The bitterling *Acheilognathus longipinnis* is listed on the IUCN red species and is a domestic species with national monument in Japan. However, this bitterling has not been found in the Yodo River system since 2006 [32]. The Kizu River joins the Yodo River. Therefore, the Yodogawa River Bureau and Osaka-fu prefecture implemented the re-introduction of the bitterling *Acheilognathus longipinnis* in the autumn of 2009 as an experimental trial. Mature adults (five hundred individuals) were released into embayments of the Yodo River in autumn 2009 during the spawning season. A total of 133 juveniles were observed in the embayments, but no juveniles were found in the spring of 2011. The bitterling restoration project continues, and this paper could be support of this restoration project.

Mussels were negatively related to the mud depth (Table 1; Figure 6b). Mussels also were more clearly observed in TPs than APs, especially the TPs located in the lower area of bars with a flooding frequency of 8 days/year and those located in the upper area of bars with flooding frequencies of 16–22 days/year than in APs (Figure 7b). Further, these TPs had an average mud depth of 1 cm (Figure 8c). Mussels require an appropriate depth of sediments under stable conditions because they live buried in the riverbed [21], and they filter the surrounding water to obtain food and oxygen [33]. However, deep mud may influence negative relationships with mussels, as shown in Figure 6c. Here, the pond is generated by erosion during large flood events, and is sustained by frequent intermediate flood events. In the case of TPs, the depth of the mud is likely to deepen by the deposition of suspended load by frequent intermediate flood events. Therefore, in the case of BT-TP, a flooding frequency of 16–22 days exhibits considerable mud depth as compared to that obtained at a flooding frequency of 1–8 days. If the mud depth increases, mussels buried in the mud will find it difficult to breathe owing to the reduction in DO. The authors of Reference [21] explained that the survival rates of mussels were low and that the growth was approximately zero in infrequently inundated water bodies (backwater, wetlands) by conducting field-earing experiments in the case of the Kiso River, because infrequently inundated water bodies exhibited hypoxic conditions in substratum-water. Further, other studies [29,30] indicated that the sand size of substrate is more suitable than that of silt (fine size) or gravel (larger size). However, fine substrate sediments have been reported to increase the mortality rate in some freshwater mussel species [34]. We assumed that the mud depth until 1 cm could maintain suitable DO for mussel breathing; if the depth is increased, it could become difficult for mussels to live in the Kizu River.

Further, we detected that, although the same TP with the same flooding frequency, they showed different habitat conditions and abundance of mussels. The mud depth of BH-TP was lower than that observed in BT-TP by considering the same flooding frequency. Thus, the HSI of mussels belonging to BH-TP exhibited higher values as compared to that exhibited by BT-TP with a flooding frequency of 16–22 days. However, we were unable to explain the reason for this difference. We inferred that ponds located on the upper area in the bar have much more powerful river strength than ponds located in the lower area in the flood event. Further, the accumulation of chlorophyll and mud in ponds is prevented by tractive force. In future studies, we should perform various other comparative analyses and include hydraulic aspects.

4.2. Relationships between Habitats Structures and FVSI

FVSI reflects the channel configuration in the Kizu River. The concave reaches with an FVSI < -0.35 had a lot of ponds with frequent flooding frequency more than 22 days/year (Figure 11) and only active ponds (Figure 10b). This indicates the significantly concave reaches with FVSI < -0.35 had low habitat suitability for bitterlings and mussels, because they have few suitable habitat structures (terrace ponds) with flooding frequency (8–22 days/year). These reaches were related with wandering, bifurcated, and braided channels in the Kizu River (Figure 5). Their floodplain is frequently inundated, and thus active channel was wide and area of vegetation was small. Hence, active ponds or backwaters on active channels are mainly located on these reaches. Although these channel types may have low potential of bitterlings and mussels, they may have high potential for other fish or aspects of lotic habitats. For example, Sukhodolov et al. and, Payne and Lapointe [15,16] noted that braided channels supply suitable conditions for fish shelters in terms of depth and velocity. Choi et al. [13] also examined that the bifurcated channel exhibits higher habitat diversity for both lentic and lotic habitats as compared to that exhibited by the single or slightly wandering channels. Since this study focused on lentic habitat with bitterlings and mussels, channel types with FVSI < -0.35 tended to have lower potential than others.

On the other hand, the convex channel reaches with an FVSI > 0.05 had a lot of terrace ponds with infrequent flooding frequency (1 day/year). Although habitat with infrequent flooding was suitable for bitterlings and mussels in the Kizu River, less flooding about 1 day/year showed the low suitability for bitterlings and mussels (Figure 7). However, unfortunately, there was no clear difference of environmental conditions between flooding frequency of 1 day and others (8, 16, 22, 45, 71, 185, and 365 days), and between active ponds and terrace ponds in the same flooding frequency 1 day.

Therefore, the intermediate channel types tended to be suitable excluding reaches of significant concave floodplain vertical shape (FVSI < -0.35) and convex shape (FVSI > 0.05). These suitable reaches were related to single or slightly wandering channels in the Kizu River (Figure 5), and these reaches were detected more after the 1990s (Figure 9). Similarly, studies related to the potential of recent channel conditions of the Kizu River were also noted by Reference [35]. They classified riffle structures into four types (traverse, converge, diverge I and II types), and surveyed the biomass and taxonomic richness of invertebrates per these riffle types. Further, the areas of each riffle type were measured using aerial photos that were obtained over 65 years. Their results exhibited that traverse and converge types of riffle had more biomass and taxonomic richness of invertebrates as compared to the diverge types; further, the areas of these riffle types increased after the 1970s. They discussed that river conditions from the 1980s to the 2010s may have had more biomass and richness than before. Thus, we recommend establishing countermeasures to prevent an increase in single or slightly wandering channel types to an FVSI of more than 0.05 to include many ponds with a flooding frequency of 1 day/year rather than restoring the previous channel conditions.

4.3. Application of FVSI to Sediment Management

Dam constructions cause the regular flow and the reduction of sediment transport. Stabilized flow and low sediment volume results in channel changes from braided (mobile) channel to single or single-thread (stabilized) channels [36]. In the Kizu River, these channel changes occurred over 60 years [13] with a reduction of sediment transport (Figure 2). A stabilized channel shows channel narrowing and incision with vegetation encroachment [2]. The encroachment of riparian vegetation into high level of floodplain by reduction in flood scour and sediment deposition is related to the high value of floodplain vertical shape index. As sediment deposition and vegetation expansion lead to an increase in the relative elevation of floodplains, FVSI was calculated by difference between elevation of floodplain and water level in main channel increases. Thus, positive values (convex shape) of FVSI meant stabilized channel such as single or semi wandering channels, and negative values (concave shape) meant more mobile channels such as wandering, bifurcated and braided channels in the Kizu River (Figure 5). This means that a greater reduction of sediment transport could further increase the FVSI. Therefore, sediment supplies need to maintain the current channel condition, not the more stabilized channel, for bitterlings and mussels. If we want to maintain single or slightly wandering channels with FVSI less than 0.05, we could consider a suitable period as between 1990s and 2000s (Figure 9), and suggest the volume of sediment transport at the time (20,000–50,000 m^3/y) of sediment management. However, this is an indirect and long-term measurement. The consideration of various external and internal conditions should precede before the management.

On the other hand, the sediment management such as replenishment directly leads to improved habitat quality. The sediment excavated at the upstream of dam was moved to downstream by dump truck or other transportation [5]. Then, the riverbed was reformed by magnitude or duration of flushing flow or natural flood. Kondolf [2] stated that sediment supply contributed to enhancing the available spawning habitat for chinook salmon. Wood and Armitage [6] said that replenishment of coarse sediment increased the lotic habitat quality for invertebrates and fishes. Additionally, it is possible to convey sediment from convex floodplain vertical shape (FVSI greater than 0.05 having low potential of bitterlings and mussels) to significant concave floodplain vertical shape (FVSI less than −0.35 having low potential of bitterlings and mussels). The area of convex floodplain vertical shape requires a reduction of sediment deposition on the floodplain for changing the movable channels, and significant concave floodplain vertical shape requires sediment deposition for changing the stabilized channel. These processes could be predicted through simulations or experiments. The experiments and simulations of the sediment replenishment had been undertaken in tributary of the Kizu River [5]. However, ecological evaluation of the sediment management has not yet been properly performed. This study tried to support ecological connectivity between species abundance and sediment management based on relationships among animal communities, habitat structures with flooding frequency, and channel characteristics.

5. Conclusions

The reach-scale channel characteristics helped to determine a suitable target image for ecological sediment management on a catchment scale based on the multiscale relationships. The bitterlings and mussels were more abundant in terrace ponds than in active ponds, especially terrace ponds with a flooding frequency of 8–22 days/year. These suitable pond types tended to be located on the reaches with floodplain vertical shape index between −0.35 and 0.05 under the current channel conditions in the Kizu River. Thus, we recommend establishing long-term countermeasures of required sediment volume to maintain current conditions at sediment supply downstream. In terms of short-term countermeasures, we could suggest the target conditions at experiments and simulation of sediment replenishment. Although this study is based on a simple relationship analysis and requires considerably detailed correlation analysis, it could indicate an eco-geomorphic framework for river management based on the hierarchical interrelationships. In future, we will supplement the biological characteristics of bitterlings and mussels and various habitat conditions based on the hydraulic interpretation.

Author Contributions: Conceptualization, M.C. and Y.T.; Methodology, Y.T.; Software, K.I.; Formal Analysis, M.C.; Investigation, K.I.; Resources, K.I.; Writing-Original Draft Preparation, M.C.; Writing-Review & Editing, M.C.; Supervision, K.J.; Project Administration, K.J.; Funding Acquisition, K.J.

Acknowledgments: We would like to thank Ikeda and Nishii in ASIA AIR Survey CO. Ltd. and the Yodogawa River Bureau for providing historical aerial photos, cross-section data, and historical hydraulic data of the Kizu River.

References

1. Kondolf, G.M.; Gao, Y.; Annandale, G.W.; Morris, G.L.; Jiang, E.; Zhang, J.; Cao, Y.; Carling, P.; Fu, K.; Guo, Q.; et al. Sustainable sediment management in reservoirs and regulated rivers: Experiences from five continents. *Earth's Future* **2014**, *2*, 256–280. [CrossRef]
2. Kondolf, G.M. Hungry Water: Effects of Dams and Gravel Mining on River Channels. *Environ. Manag.* **1997**, *21*, 533–551. [CrossRef]
3. Auel, C.; Kobayashi, T.; Takemon, Y. Effects of sediment bypass tunnels on sediment grain size distribution and benthic habitats. *Int. J. River Basin Manag.* **2017**, *15*, 433–444. [CrossRef]
4. Kantoush, S.A.; Sumi, T.; Kubota, A. Geomorphic response of rivers below dams by sediment replenishment technique. In *River Flow 2010*; Dittrich, A., Koll, K., Aberle, J., Geisenhainer, P., Eds.; Bundesanstalt für Wasserbau: Karlsruhe, Germany, 2010; pp. 1155–1163.
5. Ock, G.; Sumi, T.; Takemon, Y. Sediment replenishment to downstream reaches below dams: Implementation perspectives. *Hydrol. Res. Lett.* **2013**, *7*, 54–59. [CrossRef]
6. Wood, P.J.; Armitage, P.D. Biological effects of fine sediment in the lotic environment. *Environ. Manag.* **1997**, *21*, 203–217. [CrossRef]
7. Leopold, L.B.; Wolman, M.G. *River Channel Patterns: Braided, Meandering and Straight*; US Geological Survey Professional Paper; U.S. Government Printing Office: Washington, DC, USA, 1957; Volume 282B, pp. 39–85.
8. Schumm, S.A. Patterns of alluvial rivers. *Annu. Rev. Earth Planet. Sci.* **1985**, *13*, 5–27. [CrossRef]
9. Rosgen, D.L. A classification of natural rivers. *Catena* **1994**, *22*, 169–199. [CrossRef]
10. Brierley, G.; Fryirs, K. *Geomorphology and River Management: Applications of the River Styles Framework*; Blackwell: Oxford, UK, 2005.
11. Wheaton, J.M.; Brasington, J.; Darby, S.E.; Merz, J.; Pasternack, G.B.; Sear, D.; Vericat, D. Linking geomorphic changes to salmonid habitat at a scale relevant to fish. *River Res. Appl.* **2010**, *26*, 469–486. [CrossRef]
12. Wyrick, J.R.; Pasternack, G.B. Geospatial organization of fluvial landforms in a gravel-cobble river: Beyond the riffle-pool couplet. *Geomorphology* **2014**, *213*, 48–65. [CrossRef]
13. Choi, M.; Takemon, Y.; Yu, W.; Jung, K. Ecological evaluation of reach scale channel configuration based on habitat structures for river management. *J. Hydroinform.* **2018**, *20*, 622–632. [CrossRef]
14. Frothingham, K.M.; Rhoads, B.L.; Herricks, E.E. A multiscale conceptual framework for integrated eco-geomorphological research to support stream naturalization in the agricultural Midwest. *Environ. Manag.* **2002**, *29*, 16–23. [CrossRef]

15. Sukhodolov, A.; Bertoldi, W.; Wolter, C.; Surian, N.; Tubino, M. Implication of channel processes for juvenile fish habitats in Alpine rivers. *Aquat. Sci.* **2009**, *71*, 338–349. [CrossRef]
16. Payne, B.A.; Lapointe, M.F. Channel morphology and lateral stability: Effects on distribution of spawning and rearing habitat for Atlantic salmon in a wandering cobble-bed river. *Can. J. Fish. Aquat. Sci.* **1997**, *54*, 2627–2636. [CrossRef]
17. Beechie, T.J.; Liermann, M.; Pollock, M.M.; Baker, S.; Davies, J. Channel pattern and river-floodplain dynamics in forested mountain river systems. *Geomorphology* **2006**, *78*, 124–141. [CrossRef]
18. Tockner, K.; Ward, J.V.; Arscott, D.B.; Edwards, P.J.; Kollmann, J.; Gurnell, A.M.; Petts, G.E.; Maiolini, B. The Tagliamento River: A model ecosystem of European importance. *Aquat. Sci.* **2003**, *65*, 239–253. [CrossRef]
19. Arscott, D.B.; Tockner, K.; van der Nat, D.; Ward, J.V. Aquatic Habitat Dynamics along a Braided Alpine River Ecosystem (Tagliamento River, Northeast Italy). *Ecosystems* **2002**, *5*, 802–814. [CrossRef]
20. Kizu River Research Group. *Integrated Research of the Kizu River II*; Kizu River Research Group: Kyoto, Japan, 2003. (In Japanese)
21. Negishi, J.N.; Sagawa, S.; Kayaba, Y.; Sanada, S.; Kume, M.; Miyashita, T. Mussel responses to flood pulse frequency: The important of local habitat. *Freshw. Biol.* **2012**, *57*, 1500–1511. [CrossRef]
22. Takemon, Y.; Kobayashi, S.; Choi, M.; Terada, M.; Takebayashi, H.; Sumi, T. River Habitat Evaluation based on Cross-sectional Bed Profile and Frequency Distribution of Relative Elevation. *Adv. River Eng.* **2013**, *19*, 519–524. (In Japanese)
23. Egashira, S.; Jin, H.; Takebayashi, H.; Nida, B.; Nagata, T. Bed variation and sediment budget in the downstream reach of Kizu River. *Annu. J. Hydraul. Eng.* **2002**, *44*, 777–782. (In Japanese) [CrossRef]
24. Jowett, I.G.; Davey, A.J.H. A Comparison of Composite Habitat Suitability Indices and Generalized Additive Models of Invertebrate Abundance and Fish Presence-Habitat Availability. *Trans. Am. Fish. Soc.* **2007**, *136*, 428–444. [CrossRef]
25. Bovee, K.D. *Development and Evaluation of Habitat Suitability Criteria for Use in the Instream Flow Incremental Methodology*; National Ecology Center, U.S. Fish and Wildlife Service: Washington, DC, USA, 1986.
26. *Survey Report on the Kizu River*; Yodogawa River Bureau, ASIA AIR SURVEY CO., LTD.: Osaka, Japan, 2007. (In Japanese)
27. *Survey Report on the Kizu River*; Yodogawa River Bureau, ASIA AIR SURVEY CO., LTD.: Osaka, Japan, 2009. (In Japanese)
28. *Survey Report on the Kizu River*; Yodogawa River Bureau, ASIA AIR SURVEY CO., LTD.: Osaka, Japan, 2010. (In Japanese)
29. Johnson, P.D.; Brown, K.M. The importance of microhabitat factors and habitat stability to the threatened Louisiana pearl shell, Mrgaritifera hembeli (Conrad). *Can. J. Zool.* **2000**, *78*, 271–277. [CrossRef]
30. Yoshihiro, B.A.; Takashi, M. Habitat Characteristics Influencing Distribution of the Freshwater Mussel Pronodularia japanensis and Potential Impact on the Tokyo Bitterling, Tanakia tanago. *Zool. Sci.* **2010**, *27*, 912–916.
31. Terada, M.; Takemon, Y.; Sumi, T. A study on habitat evaluation of tamari for bitterling and mussel. In Proceedings of the JSCE-Kansai, Uji, Japan, 26 August 2011; Volume II-12. (In Japanese)
32. *Global Re-introduction Perspectives: 2011*; IUCN/SSC Re-Introduction Specialist Group: Calgary, AB, Canada, 2011.
33. Reichard, M.; Liu, H.; Smith, C. The co-evolutionary relationship between bitterling fishes and freshwater mussels: Insights from interspecific comparison. *Evolut. Ecol. Res.* **2007**, *9*, 239–259.
34. Ellis, M.M. Erosion silt as a factor in aquatic environments. *Ecology* **1936**, *17*, 29–42. [CrossRef]
35. Kobayashi, S.; Takemon, Y. Historical Changes of Riffle Morphology for Benthic Invertebrate Habitats in the Kizu River. *Annu. Disaster Prev. Res. Inst. Kyoto Univ.* **2013**, *56B*, 681–689. (In Japanese)
36. Surian, N.; Rinaldi, M. Morphological response to river engineering and management in alluvial channels in Italy. *Geomorphology* **2003**, *50*, 307–326. [CrossRef]

9

Estimating Sediment Flushing Efficiency of a Shaft Spillway Pipe and Bed Evolution in a Reservoir

Ching-Nuo Chen [1,*] and Chih-Heng Tsai [2]

1 International Master Program of Soil and Water Engineering, National Pingtung University of Science and Technology, Pingtung 91201, Taiwan
2 Department of Recreation and Healthcare Management, Chia Nan University of Pharmacy and Science, Tainan 71710, Taiwan; jht581212@gmail.com
* Correspondence: ginrochen@mail.npust.edu.tw

Abstract: Control of reservoir sedimentation in order to ensure their sustainable use has drawn attention among water engineers and water resource managers. Several methods have been proposed, but most of the developed methodologies are incapable of modelling bed evolutions, while at the same time, compute sediment flushing efficiency. In this study a two-dimensional bed evolution model is proposed to estimate sediment distribution, bed evolution and sediment flushing efficiency of reservoirs. A-Gong-Dian reservoir, in southern Taiwan, is used as an illustrative example. Typhoon events were used to verify the proposed model. Simulations were conducted for one and two-day storm events under return periods, 2, 5, 10, 25, 50, 100, and 200-year. The results indicated that the average sediment flushing efficiency of the shaft spillway under one and two-day storms were close, 58.50% and 59.39%, respectively. These results were similar to observed laboratory tests experiments, where an efficiency of 65.34% was obtained. This study suggests that the applied model could be adopted to ensure the sustainable use of reservoirs, and also to find an optimal area for the location of a shaft spillway pipe. Therefore, the proposed model could serve as a reference to the reservoir management personnel.

Keywords: two-dimensional bed evolution model; sediment flushing of empty storage; shaft spillway pipe; sediment flushing efficiency

---

1. Introduction

Reservoirs are often affected by accelerated sediment deposition rates and has shortened the life of reservoirs by more than 65% in China alone [1]. As a result, the economic value of such projects has severely declined. Not only do they influence the life of reservoirs, they also pose safety hazards, as illustrated by [2]. Their sustainability is strongly dependent on how well the rate of sediment deposition is reduced and on the techniques of managing the reservoirs. Several techniques are available for their management, amongst which are mechanical excavation, dredging (conventional dredging, dry excavation), and hydraulic desilting. For an exhaustive review of the different techniques, the reader is referred to [3], who explored sustainable sediment management in reservoirs based on experience from five continents. Mechanical excavation and dredging boats, however, are associated with higher costs when compared to hydraulic desilting, and are often plagued by subsequent disposal problems. Hydraulic desilting employs stream power and hydraulics to cut down sediment deposits downstream. Flushing out sediments in reservoirs has been shown to cut costs [4], despite the large amount consumed by the flushing operations. Emamgholizadeh and Samadi [5] classify flushing into two, complete (also termed empty) and partial drawdown flushing. These, in turn, include hydro-suction, sediment sluicing, sediment bypass, density current venting, and hydraulic flushing through the reservoir, used independently or in combination [6]. The efficiency of sediment flushing depends on the geometry of the reservoir, sediment particle size, characteristics of sediments deposited, flow discharge, and flow depth. Several authors have argued that the efficiency of sediment flushing is influenced by the ratio of storage volume to incoming runoff [7], which should be less than 0.05 [8] for this technique to work. Moreover, we did not evaluate this threshold in this study, since the reservoir under investigation already applied hydraulic sediment flushing. Madadi, et al. [9] managed to improve flushing efficiencies by up to 280% through reconfigured reservoir bottom outlets in laboratory experiments.

Effective management of reservoirs system require a model that can predict future behaviour and response to perturbation [10], and all of the models are developed through experiments and depending on the status quo, they may grow in complexity to include conceptual frameworks, computer calculations, numerical simulations, and physical scale modelling [11]. Physical models have been applied to study the process and efficiency of sediment flushing in a reservoir. Although they have been successfully used to understand and reproduce to some extent complex physical processes that occur in nature, and have contributed significantly in hydraulic construction designs, they are relatively costly and time consuming [12]. More recently photogrammetry-based surveys using unmanned aerial systems have been used to evaluate flushed sediments [13–15]. Moreover, such techniques can only compute the amount of flushed sediments only when the reservoir is dry (i.e., empty), and subsequent images are necessary to compute the Digital Elevation Models (DEMs) of difference, from which the flushing efficiencies may be computed. Network-based programming techniques have also been employed in multi-reservoir systems [16], though their core emphasis is on determining empty flushing of sediments.

Given the recent advances in computational power, multi-dimensional models have increased the capability of assessing sedimentation problems and the multi-dimensional models have been extensively adopted in engineering application and analysis. For models to be adopted, they should reflect the physical characteristics of the reservoir and complexity in question. Numerical sediment transport models are available in one, two, and three dimensions. The widely used models, however, are one- (1D) and two-dimensional (2D) models when compared to the high computer intensive three-dimensional (3D) models. Examples of 1D sediment transport models are HEC-6 [17], HEC-RAS [18], and FLUVIAL-12 [19,20]. Castillo, Carrillo, and Álvarez [7] employed four complementary methods, which included 1D model, to determine sedimentation and flushing in a reservoir These models are capable of simulating longitudinal flows in rivers, moreover, they run short in the simulation of sediment transport and bed evolution in reservoirs. To be applied in sediment flushing, several assumptions should be made, thus, compromising the accuracy and efficiency in reservoir management. In such situations, reservoirs are narrow in shape, flow highly channelized, while closely following the thalweg [12]. However, most reservoir pools are wide and have no single clear flow direction, and they often constitute complex topography and geometry. As a result, multi-dimensional models are used. Olsen [21] used a depth-averaged 2D model to study the flushing process in a water reservoir in Nepal. Besides two-dimensional models, three-dimensional models have also been applied to study sediments in reservoirs. Olsen and Skoglund [22] applied a 3D model to calculate the sediment deposition in a hydropower reservoir, and also in a sand trap. Fang and Rodi [23] used a 3D model to simulate flow and sediment transport in the Three Gorges Project (TGP) reservoir in Yangtze River. Khosronejad, et al. [24] used a three-dimensional finite volume model to study the effects of various parameters on the quantity of sediment that was released from a reservoir in the reservoir flushing process.

Although the above models could estimate sediment erosion and deposition, bed evolution in a reservoir, and the efficiency of flushing, we have not found a model that is capable of combining all of these key reservoir management strategies in a single package. In addition, the above stated models require suspended sediment concentration, and sediment yield hydrograph into the reservoir, which are not easily obtained. Consequently, rating curves of discharge and suspended sediment transport rate are used, and these are associated with high errors [25]. Incorrect estimation of sediment inflow into reservoirs especially during flood events, will eventually lead to inefficient flushing of sediments and to misleading bed evolution in the reservoirs. It is therefore imperative to develop models that are highly efficient in estimating inflow hydrographs and sedigraphs, in turn, correctly estimating the amount of sediments to be flushed, while estimating the resultant bed evolution. A two-dimensional bed evolution model having these capabilities is developed and applied in this study. The upstream boundary condition hydrographs of inflow discharge and suspended sediment concentration for the 2D dimensional bed evolution model were calculated by the Physiographic Soil Erosion and Deposition (PSED) [26]. The PSED model can accurately estimate discharge hydrographs, concentration of suspended sediment hydrograph and suspended sediment transport rate from a watershed.

2. Numerical Model

The depth-averaged two-dimensional bed evolution model is divided into three parts: (1) water flow calculations; (2) sediment transport calculations; and, (3) bed elevation variation calculations.

2.1. Governing Equations for Water

The depth-averaged continuity and momentum equations are given below:

$$\frac{\partial h}{\partial t}+\frac{\partial (uh)}{\partial x}+\frac{\partial (vh)}{\partial y}=0 \tag{1}$$

$$\frac{\partial (uh)}{\partial t}+\frac{\partial (uuh)}{\partial x}+\frac{\partial (uvh)}{\partial y}=\frac{\partial}{\partial x}\left(2\varepsilon h\frac{\partial u}{\partial x}\right)+\frac{\partial}{\partial y}\left[\varepsilon h\left(\frac{\partial u}{\partial y}+\frac{\partial v}{\partial x}\right)\right]-\frac{\tau_{bx}}{\rho}-gh\frac{\partial H}{\partial x} \tag{2}$$

$$\frac{\partial (vh)}{\partial t}+\frac{\partial (vuh)}{\partial x}+\frac{\partial (vvh)}{\partial y}=\frac{\partial}{\partial x}\left[\varepsilon h\left(\frac{\partial u}{\partial y}+\frac{\partial v}{\partial x}\right)\right]+\frac{\partial}{\partial y}\left(2\varepsilon h\frac{\partial v}{\partial y}\right)-\frac{\tau_{by}}{\rho}-gh\frac{\partial H}{\partial y} \tag{3}$$

in which t is time; x and y are horizontal Cartesian coordinates; h is the depth; u and v are depth-average flow velocities in x and y directions; H is water surface elevation; g is gravitational acceleration; ρ is density of flow; ε is the depth-average kinematic eddy viscosities of water; and, τ_{bx} and τ_{by} are bed shear stresses τ_b in x and y directions, $\tau_b=\rho ghS_f$, S_f is the friction slope.

The depth-average kinematic eddy viscosities of water can be approximated and expressed as [27]:

$$\varepsilon=\frac{\kappa}{6}u_*h \tag{4}$$

in which κ is the von Karman constant, and $\kappa=0.4$ is chosen in this study. u_* is the shear velocity and $u_*=\sqrt{ghS_f}$.

2.2. Governing Equations for Sediment Transport with Source Terms

The convective-diffusive equation of suspended sediment, can be expressed as [28]:

$$\frac{\partial (Ch)}{\partial t}+\frac{\partial (uCh)}{\partial x}+\frac{\partial (vCh)}{\partial y}=\frac{\partial}{\partial x}\left(\varepsilon\frac{\partial C}{\partial x}\right)+\frac{\partial}{\partial y}\left(\varepsilon\frac{\partial C}{\partial y}\right)+q_{se}-q_{sd} \tag{5}$$

where C is the depth-averaged volumetric concentration of suspended sediment; q_{se} and q_{sd} are the entrainment and deposition terms of river bed, respectively. According to Itakura and Kishi [29], the entrained rate of channel bed can be expressed as:

$$q_{se}=0.008\sqrt{sgd}\left[0.14\frac{\rho}{\rho_s}\left(14\sqrt{\tau_*}-\frac{0.9}{\sqrt{\tau_*}}\right)-\frac{\omega_s}{\sqrt{sgd}}\right] \tag{6}$$

in which $s = (\rho_s - \rho)/\rho$ is the submerged specific gravity of the sediment; ρ_s is density of the sediment; d is diameter of the sediment; ω_s is fall velocity of the sediment; and, τ_* is non-dimensional bed shear stress, $\tau_* = {u_*}^2/sgd$.

The deposition rate of suspended sediment can be expressed as:

$$q_{sd} = \omega_s\, C_a \tag{7}$$

where, C_a is the concentration of sediment near the channel bed. C_a can be estimated from the volumetric concentration of suspended sediment obtained at 0.05 depth from the channel bed. Using the exponential law, the volumetric concentration of suspended sediment may be expressed as [30]:

$$C_a = \frac{P_e}{[\,1 - \exp\,(-P_e)\,]}\,C \tag{8}$$

where P_e is the Peclet's number, which may be expressed as $\omega_s h/\varepsilon$.

An extensively used bed load transport formula is the Meyer Peter and Muller formula (MPM) [31]. Moreover, Wong and Parker [32] amended the MPM formula, and more accurate estimates of bed load transport rate were obtained. Nonetheless, since the original MPM formula is relatively easy to apply when establishing a numerical model, bed load transport was calculated using the original MPM formula in this study.

$$\left(\frac{k_n}{k'}\right)^{\frac{3}{2}} \gamma h S_f = 0.047(\gamma_s - \gamma)d + 0.25\left(\frac{\gamma}{g}\right)^{1/3}\left(\frac{\gamma_s - \gamma}{\gamma}\right)^{2/3} {q_b}^{2/3} \tag{9}$$

where γ and γ_s are the specific weight of water and the specific weight of sediment, respectively; q_b is the bed load transport rate per unit width of bed; k_n is Strickler's roughness coefficient, which can be represented as the reciprocal of Manning's roughness coefficient; and, $k' = 26/d_{90}^{1/6}$, d_{90} is the size of sediment in the unit of meter for which 90% of the material is finer.

2.3. Governing Equations for Bed Variation

The bed evolution due to sediment transport rate is not equal throughout an alluvial river. The continuity equation for bed elevation variations can be written as [33,34]:

$$\frac{\partial z}{\partial t} + \frac{1}{1-\lambda}\left[\frac{\partial q_{bx}}{\partial x} + \frac{\partial q_{by}}{\partial y} + (q_{se} - q_{sd})\right] = 0 \tag{10}$$

where z is the channel bed elevation; λ is the porosity, $\lambda = 0.245 + 0.0864/{d_m}^{0.21}$, d_m is the mean sediment diameter [35]; and, q_{bx} and q_{by} are the components of q_b in x and y directions, respectively.

2.4. Numerical Scheme

MacCormack explicit finite-difference method [36] is adopted and is divided into predictor and corrector steps. The forward finite-difference is used to discretize the predictor step while the backward finite-difference is used to discretize the corrector step.

The forward finite-difference is used to compute the water depth h by the continuity equation (Equation (1)) and the u and v are computed by the momentum equations (Equations (2) and (3)). The predicted value may be written as:

$$h^*_{i,j} = h^n_{i,j} + \Omega'_{hf} \tag{11}$$

$$u^*_{i,j} = u^n_{i,j}\frac{h^n_{i,j}}{h^{n+1}_{i,j}} + \Omega'_{uf} \tag{12}$$

$$v_{i,j}^{*} = v_{i,j}^{n} \frac{h_{i,j}^{n}}{h_{i,j}^{n+1}} + \Omega'_{vf} \tag{13}$$

The superscript * denotes the predicted value; the superscript n and $n + 1$ refers to the variables at the known and unknown time levels; the subscript i and j denote the grid in x- and y-directions; and, Ω'_{hf}, Ω'_{uf} and Ω'_{vf} are the functions of the known value of variables h, u, v at the time level n.

The backward finite-difference is used to calculate the water depth h by the continuity equation (Equation (1)) and the u and v are computed by the momentum equations (Equations (2) and (3)). The corrected value may be written as:

$$h_{i,j}^{**} = h_{i,j}^{n} + \Omega'_{hb} \tag{14}$$

$$u_{i,j}^{**} = u_{i,j}^{n} \frac{h_{i,j}^{n}}{h_{i,j}^{n+1}} + \Omega'_{ub} \tag{15}$$

$$v_{i,j}^{**} = v_{i,j}^{n} \frac{h_{i,j}^{n}}{h_{i,j}^{n+1}} + \Omega'_{vb} \tag{16}$$

The superscript ** denotes the corrected value; Ω'_{hb}, Ω'_{ub}, and Ω'_{vb} are the functions of the known value of variables h, u, v at the time level n.

The value of the variables at the unknown time level could be calculated by the predicted and corrected values that may be written as:

$$h_{i,j}^{n+1} = \frac{1}{2}\left(h_{i,j}^{*} + h_{i,j}^{**}\right) \tag{17}$$

$$u_{i,j}^{n+1} = \frac{1}{2}\left(u_{i,j}^{*} + u_{i,j}^{**}\right) \tag{18}$$

$$v_{i,j}^{n+1} = \frac{1}{2}\left(v_{i,j}^{*} + v_{i,j}^{**}\right) \tag{19}$$

Bed Evolution Model—MacCormack Explicit Finite—Difference Method

The explicit finite-difference method is used to discretize the suspended sediment concentration convection-diffusion equation (Equation (5)) and the continuity equation for bed elevation variations (Equation (10)) to calculate the volumetric concentration of suspended sediments and bed elevation. The volumetric concentration of suspended sediments C is calculated by:

$$C_{i,j}^{n+1} = \frac{C_{i,j}^{n}\, h_{i,j}^{n}}{h_{i,j}^{n+1}} + \Omega'_{c} \tag{20}$$

where Ω'_c is composed of the known value of variables h, u, v, C at the time level n.

The bed elevation may be written as:

$$z_{i,j}^{n+1} = z_{i,j}^{n} + \Omega'_{z} \tag{21}$$

where Ω'_z is composed of the known value of variables h, u, v, C at the time level n.

3. Study Area

A-Gong-Dian Reservoir (Figure 1) is used as an illustrative example in this study. This reservoir, located in Kaohsiung City, (southern Taiwan), collects water from Joushui River and Wanglai River. The total watershed area is 29.58 with 12.81 km^2 (43%) from Joushui River watershed and 16.77 km^2 (57%) from Wanglai River watershed. The length of the dam is 2.38 km, making it the longest dam in Taiwan. Its major purpose is flood control, while other uses, such as irrigation, water supply, and tourism benefit.

The elevation of the dam top, design water level, and maximum water level are 42, 37, and 40 m, respectively. The reservoir was completed in 1953. However, since its completion, large amounst of green-grey clay and yellow silty clay from the upstream watersheds of Joushui and Wanglai rivers have been washed into the reservoir and severe sedimentation has been observed [37]. The effective reservoir capacity was slashed from 20.5 to 5.9 million cubic meters in 1996. In order to revamp the reservoir, the A-Gong-Dian reservoir improvement project was implemented in 1997. Large-scale sediment flushing of the reservoir was executed and 11.6 million cubic meters of sediments were dredged, and the reservoir reached empty storage. The reservoir improvement project involved dam improvement, conduit spillway reconstruction, water intake tower reconstruction, trans-basin waterway, etc. It was finally completed in 2005, and re-opened in June 2006. The shaft spillway pipe (Figure 2) has been operated ever since, for the period 1 June to 10 September annually, which corresponds to the wet season in this region.

Although its design capacity storage is 20.5 million cubic meters, currently, the effective storage capacity is 16.69 million cubic meters, and the total water storage is 45 million cubic meters. The reservoir adopts a shaft spillway pipe having a 2.8 m diameter to reduce the pipe top elevation to 27 m. Based on hydraulic model test, flow discharge of the shaft spillway pipe can be expressed by Equations (22) and (23) [37]:

$$\text{Free overfall } Q_{\text{out}} = 34.12(H_s - 27)^{1.5},\ H_s \leq 28.57 \text{ m} \tag{22}$$

$$\text{Pipe flow } Q_{out} = 17.63(H_s - 14)^{0.5},\ H_s > 28.57 \text{ m} \tag{23}$$

where H_s is water level in the reservoir, Q_{out} is the releasing discharge. The maximum releasing discharge of pipe flow for the shaft spillway pipe is 89.90 m^3/s when the water level reaches the 40-m design maximum flood retention level.

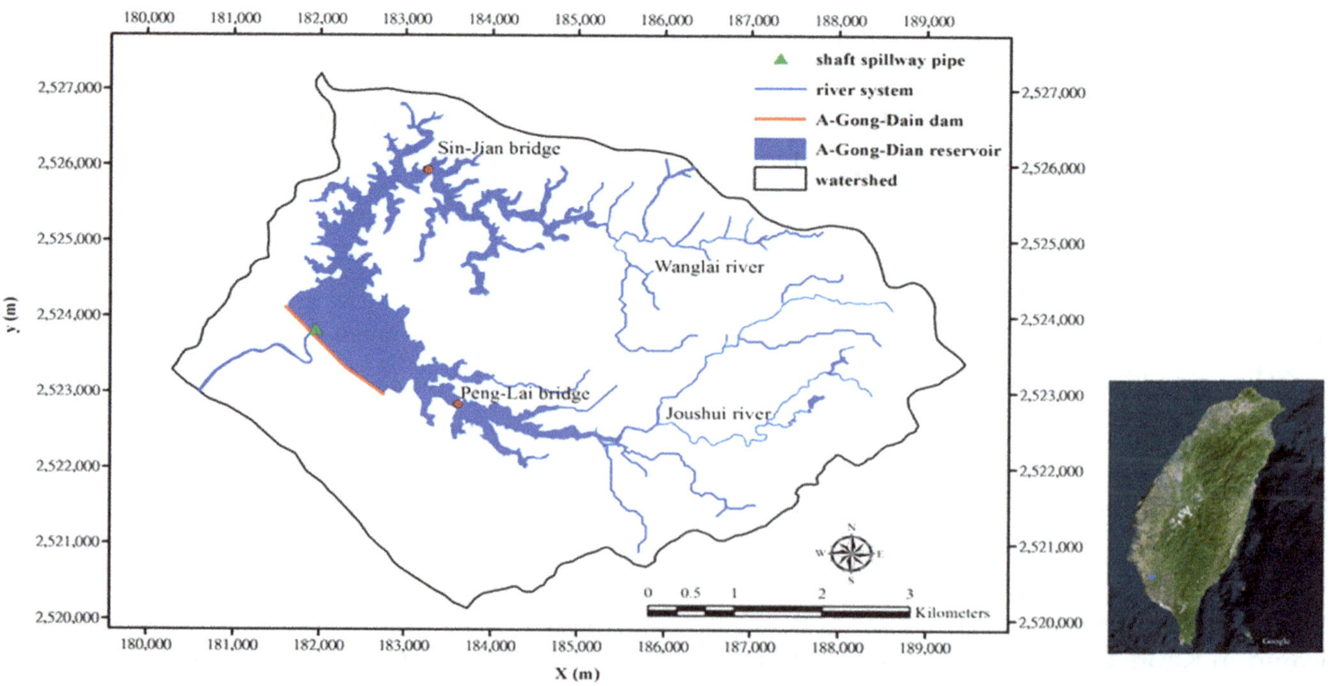

Figure 1. A-Gong-Dian Reservoir watershed and its river system.

Figure 2. Shaft spillway pipe.

4. Methodology

In order to understand the sedimentation pattern, distribution of sediments into the A-Gong-Dian reservoir, bed evolution within the reservoir, and the sediment flushing efficiency of an empty storage operation of the shaft spillway pipe during a flood season, a depth-averaged two-dimensional bed evolution model was developed and applied. The upstream boundary condition hydrographs of inflow discharge and suspended sediment concentration for the two-dimensional bed evolution model were calculated by the PSED model [26,38].

The PSED model can accurately estimate discharge hydrographs, sediment hydrograph, suspended sediment transport rate, and sediment yield from a watershed. In this model, GIS is used to partition the river catchment into computed river cells and land cells, according to the spatial distribution of the physiographical characteristics, such as topography, landform, and vegetation distribution in the watershed. Due to the complex nature of these earth features, a large amount of data is often generated. With the assistance of GIS, the PSED model can handle enormous hydrologic and physiographic datasets, simulating the physical erosion process without the need for simplification.

Computations were done for the storage region between Sin-Jian Bridge of the Wanglai river and the Peng-Lai Bridge of the Joushui river, which is about 5 km long. Cross-sections were measured by the Southern Water Resources Bureau in 2008, and the measured elevations are shown in Figure 3. The study area was discretised into 26,251, squared, $\Delta x = \Delta y = 10$ m, computational cells and the time step was 0.05 s (Δt = 0.05). Conducted field experiments have shown the average sediment particle size to be 0.015 mm, while the Manning's roughness coefficient was 0.028. The initial water depth condition was 0.1 m, while velocity and sediment concentration were 0 for each computed grid.

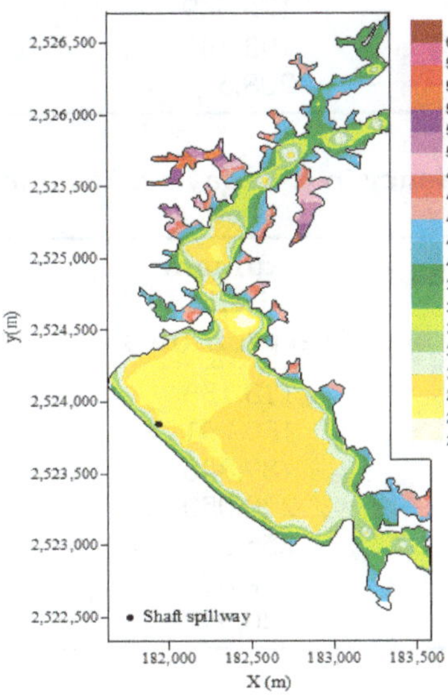

Figure 3. Bed elevation in storage area of A-Gong-Dian reservoir.

The hydrograph boundary conditions of inflow discharge and suspended sediment concentration for the typhoon event and design rainfall of one- and two-day storms under various return periods (2, 5, 10, 25, 50, 100, 200-year) for the depth-averaged two-dimensional bed evolution model were simulated and calculated by the PSED model.

5. Results and Discussions

5.1. Analysis of the Sediment Flushing Efficiency of an Empty Storage

Reducing sediment deposition to increase flood control volume is crucial for Agongdian reservoir. Therefore, sediment flushing of empty storage by a shaft spillway pipe is operated during the wet period to decrease the sediment deposition within the reservoir bed. To understand the sediment flushing efficiency of empty storage by a shaft spillway pipe under one-day and two-day storms of various return periods, the operations for sediment flushing of the empty storage were simulated using the depth-averaged two-dimensional bed evolution model, and the results are shown in Tables 1 and 2.

The results indicate that the average sediment flushing efficiency for one-day and two-day storms are close, 58.50% and 59.39%, respectively. For the simulated storms, the lowest efficiencies are 58.16%, 58.62% (200-year return period), and the highest efficiencies are 58.98%, 60.19% (10-year, 2-year return period). The simulation results are close to 65.34% [39], which agree well with the efficiency obtained from the hydraulic model test in the laboratory. The results show that the two-dimensional bed evolution model can reasonably simulate sediment flushing of empty storage by a shaft spillway pipe, as sediment flushing efficiency reached up to 60%.

Table 1. Empty flushing efficiency in one-day storms under different return periods.

Return Period (Year)	Sediment Yield of the Watershed (m^3)	Sediment Deposited in the Reservoir (m^3)	Sediment Flushing by Shaft Spillway (m^3)	Sediment Flushing Efficiency (%)
2	201,341	83,154	118,187	58.70
5	280,522	116,875	163,647	58.34
10	331,857	136,121	195,736	58.98
25	375,159	156,963	218,196	58.16
50	427,718	176,549	251,169	58.72
100	464,325	193,141	271,184	58.40
200	498,867	208,575	290,292	58.16

Table 2. Empty flushing efficiency in two-day storms under different return periods.

Return Period (Year)	Sediment Yield of the Watershed (m^3)	Sediment Deposited in the Reservoir (m^3)	Sediment Flushing by Shaft Spillway (m^3)	Sediment Flushing Efficiency (%)
2	284,238	113,155	171,083	60.19
5	397,096	159,791	237,305	59.76
10	471,082	189,846	281,236	59.70
25	541,196	220,050	321,146	59.34
50	625,933	255,506	370,427	59.18
100	701,485	287,889	413,596	58.96
200	775,718	320,992	454,726	58.62

5.2. Analysis of Sediment Delivery Behaviour in Reservoirs

Sediment flushing efficiency is related to the life and function of the reservoir. Therefore, this topic of great importance for reservoir management. In addition, it is necessary to carry out the analysis from sediment transport into the reservoir to bed evolution during the process of flooding and sediment flushing. Management of reservoirs does not only entail improving the flushing efficiencies, but in depth investigations are needed to understand the whole process from catchment erosion, to sediment transport and bed evolution, and finally to the removal of the sediments. To understand the whole complex process of sediments transport during heavy storms or floods in a reservoir, the severe storm event typhoon Morakot, which hit Taiwan in 6 to 10 August 2009, was numerically simulated to estimate the variations of water depth, suspended sediment concentration, sediment delivery, and bed evolution. Results are shown in Figures 4–6. Figure 4 show water depth variation in the reservoir at 36, 60, 72, 84, 96, and 120-h, while Figure 5 show distribution of suspended sediment concentration, and Figure 6, bed evolution.

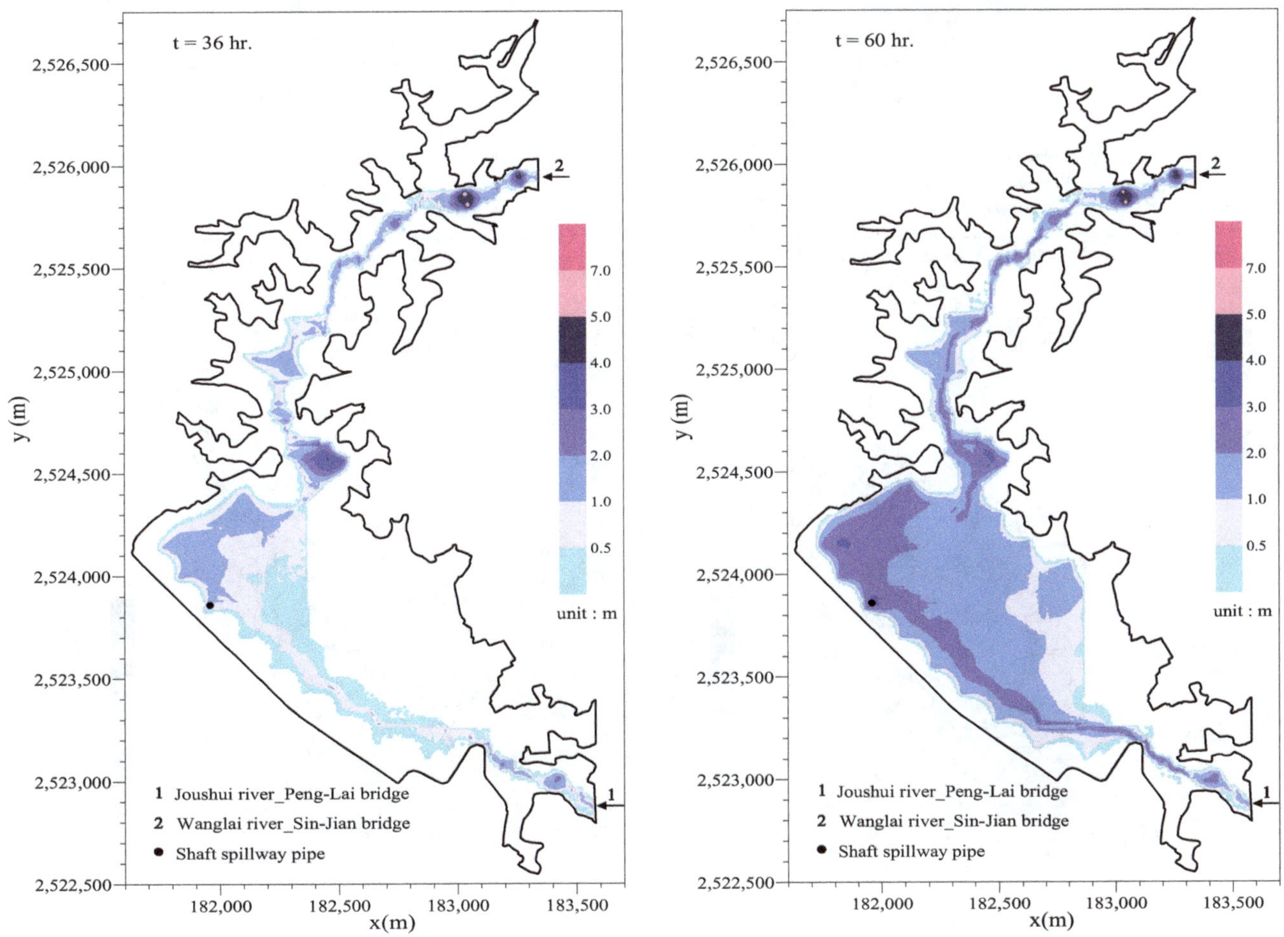

(a) (b)

Figure 4. *Cont.*

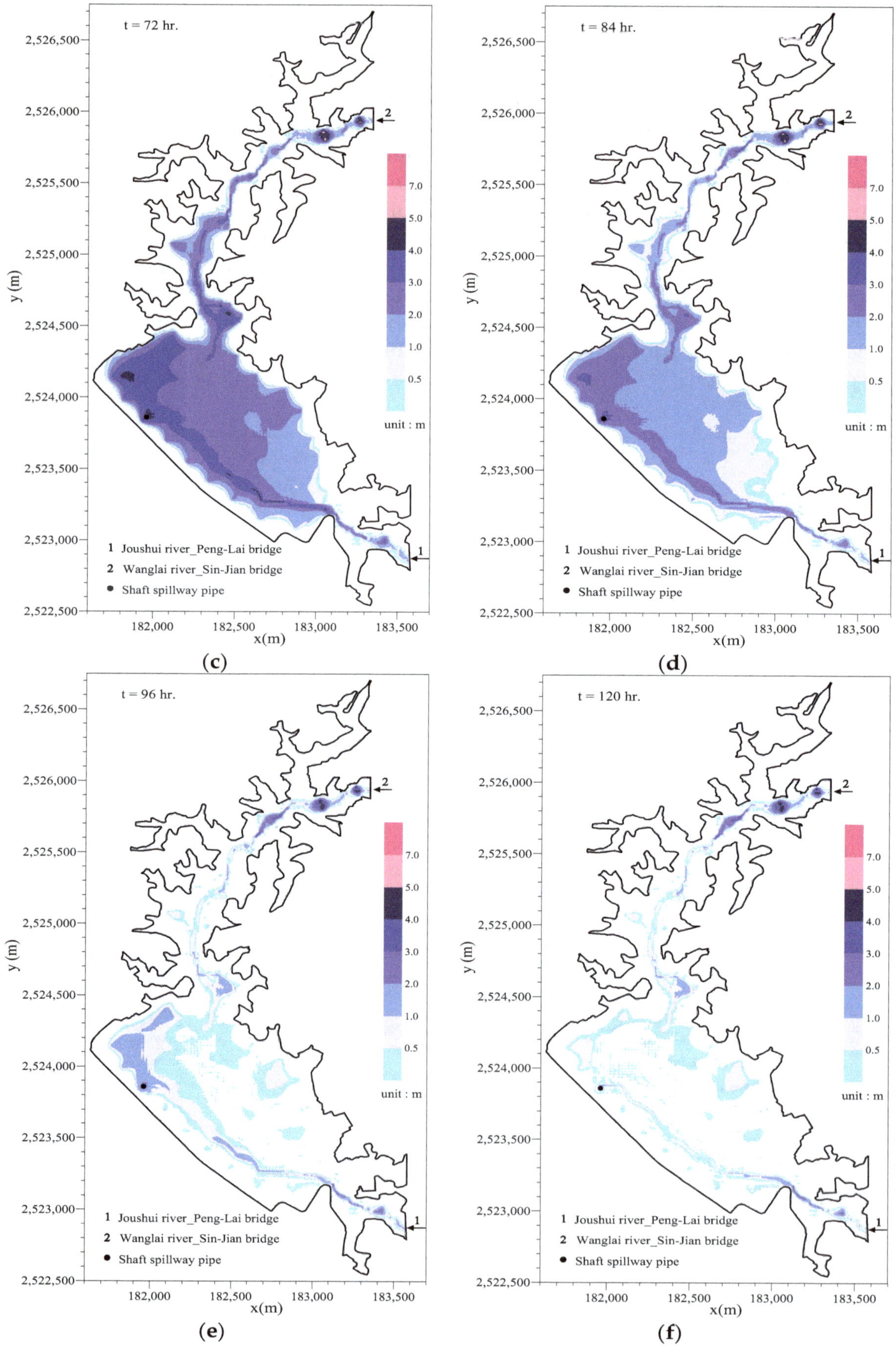

Figure 4. (**a**) Water depth in the reservoir at t = 36 h; (**b**) Water depth in the reservoir at t = 60 h; (**c**) Water depth in the reservoir at t = 72 h; (**d**) Water depth in the reservoir at t = 84 h; (**e**) Water depth in the reservoir at t = 96 h; and, (**f**) Water depth in the reservoir at t = 120 h.

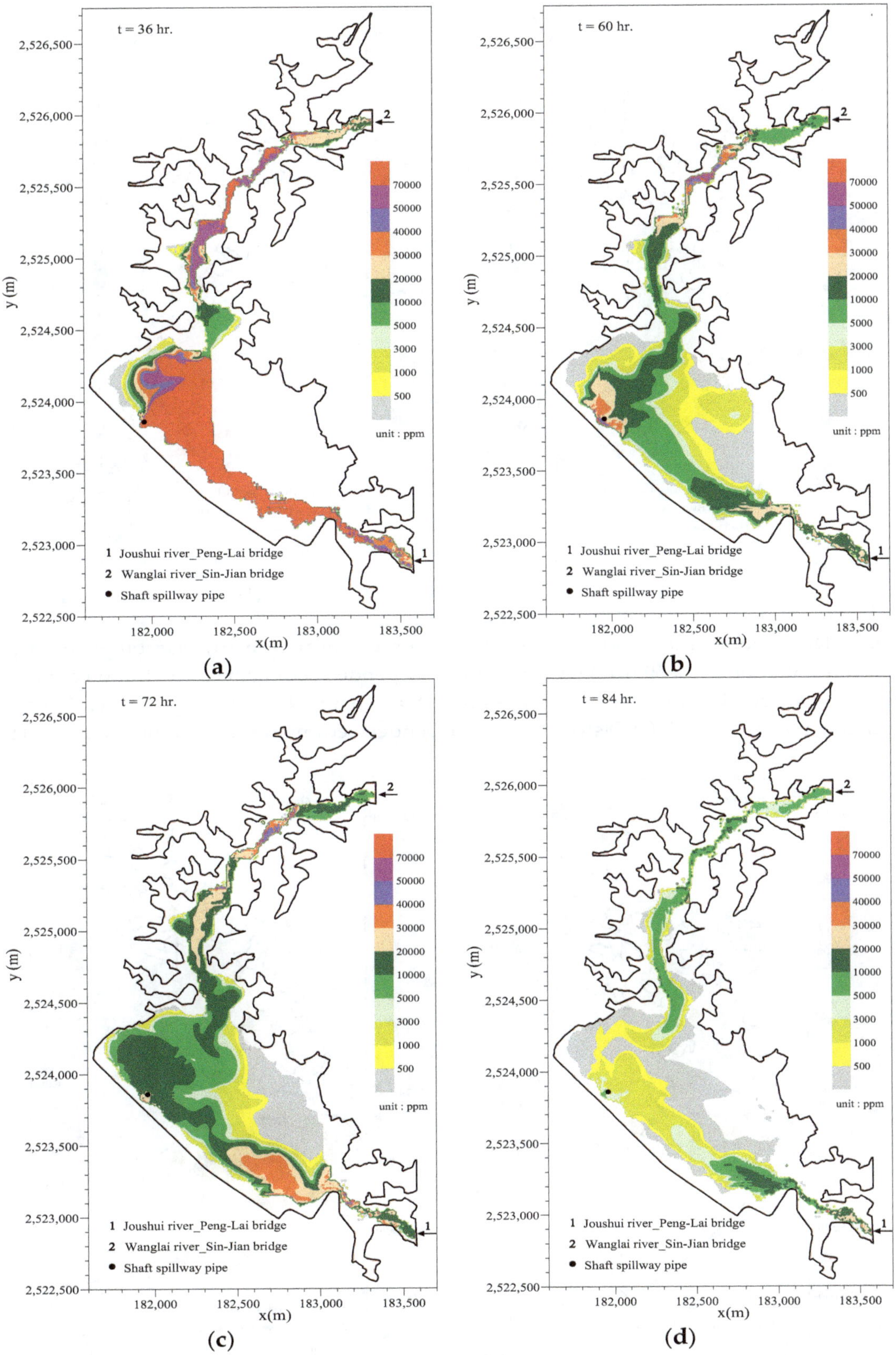

Figure 5. *Cont.*

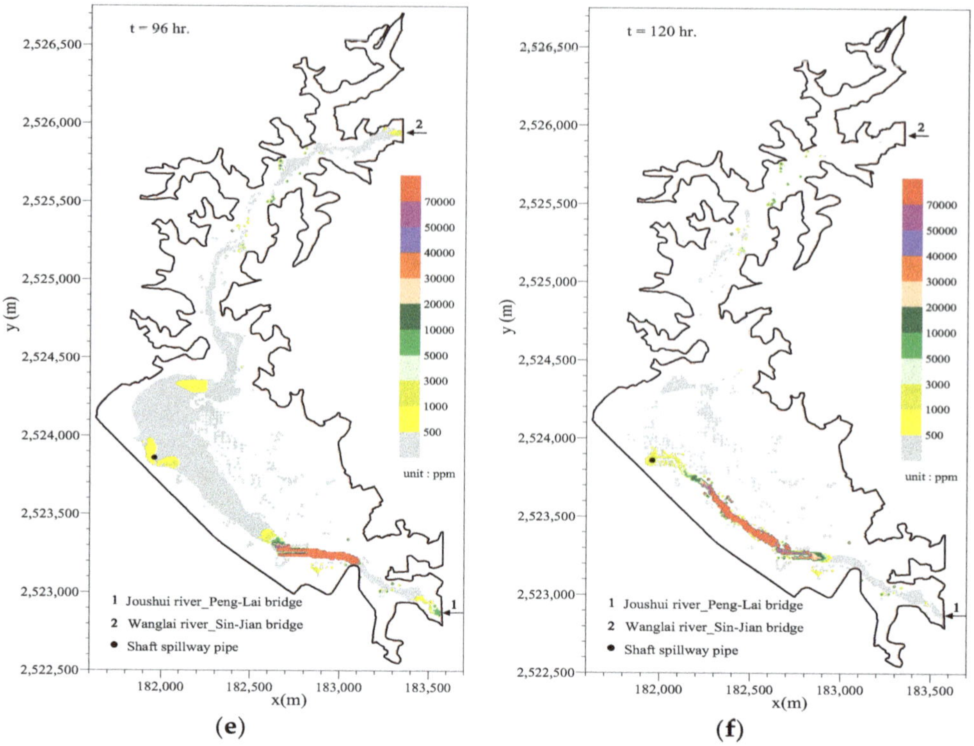

Figure 5. (**a**) Distribution of suspended sediment concentration at t = 36 h; (**b**) Distribution of suspended sediment concentration at t = 60 h; (**c**) Distribution of suspended sediment concentration at t = 72 h; (**d**) Distribution of suspended sediment concentration at t = 84 h; (**e**) Distribution of suspended sediment concentration at t = 96 h; and, (**f**) Distribution of suspended sediment concentration at t = 120 h.

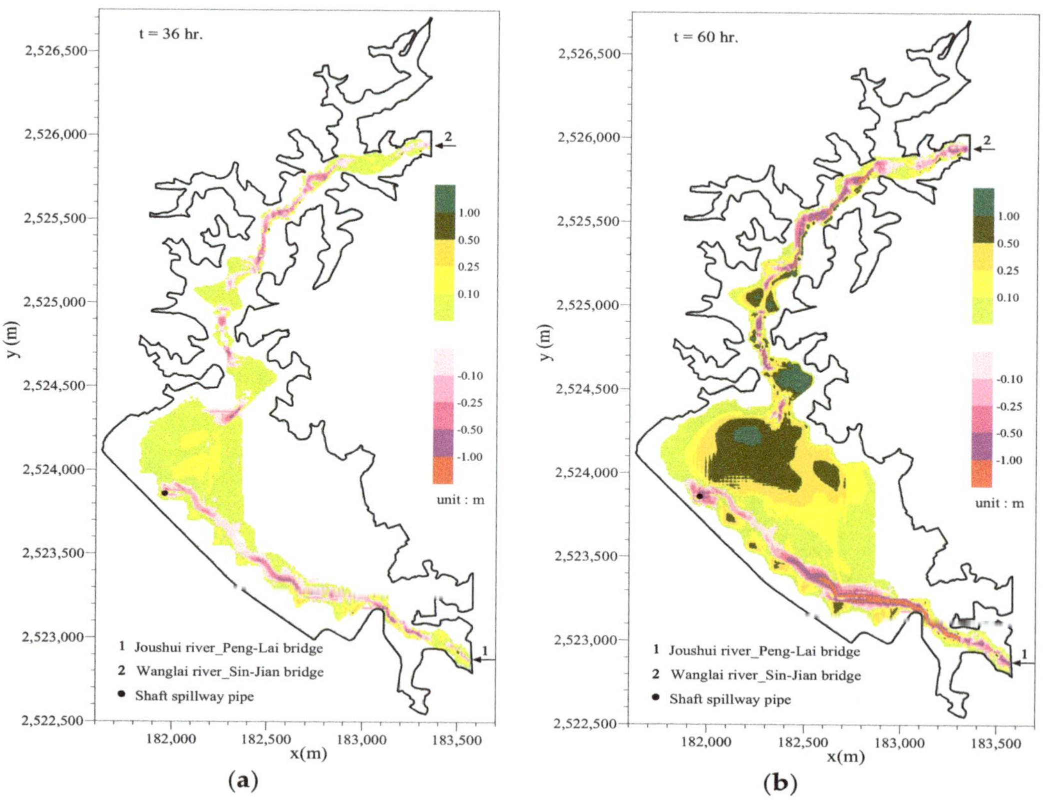

Figure 6. *Cont.*

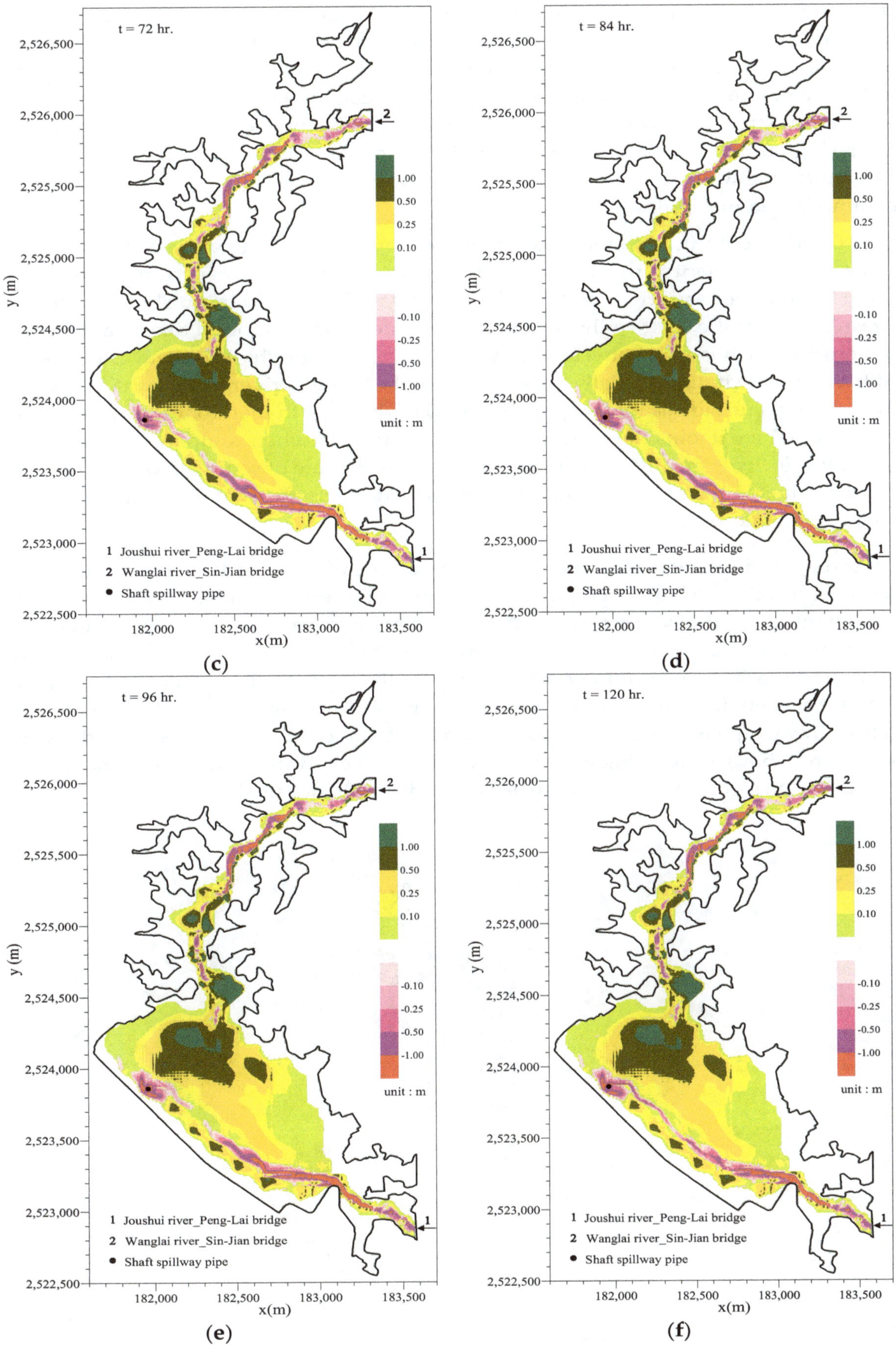

Figure 6. (**a**) Bed evolution in the reservoir at t = 36 h; (**b**) Bed evolution in the reservoir at t = 60 h; (**c**) Bed evolution in the reservoir at t = 72 h; (**d**) Bed evolution in the reservoir at t = 84 h; (**e**) Bed evolution in the reservoir at t = 96 h; and, (**f**) Bed evolution in the reservoir at t = 120 h.

Water depth is shown to initially increase in Figure 4 while flow entered the reservoir, and after t = 72-h it began to gradually decrease as the inflow discharge became less than the releasing discharge of the shaft spillway. Figure 5 shows the distribution of suspended sediment concentration. The results show muddy water into the reservoir and convection-diffusion of sediment in the reservoir. In addition, the high bed shear stress that was caused by the increased velocity entrained more sediments into the flow, thus the high turbid flow in the reservoir. Sediment concentration reached more than 70,000 ppm and the high turbid flow was released by the shaft spillway pipe, and the siltation in the reservoir was due to the unreleased turbid flow.

Figure 6 shows both the erosion and the deposition depth during typhoon Morakot. Erosion is seen to be more dominant at the thalweg (>0.5 m), and is more apparent at the erosion ditch that is generated between Joushui river and the shaft spillway. This could be attributed to the high flow velocity, resulting from the shaft spillway pipe, and the steep slopes around this area. Moreover, a not significant erosion ditch is seen between the Wanglai river and the shaft spillway pipe. Deposition is prevalent from the 60th hour, especially downstream of the Wanglai river, greater than 0.5 m, and is not severe from the Joushui side as it hardly reaches 0.25 m. It is worth mentioning that the downstream of Wanglai river where it joins the reservoir is wider (Figure 2) when compared to the Joushui river, hence the flow velocity is greatly reduced and it increases the rate of sediment deposition. Part of the sediments deposited in the reservoir could not be flushed out by the shaft spillway pipe, resulting in erosion-deposition interplay.

Final bed evolution under the different return periods of one and two-day storms, with a total simulation time of 120 h is shown in Figures 7–13. Similar patterns of erosion and deposition are seen with the different return periods; moreover, these intensify with increasing the return period. There is a significant erosion ditch (>1 m) from the Joushui river down to all around the shaft spillway pipe. The maximum deposition area in the north-eastern of the shaft spillway pipe is mainly due to the non-significant erosion ditch that is seen between the Wanglai river and the shaft spillway pipe. Hence, if the shaft spillway pipe were shifted to the applicable distance in the north-eastern direction, it will improve the probability of forming an erosion ditch between Wanglai river and the shaft spillway. There would be two significant erosion ditches formed to enhance flushing.

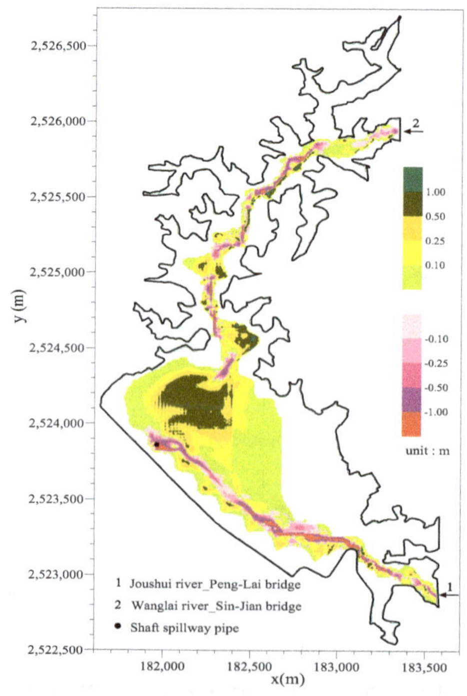

Figure 7. Bed evolution in the reservoir during a two-year return period, t = 120 h.

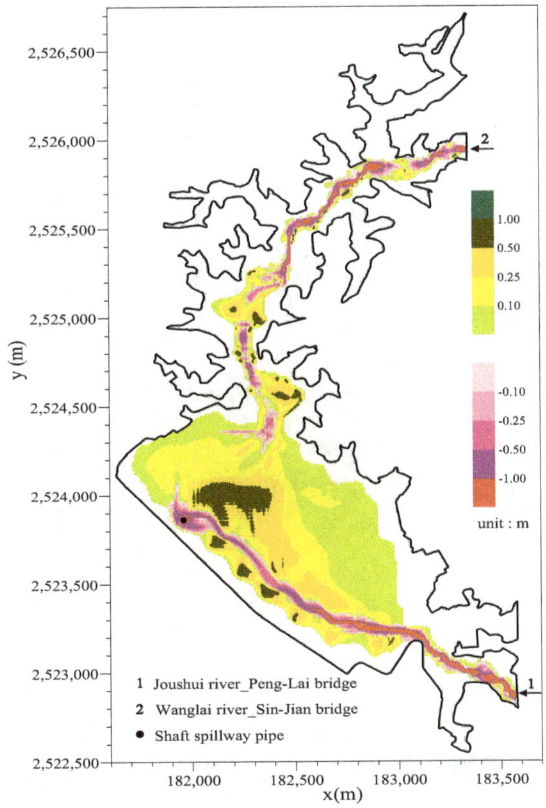

Figure 8. Bed evolution in the reservoir during a five-year return period, $t = 120$ h.

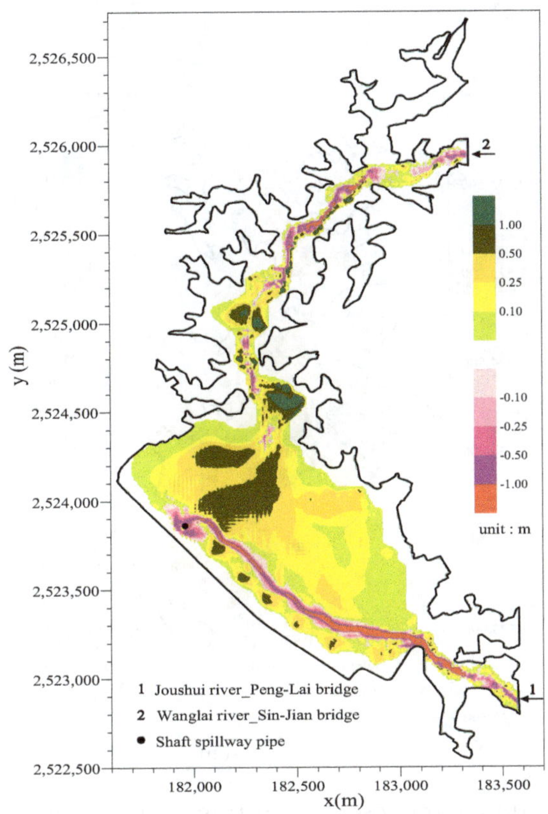

Figure 9. Bed evolution in the reservoir during a 10-year return period, $t = 120$ h.

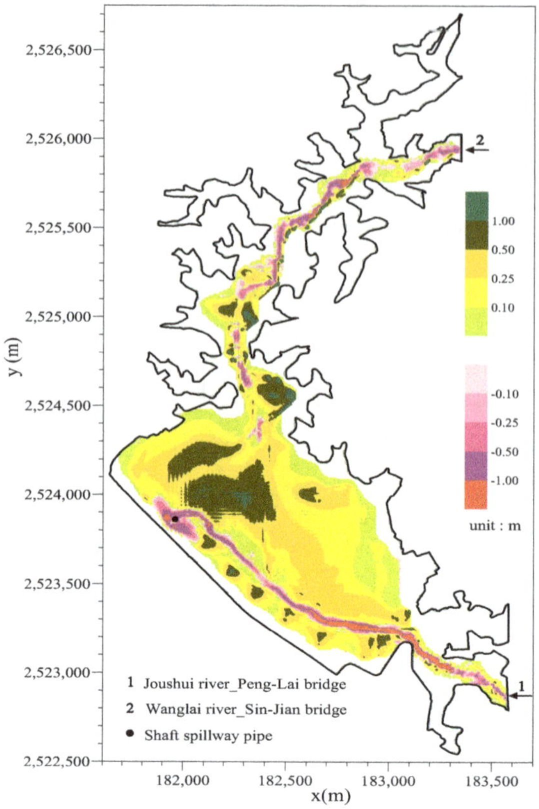

Figure 10. Bed evolution in the reservoir during a 25-year return period, t = 120 h.

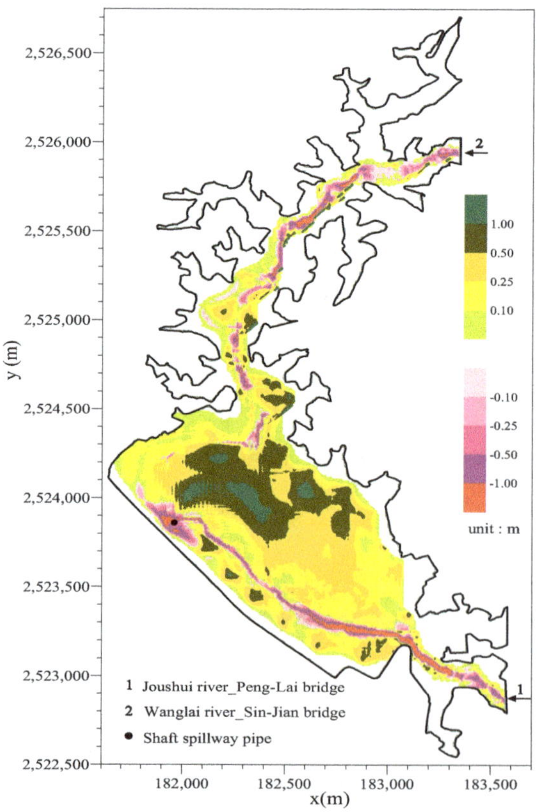

Figure 11. Bed evolution in the reservoir during a 50-year return period, t = 120 h.

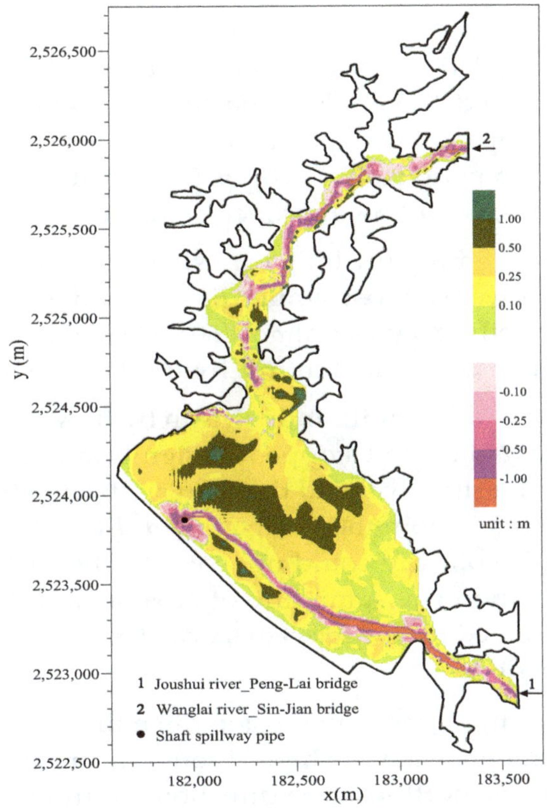

Figure 12. Bed evolution in the reservoir during a 100-year return period, t = 120 h.

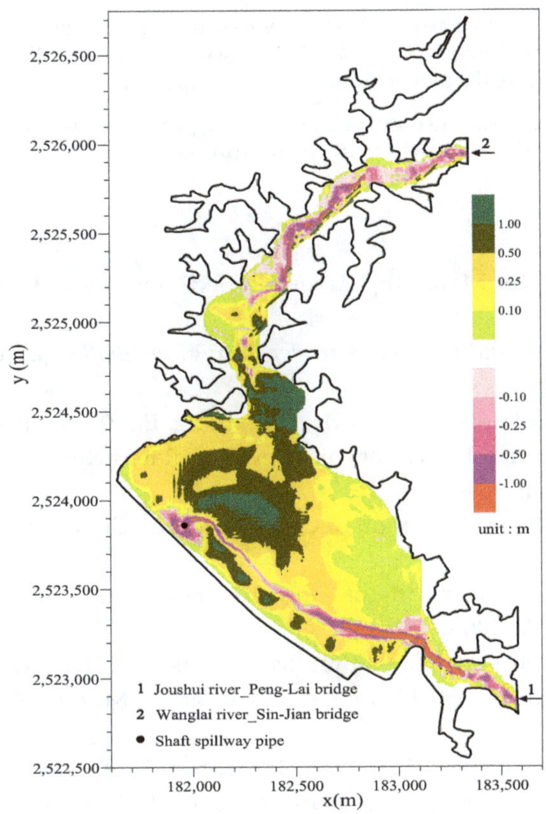

Figure 13. Bed evolution in the reservoir during a 200-year return period, t = 120 h.

6. Conclusions

A model is developed and applied to the simulation of sediment erosion/deposition and sediment distribution within a reservoir, and to simulate the flushing efficiency of a shaft spillway pipe. Hydrograph boundary conditions of inflow and suspended sediment concentration of one-day and two-day storms of different return periods, (2, 5, 10, 25, 50, 100, 200-year), were computed by the 1D, PSED model. Based on the results, the following conclusions may be drawn.

The average efficiency of the empty flushing by a shaft spillway pipe under one- and two-day storms of various return periods were almost similar, 58.49% and 59.39%, respectively. These results were found to be adequate as the efficiency was almost 60%. The overall simulation results are close to 65.34%, which is the efficiency that is obtained from a hydraulic model test in a laboratory.

Bed evolution in the reservoir was significantly driven by flow velocity under empty flushing by the shaft spillway pipe after the rainfall had stopped. A significant erosion ditch was generated after 96 h of simulation time between Joushui river and the shaft spillway pipe. At the end of the simulation time, no erosion ditch was developed from the Wanglai river due to the relatively wider cross section and low flow velocity. The shaft spillway could not completely flush out the sediments that were deposited; as a result, bed evolution was a mixture of erosion and deposition. Similar patterns of erosion and deposition were observed for the selected storm, typhoon Morakot, and the design rainfall events under different return periods.

A common observation was the significant erosion ditch that formed between the Wanglai river and the shaft spillway pipe that enhanced flushing. From our findings, we propose a relocation of the shaft spillway pipe towards the north-eastern direction. Shifting the pipe to this direction could improve the probability of forming the erosion ditch between Wanglai river and the shaft spillway. Two significant erosion ditches would improve the empty flushing efficiency even further.

Acknowledgments: Observed and hydraulic model test data for this study was provided by A-Gong-Dian Reservoir Management Centre, Southern Region Water Resources office, Water Resources Agency, Ministry of Economic Affairs. The authors gratefully acknowledge their support.

Author Contributions: Ching-Nuo Chen conceived and designed the study and further wrote the manuscript while Chih-Heng Tsai contributed significantly with data analysis.

References

1. Wang, Z.-Y.; Hu, C. Strategies for Managing Reservoir Sedimentation. *Int. J. Sediment Res.* **2009**, *24*, 369–384. [CrossRef]
2. Schleiss, A.J.; Boes, R.M. *Dams and Reservoirs under Changing Challenges*; CRC Press: Boca Raton, FL, USA, 2011.
3. Kondolf, G.M.; Gao, Y.; Annandale, G.W.; Morris, G.L.; Jiang, E.; Zhang, J.; Cao, Y.; Carling, P.; Fu, K.; Guo, Q. Sustainable Sediment Management in Reservoirs and Regulated Rivers: Experiences from Five Continents. *Earth Future* **2014**, *2*, 256–280. [CrossRef]
4. Chang, F.-J.; Lai, J.-S.; Kao, L.-S. Optimization of Operation Rule Curves and Flushing Schedule in a Reservoir. *Hydrol. Process.* **2003**, *17*, 1623–1640. [CrossRef]
5. Emamgholizadeh, S.; Samadi, H. Desilting of Deposited Sediment at the Upstream of the Dez Reservoir in Iran. *J. Appl. Sci. Environ. Sanit.* **2008**, *3*, 25–35.
6. Chaudhry, M.A.; Habib-ur-Rehman, M.; Akhtar, M.N.; Hashmi, H.N. Modeling Sediment Deposition and Sediment Flushing through Reservoirs Using 1-D Numerical Model. *Arab. J. Sci. Eng.* **2014**, *39*, 647–658. [CrossRef]
7. Castillo, L.G.; Carrillo, J.M.; Álvarez, M.A. Complementary Methods for Determining the Sedimentation and Flushing in a Reservoir. *J. Hydraul. Eng.* **2015**, *141*, 05015004. [CrossRef]
8. Basson, G.R.; Rooseboom, A. Sediment Pass-through Operations in Reservoirs. In Proceedings of the International Conference on Reservoir Sedimentation, Fort Collins, CL, USA, 9–13 September 1996; pp. 1107–1130.

9. Madadi, M.R.; Rahimpour, M.; Qaderi, K. Improving the Pressurized Flushing Efficiency in Reservoirs: An Experimental Study. *Water Resour. Manag.* **2017**, *31*, 4633–4647. [CrossRef]
10. Morris, G.L.; Fan, J. *Reservoir Sedimentation Handbook: Design and Management of Dams, Reservoirs, and Watersheds for Sustainable Use*; McGraw-Hill: New York, NY, USA, 1998.
11. United States Society on Dams. *Modeling Sediment Movement in Reservoirs*; United States Society on Dams: Denver, CO, USA, 2015.
12. Basson, G.R.; Rooseboom, A. *Dealing with Reservoir Sedimentation: Guidelines and Case Studies*; International Commission on Large Dams: Paris, France, 1996.
13. Pagliari, D.; Rossi, L.; Passoni, D.; Pinto, L.; De Michele, C.; Avanzi, F. Measuring the Volume of Flushed Sediments in a Reservoir Using Multi-Temporal Images Acquired with Uas. *Geomat. Nat. Hazards Risk* **2017**, *8*, 150–166. [CrossRef]
14. Su, T.-C.; Chou, H.-T. Application of Multispectral Sensors Carried on Unmanned Aerial Vehicle (Uav) to Trophic State Mapping of Small Reservoirs: A Case Study of Tain-Pu Reservoir in Kinmen, Taiwan. *Remote Sens.* **2015**, *7*, 10078–10097. [CrossRef]
15. Guertault, L.; Camenen, B.; Peteuil, C.; Paquier, A. Long Term Evolution of a Dam Reservoir Subjected to Regular Flushing Events. *Adv. Geosci.* **2014**, *39*, 89–94. [CrossRef]
16. Chou, F.N.F.; Wu, C.W. Assessment of Optimal Empty Flushing Strategies in a Multi-Reservoir System. *Hydrol. Earth Syst. Sci. Discuss* **2016**, *2016*, 1–49. [CrossRef]
17. U.S. Army Corps of Engineers. *Hec-6 Scour and Deposition in Rivers and Reservoir, User's Manual*; Hydrologic Engineering Centre: Davis, CA, USA, 1991.
18. U.S. Army Corps of Engineers. *Hec-Ras (Hydrologic Engineering Center River Analysis System) Manual*; Hydrologic Engineering Centre: Davis, CA, USA, 2005.
19. Chang, H.H. *Fluvial Processes in River Engineering*; Wiley: Hoboken, NJ, USA, 1988.
20. Chang, H.H.; Harrison, L.L.; Lee, W.; Tu, S. Numerical Modeling for Sediment-Pass-through Reservoirs. *J. Hydraul. Eng.* **1996**, *122*, 381–388. [CrossRef]
21. Olsen, N.R.B. Two-Dimensional Numerical Modelling of Flushing Processes in Water Reservoirs. *J. Hydraul. Res.* **1999**, *37*, 3–16. [CrossRef]
22. Olsen, N.R.B.; Skoglund, M. Three-Dimensional Numerical Modeling of Water and Sediment Flow in a Sand Trap. *J. Hydraul. Res.* **1994**, *32*, 833–844. [CrossRef]
23. Fang, H.-W.; Rodi, W. Three-Dimensional Calculations of Flow and Suspended Sediment Transport in the Neighborhood of the Dam for the Three Gorges Project (Tgp) Reservoir in the Yangtze River. *J. Hydraul. Res.* **2003**, *41*, 379–394. [CrossRef]
24. Khosronejad, A.; Rennie, C.D.; Neyshabouri, A.A.S.; Gholami, I. Three-Dimensional Numerical Modeling of Reservoir Sediment Release. *J. Hydraul. Res.* **2008**, *46*, 209–223. [CrossRef]
25. Wang, Y.M.; Tfwala, S.S.; Chan, H.C.; Lin, Y.C. The Effects of Sporadic Torrential Rainfall Events on Suspended Sediments. *Arch. Sci. J.* **2013**, *66*, 211–224.
26. Chen, C.-N.; Tsai, C.-H.; Tsai, C.-T. Simulation of Sediment Yield from Watershed by Physiographic Soil Erosion–Deposition Model. *J. Hydrol.* **2006**, *327*, 293–303. [CrossRef]
27. Tsai, C.H.; Tsai, C.T. Simulation of Two-Dimensional Gradually-Varied Unsteady Flow in Curved Channels. *J. Chin. Inst. Civ. Hydraul. Eng.* **1995**, *7*, 461–473.
28. Tsai, C.H. Numerical Simulation of Flow and Bed Evolution in Alluvial River with Levees. Ph.D. Thesis, National Cheng-Kung University, Tainan, Taiwan, 2000.
29. Itakura, T.; Kishi, T. Open Channel Flow with Suspended Sediments. *J. Hydraul. Div.* **1980**, *106*, 1325–1343.
30. Shimizu, Y.; Yamaguchi, H.; Itakura, T. Three-Dimensional Computation of Flow and Bed Deformation. *J. Hydraul. Eng.* **1990**, *116*, 1090–1108. [CrossRef]
31. Meyer-Peter, P.E.; Muller, R. Formulas for Bedload Transport. In Proceedings of the 2nd International Association for Hydraulic Research, Stockholm, Sweden, 7–9 June 1948; pp. 39–64.
32. Wong, M.; Parker, G. Reanalysis and Correction of Bed-Load Relation of Meyer-Peter and Müller Using Their Own Database. *J. Hydraul. Eng.* **2006**, *132*, 1159–1168. [CrossRef]
33. Paola, C.; Voller, V.R. A Generalized Exner Equation for Sediment Mass Balance. *J. Geophys. Res. Earth Surf.* **2005**, *110*, 2156–2202. [CrossRef]
34. Parker, G.; Paola, C.; Leclair, S. Probabilistic Exner Sediment Continuity Equation for Mixtures with No Active Layer. *J. Hydraul. Eng.* **2000**, *126*, 818–826. [CrossRef]

35. Komura, S.; Simons, D.B. River-Bed Degradation Below Dams. *J. Hydraul. Div.* **1967**, *93*, 1–14.
36. Bhallamudi, M.S.; Chaudhry, H.M. Computation of Flows in Open-Channel Transitions. *J. Hydraul. Res.* **1992**, *30*, 77–93. [CrossRef]
37. Southern Water Resources Bureau. *Effect Assessment of Empty Storage Operation for Sediment Prevention Observation Program of Agongdian Reservoir*; Water Resources Agency, MOEA: Taichung, Taiwan, 2010.
38. Chen, C.-N.; Tsai, C.-H. The Impact of Climate Change on Inundation Potential. *Int. J. Environ. Sci. Dev.* **2013**, *4*, 496–500. [CrossRef]
39. Water Resources Planning Institute. *Hydraulic Model Studies on the Functions and Operations of Silting Prevention in a-Kung-Tien Reservior*; Water Resources Agency, Ministry of Economic Affairs: Taichung, Taiwan, 2003.

10

Piecewise-Linear Hedging Rules for Reservoir Operation with Economic and Ecologic Objectives

Yueyi Liu [1], Jianshi Zhao [2] and Hang Zheng [1,*]

[1] School of Environment and Civil Engineering, Dongguan University of Technology, Dongguan 523808, China; liuyueyi@dgut.edu.cn
[2] State Key Laboratory of Hydro-Science and Engineering, Department of Hydraulic Engineering, Tsinghua University, Beijing 100086, China; zhaojianshi@tsinghua.edu.cn
* Correspondence: zhengh@dgut.edu.cn

Abstract: The construction of dams and operation of reservoirs have a significant impact on the interruption of aquatic and riparian ecological systems, by altering natural stream flows in river courses. Recently, the ecological requirements are included as an additional objective in reservoir operation in order to restore the natural stream flows and reduce the negative impacts of reservoir operations on ecosystems that rely on the natural flows. The key challenge involving ecological requirements is to balance the ecological and economic objectives by operation rules, on the basis of quantitatively identifying the objective of ecological flows required to maintain the natural flow regime for ecosystem. This study develops a piecewise-linear multi-objective hedging rule (PMHR) for reservoir operations, with ecologic flow objectives represented by 33 hydrologic parameters from the indicators of hydrologic alteration (IHA). Variables of the PMHR are obtained through optimization using a vector evaluated genetic algorithm. The results show that the PMHR improves the ecological water releases without reducing economic water supplies in the case river in the context of hydrological uncertainty. It can offer technological references for improving the utility of water resource management under competitive conditions of water resources.

Keywords: natural flow regime; multi-objective model; uncertainty; genetic algorithm

---

1. Introduction

For decades, dams have been built and operated mostly for economic purposes that require a reliable water supply for human needs, such as hydropower electricity, irrigation, living and industry water supply, and navigation. Water withdrawn for increasing economic demands has led to conflicts between human water use and ecosystem water needs [1–3]. The operation of dams leads to negative effects on aquatic and riparian ecosystem systems [4–6]. The alteration of the flow regime caused by reservoir operation is recognized as a major driving factor that threatens the integrity of the river ecosystem [7–11]. Bunn and Arithington reviewed the studies focused on the relationship between hydrological flow regime and river ecosystems, and illustrate four critical negative impacts of the alteration: (1) alteration of the magnitude of flow as the determinant of physical habitat and biotic composition; (2) alteration of the timing and frequency of peak flow as key factors of the life history of aquatic species; (3) destruction of the patterns of longitudinal and lateral connectivity for riverine species; and (4) the extinguishing of extreme low flow avoidance, meant to prevent the invasion of exotic species [12].

How to balance the trade-off between humans and ecosystems has been discussed for decades regarding integrated river basin management [13–16]. Many investigations have been conducted to address the conflicts between environmental flow release and the economic water supply over the past decades. To evaluate the ecological objective for aquatic and riparian ecosystems' sustainability, four kinds of approaches are often applied, including (1) estimating the minimum flow requirement for downstream habitats to maintain the survival of specific species [17–19]; (2) determining a flow regime based on fish diversity information [20]; (3) providing a regime-based, prescribed flow duration curve that considers floods and droughts for species and morphological needs [21–23]; and (4) minimizing the degree of flow alterations in order to maintain the stability of river ecosystems and population structures of species [16,24–26]. In these studies, it is can be found that the adopted ecological objectives have been shifted from an emphasis on a minimum flow requirement or single species to the development of a regime-based comprehensive approach. With the first two methods mentioned above, the flow fluctuation, which decides the ecological integrity, is generally neglected. Therefore, ecologists currently prefer to recommend the last two methods, in which the ecological integrity can be sustained by accounting for the functional and structural requirements for aquatic and riparian ecosystems.

Suen and Eheart proposed a multi-objective model based on the ecological flow regime paradigm, which incorporates the intermediate disturbance hypothesis to address the ecological and anthropic need [16]. Their study uses parts of the Taiwan Eco-Hydrology Indicators System (TEIS) as the criteria to represent ecosystem objectives. Lane et al. proposed a multi-step re-operation methodology to address environmental flow requirements and human water use objectives [23]. This model summarizes the environmental flow requirements on the basis of empirical streamflow thresholds for the maintenance of an ecosystem with geomorphic functions. Using a one-dimensional water routing model by the Water Evaluation and Planning System (WEAP), the model develops an alternative reservoir rule curve and suggests the new timing of releases to sustain key ecological and geomorphic functions. Generally, these studies have used a monthly-based interval and have therefore failed to reflect important hydrological factors with daily variations, which may play an important role for sustaining the health of an ecosystem [7,27–29]. Chen et al. proposed a time-nested approach, in order to scale down the decision variables from a monthly to a 10-day basis, and further to a daily basis for reservoir operations [30]. They used a daily flow hydrograph as a constraint in the optimization model. However, the model was extremely complex, with 730 decision variables, leading to a high cost of calculation in the optimization process. Considering daily scale indicators in reservoir operations, such as the timing and frequency of peak flow, the rising and falling rates requires a large number of decision variables in the optimization process, and lead to a computing-heavy task.

It has been noted that involving a set of fixed operation rules in reservoir modelling can reduce the number of decision variables as well as the computational costs [31]. Standard operating policy (SOP) and hedging rules (HR) are the most commonly used rules in reservoir operations. SOP, which is the traditional and simplest operating rule, releases water as close to the delivery targets as possible in order to meet the demands of the current stage [32]. By contrast, HR prefers to keep a certain proportion of available water in the reservoir, in order to minimize the possible losses caused by water shortages in the future. Usually, SOP is recommended only if the objective function is linear, while the hedging rules have been proven to be a more efficient way to optimize the reservoir operations when the supply benefits are nonlinear [33,34]. In real operations, the trade-off analysis of multiple objectives is often complicated with nonlinear systems and uncertain inflow conditions [35–37]. Therefore, HR is often used to cope with uncertain future inflows among different nonlinear objectives in practice [38]. Taghian et al. employed HR to achieve the optimal water allocation, in order to reduce the intensity of severe water shortages [39]. A hybrid model was developed to simultaneously optimize both the conventional rule curve and the hedging rule. Shiau developed a parameterization–simulation–optimization approach for optimal hedging for a water supply reservoir by considering the balance between beneficial release and carryover storage value. The proposed methodology was applied to the Shihmen Reservoir in northern Taiwan to illustrate the effects of derived optimal hedging on reservoir performance in terms of shortage-related indices and hedging uncertainty [40]. Yin et al. used a typical approach that uses three limit curves to divide the reservoir storage and the reservoir inflow into several zones [24]. A series of water supply rules and water releasing rules were derived by the actual reservoir storage and the reservoir inflows to provide the water supply and to satisfy the ecological needs of seasonally variables, such as base flows, dry season flow recessions, high-flow pulses, etc. However, daily interval variations were not exactly considered in the above-mentioned studies.

Accounting for the daily variation of downstream flows, Yang et al. proposed a set of linear rules to generate daily reservoir releases, with the aim of satisfying the needs of ecological assessment [20]. This method can explicitly engage flow regime variation based on real-time inflow information [41], while short-term inflow information becomes crucial for reservoir operation when daily scale ecologic indicators are considered. However, the proposed operation rules were based on traditional operation curves for flood control and ecologic objectives, without further considerations regarding the balance between consumptive water supply and ecological flow needs. Moreover, the ecologic objective in this paper was calculated by seven indicators, including Q_3 and Q_5, the average discharges in March and May, respectively; $Q_{3day\ min}$ and $Q_{7day\ min}$, which represent the annual minimum average of three- and seven-day discharges; $Q_{3day\ max}$, which is the annual maximum average 3-day flow; and D_{min}, representing the Julian date of the annual minimum daily flow. Further information, such as the changing trends of daily flow and the statistics of daily flow reversals, are required to establish a more comprehensive function for the ecological objective.

The purpose of this study is to develop practical hedging rules for reservoir operation with economic and ecologic objectives, which will be able to reflect daily interval variations and engage the real-time daily inflow forecast and storage information simultaneously into decision-making. In this paper, piecewise-linear hedging rules are proposed in order to generate the daily release for ecological and economic objectives, through which the parameters are optimized based on historical and synthetic streamflow series. In additional, the objective of ecological flow requirements is quantified by 33 hydrologic parameters from Indicators of Hydrologic Alteration (IHA). The proposed methodology is applied to a realistic reservoir as a performance test. The rest of this paper is organized as follows. Section 2 briefly introduces the structure of the model. Section 3 presents a real world case study of Baiguishan reservoir, China. Section 4 compares the proposed HRs with SOP and analyzes the value of forecast information. Finally, conclusions are drawn in Section 5.

2. Methodology

The framework of the method for optimizing the reservoir's target water levels and flows released to the economic and ecological objectives is shown in Figure 1. It is composed of the following three steps: (1) generating daily water release based on daily inflow and pre-defined hedging rules, (2) multi-objective optimization for releasing water to economic and ecologic objectives, (3) Monte Carlo simulation and probability analysis to identify the variations of the optimal water release in various hydrological years.

The aim of this framework is to identify a set of practical operating rules for water release. A set of pre-defined, piecewise-linear hedging rules are used to generate water release under the conditions of various inflow and water levels of the reservoirs. A multi-objective genetic algorithm is then applied to optimize the parameters of the rules' curves iteratively, by balancing the economic and ecological objectives of water release. The statistical characteristics of the parameters are obtained through Monte Carlo simulations, in which the historical and synthetic daily inflows are used as the inputs.

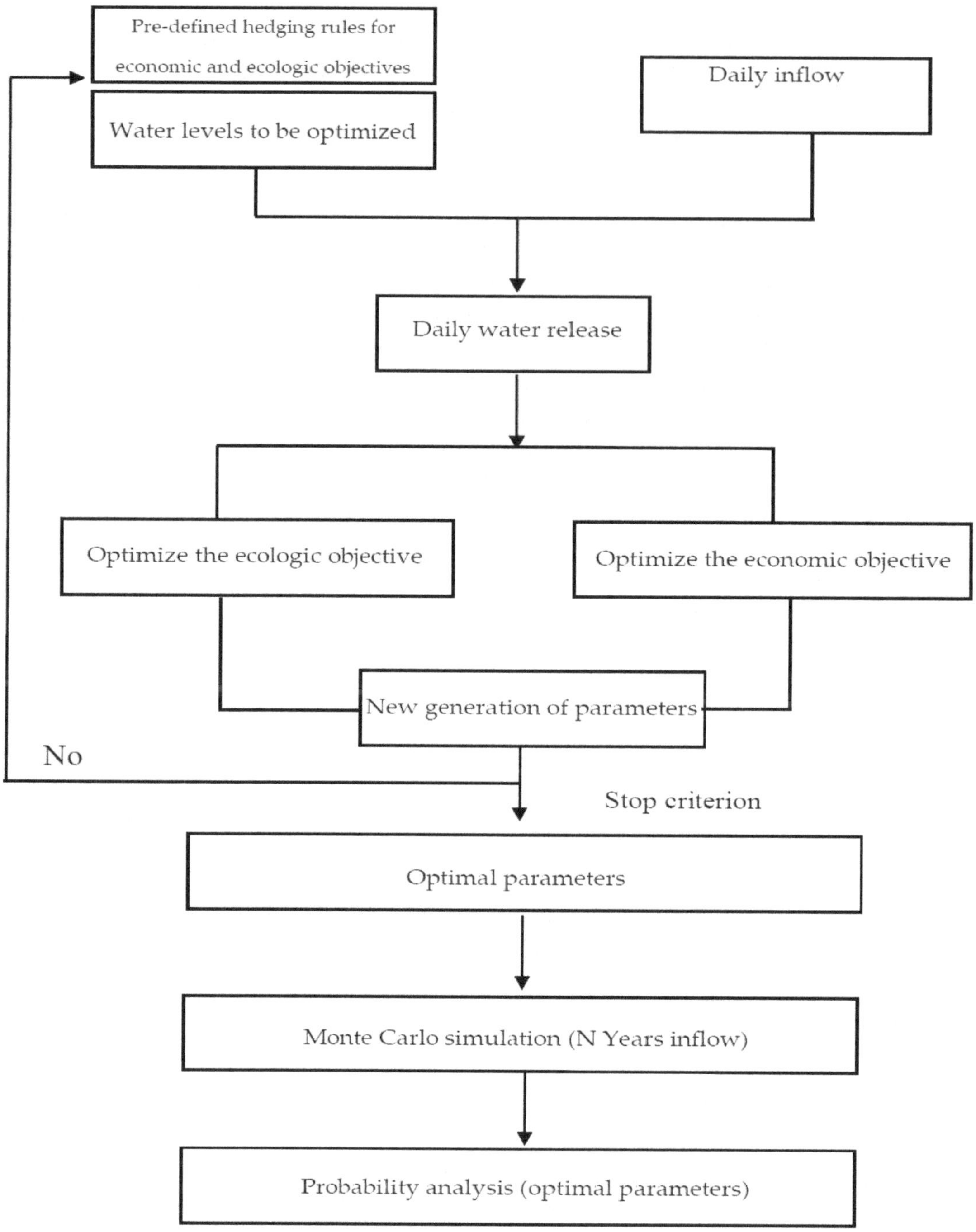

Figure 1. Process diagram for optimizing the operating rules.

2.1. Pre-Defined Piecewise-Linear Hedging Rules

In this paper, piecewise-linear hedging rules considering both economic and ecological objectives are pre-defined for the dry and wet seasons separately. As shown in Figure 2, zones 1–3 represent the rules in the dry season, and zones 4–6 stand for the rules during the wet season. Red lines show the maximum and minimum water level of different periods following the traditional operation rules currently used by the reservoir.

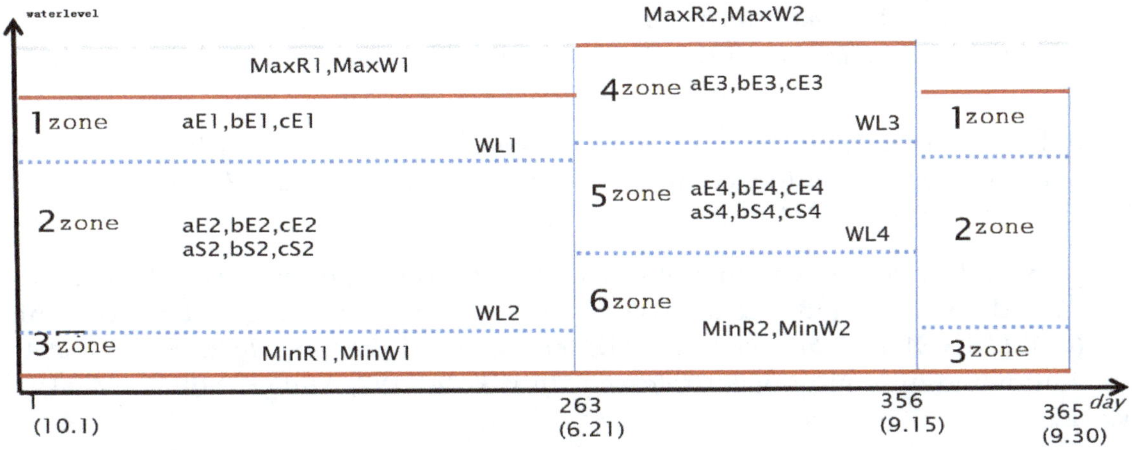

Figure 2. Release functions in different zones characterized by seasons.

For the dry season, daily water release for economic objectives can be defined as a set of piecewise-linear hedging rule functions incorporating actual storage and future inflow while considering the effect of daily inflow forecast [42], as follows:

$$\begin{cases} W_t = D_t & L_t \geq WL_1 \\ W_t = Min(D_t, Max(MinW_1, aS_1 \times I_t + bS_1 \times L_{t-1} + cS_1)) & WL_2 \leq L_t < WL_1 \\ W_t = MinW_1 & L_t < WL_2 \end{cases} \quad (1)$$

where W_t is the water release for economic objective of day t; D_t is the target release for economic objective of day t; I_t is the inflow of day t; L_t and L_{t-1} are water levels of day t and day $t-1$, respectively; WL_1 and WL_2 stand for upper and lower limits of the water level, respectively; $MinW_1$ represents the minimum release for the economic objective of day t; aS_1, bS_1, and cS_1 are the coefficients of the linear functions of hedging rules.

Specifically, when the actual water level (L_t) is higher than the upper limit (WL_1), it implies that the stored water is enough to defend the area from droughts in the future. In this case, the economic water release W_t can be described as the amount of water that is needed (D_t). When the actual water level (L_t) is between the upper limit (WL_1) and the lower limit (WL_2), it means that there is insufficient storage to defend against future droughts, and thus the water release W_t can be described as a linear function related to the forecasted inflow and current water level. When the current water level is lower than the lower limit (WL_2), it means that storage is rare and there is a huge drought risk in future. In this case, W_t is defined as a fixed minimum value, representing the minimum or basic water release to the economic system.

Similar to Equation (1), the daily water release for an ecologic objective is defined as follows:

$$\begin{cases} R_t = Min(MaxR_1, Max(MinR_1, aE_1 \times I_t + bE_1 \times L_{t-1} + cE_1)) & L_t \geq WL_1 \\ R_t = Min(MaxR_1, Max(MinR_1, aE_2 \times I_t + bE_2 \times L_{t-1} + cE_2)) & WL_2 \leq L_t < WL_1 \\ R_t = MinR_1 & L_t < WL_2 \end{cases} \quad (2)$$

where R_t represents the ecological water release of day t; aE_1, bE_1, cE_1, aE_2, bE_2, and cE_2 are the parameters of the HR, which will be obtained through the optimizations illustrated in Figure 1; $MaxR_1$ and $MinR_1$ are the maximum and minimum water release that need to be optimized, respectively.

Similarly, economic and ecological water releases in wet seasons can be defined as follows:

$$\begin{cases} W_t = D_t & L_t \geq WL_3 \\ W_t = Min(D_t, Max(MinW_2, aS_2 \times I_t + bS_2 \times L_{t-1} + cS_2)) & WL_4 \leq L_t < WL_3 \\ W_t = MinW_2 & L_t < WL_4 \end{cases} \quad (3)$$

$$\begin{cases} R_t = Min(MaxR_2, Max(MinR_2, aE_3 \times I_t + bE_3 \times L_{t-1} + cE_3)) & L_t \geq WL_3 \\ R_t = Min(MaxR_2, Max(MinR_2, aE_4 \times I_t + bE_4 \times L_{t-1} + cE_4)) & WL_4 \leq L_t < WL_3 \\ R_t = MinR_2 & L_t < WL_4 \end{cases} \quad (4)$$

where WL_3 and WL_4 are the upper and lower limit, respectively; $MaxR_2$ and $MinR_2$ are maximum and minimum water release to be optimized, respectively; and $aS2$, $bS2$, $cS2$, $aE3$, $bE3$, $cE3$, $aE4$, $bE4$, and $cE4$ are the parameters of the HR during wet seasons.

Consequently, the decision variables consist of the following three parts: (1) water level limits ($WL1$ and $WL2$ in dry seasons and $WL3$ and $WL4$ in wet seasons); (2) ecological and economic hedging coefficients ($aE1$, $bE1$, $cE1$, $aE2$, $bE2$, $cE2$, $aE3$, $bE3$, $cE3$, $aE4$, $bE4$, $cE4$ $aS2$, $bS2$, $cS2$, $aS4$, $bS4$, and $cS4$); and (3) maximum and minimum water release ($MinW1$, $MinW2$, $MinR1$, $MinR2$, $MaxW1$, $MaxW2$, $MaxR1$, $MaxR2$).

2.2. Ecological Objective

The ecological objective is to minimize reservoirs' alterations on rivers' natural flows, which have been adapted to by aquatic and riparian species over thousands years of evolution by maintaining the stability of ecosystems and population structures of species. The Indicators of Hydrologic Alteration (IHA) program, proposed by Richter, has been adopted to represent the ecological objective in this paper [4]. It contains 33 hydrologic parameters involving five groups of characteristics, including (1) the magnitude of monthly stream flow; (2) the magnitude of annual extreme flows at different time durations; (3) the timing of annual extreme flows; (4) the frequency and duration of high and low pulses; (5) the rate of change.

For each eco-hydrological indicator, a spectrum of values could be set as the target range, which could reflect the changes of environmental flow regime that a species can adapt to. If the actual value of an indicator falls in that range, it means that the alteration of the natural flow regime is acceptable by the aquatic or riparian species. Those acceptable ranges of all the indicators have been investigated by many researchers [9], among which the findings by Richter et al. are mostly widely applied through the range of variability approach (RVA) [4]. RVA is an efficient and convenient method to evaluate the degree of flow regime alteration. According to Richter and in this paper, the 25th and 75th percentiles of the historical annual value of the indicators are set as the upper and lower limits of the target range of environmental flow regime, respectively.

Here, let $f_i(r)$ represent the effect function of the i-th indicator of IHA. In addition, $iha_i(r)$ represents the value of the i-th indicator. $f_i(r)$ equals 0 if $iha_i(r)$ falls in the target range, which implies that the ecological release and flow alternation compared to the natural flow regime are acceptable. When the value of the indicator falls outside of the target range, $f_i(r)$ is calculated as the distance from the target range, as follows:

$$f_i(r) = \begin{cases} 0 & iha_{ip75} \leq iha_i(r) \leq iha_{ip25} \\ (\frac{iha_i(r) - iha_{ip25}}{iha_{ip25} - iha_{ip75}})^2 & iha_i(r) \geq iha_{ip25} \\ (\frac{iha_{ip75} - iha_i(r)}{iha_{ip25} - iha_{ip75}})^2 & iha_i(r) \leq iha_{ip75} \end{cases} \quad (5)$$

where r is the series of ecological water release generated by Equations (2) and (4); iha_{ip25} and iha_{ip75} are the upper and lower limits of the annual values of the indicators, respectively; and $iha_i(r)$ represents the value of the i-th indicator. The ecological objective can be calculated by the sum of the 33 indicators of the IHA, as

$$f_1 = \sum_{i=1}^{33} f_i(r) \quad (6)$$

The value of f_1, the range of which is greater than zero, represents the ecosystem objective to be minimized in the multi-objective optimization model.

2.3. Economic Objective

The economic objective is represented in a target-hitting form, in which the water demands of agriculture, industry, and domestic water use are considered. For equalizing the water supply shortages across time intervals throughout the year, the economic objective is employed through quantifying the water supply deficits, as follows:

$$f_2 = -g(w) = f(w) = \sum_{i=1}^{T} (\frac{d_i - w_i}{d_i})^2 \quad (7)$$

where T is the total number of time periods; d_i is the economic water demand of the i-th day, including the water demands of agriculture, industry and domestic sectors; and w_i is the water supplied in the i-th day. The value of f_2, range from 0 to 1, and represents the economic objective to be minimized in the multi-objective optimization model.

2.4. Constraints

Constraints within the optimization model are set as follows.

(1) Water balance constraint:

$$V_i = V_{i-1} + I_i \Delta t - q_i \Delta t - E_i \quad (8)$$

where V_i and V_{i-1} represent the reservoir storage available at the end of the i-th day and $i - 1$ day, respectively; I_i is the inflow of the i-th day; q_i is the total release during the i-th day; and E_i is the evaporation of the reservoir in the i-th day.

(2) Reservoir storage capacity constraint:

$$L_i^{min} \leq L_i \leq L_i^{max} \quad (9)$$

where L_i^{min} and L_i^{max} are the minimum and maximum water level limits at the end of the i-th day, respectively; and L_i is the water level of the i-th day.

(3) Initial and end storage constraint:

Considering the initial water storage of next year, here we set up initial storage and end storage equally as

$$V_0 = V_a \quad (10)$$

where V_0 is the initial storage of the reservoir and V_a is its end storage.

The multiple objectives of the model are set to meet both the ecosystem and human demand, which can be expressed as follows:

$$Obj.\ Min\{f1(r), f2(w)\} \quad (11)$$

2.5. Vector Evaluated Genetic Algorithm

This multi-objective model is solved by the vector evaluated genetic algorithm (VEGA). Different from the standard procedure of a genetic algorithm, VEGA focuses on the optimization of several objectives [43]. Through the VEGA, the initial population is divided into a number of objectives, and the elite of each group is selected for the next generation while the others are put in crossover and mutation pools. A population with a higher fitness value has a greater probability of persevering to the next generation. The procedure of VEGA is presented in Figure 3.

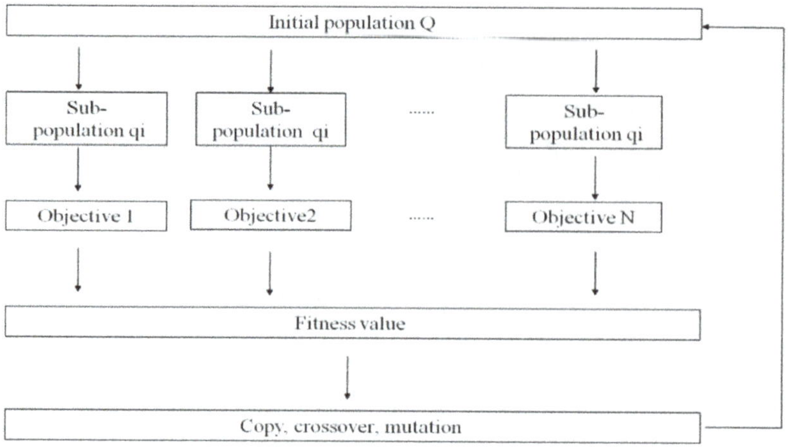

Figure 3. Procedure of the vector evaluated genetic algorithm (VEGA).

3. Results

3.1. Study Site

The Baiguishan Reservoir, which is located on the Shahe River (Figure 4) in the Huaihe River Basin in China, was selected as a case study to test the proposed model. It is a multipurpose reservoir serving as a water supply, flood control, a source of recreation, and ecologic purposes, with a total storage of 0.92 billion m^3. Historical daily inflow series from 1976–2005 were used to design the operation rules. Affected by pacific monsoons, 80% of the rainfall to the reservoir occurs in summer, and this divides the year of operation into two seasons, including the dry season (16 September–20 June) and the wet season (21 June–15 September). The maximum water level used in the current operation is 103 m and 105.9 m in the dry and wet seasons, respectively; the minimum water level is limited to 92 m.

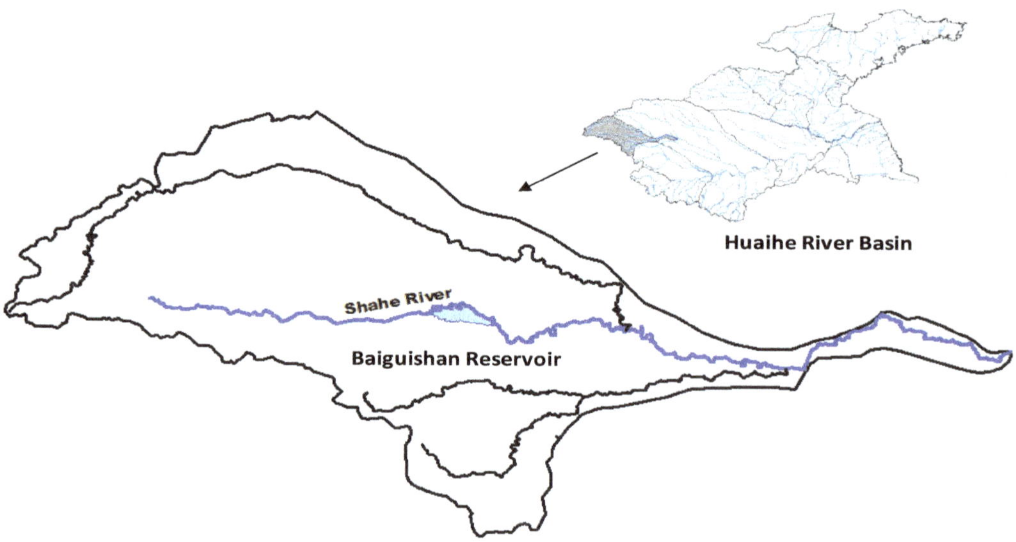

Figure 4. Location of the Baiguishan Reservoir.

3.2. Ecological Management Target Range

Generally, IHA should be calculated based on historically natural flow without intervention. In this paper, the ecological objective is defined based on the historical inflow data of the Baiguishan Reservoir instead, due to the lack of historical daily hydrologic data for the Shahe River before the

1950s—in other words, when the stream flow had not yet been altered by human beings in China. For each eco-hydrological indicator, the target range was calculated by the values of the 25th and 75th percentile of the historical series. Table 1 shows the ecological target ranges calculated by data from 1976–2005. Moreover, the expected value and the standard deviation (SD) are investigated for further discussion.

Table 1. Ecological target ranges and statistical analysis of eco-hydrologic indicators.

No.	Indicator	Unit	Ecological Target Range		Statistical Analysis	
			P25%	P75%	Expected	SD
IHA1	October	m^3/s	19.81	8.21	17.88	20.91
IHA2	November	m^3/s	20.52	4.26	12.68	11.3
IHA3	December	m^3/s	20.71	5.02	11.41	9.03
IHA4	January	m^3/s	14.59	6.6	9.05	6.33
IHA5	February	m^3/s	17.25	5.02	11.44	7.62
IHA6	March	m^3/s	21.48	2.57	13.5	9.11
IHA7	April	m^3/s	17.81	2.16	11.57	10.8
IHA8	May	m^3/s	23.1	8.2	17.36	17.06
IHA9	June	m^3/s	36.65	9.09	24.4	17.86
IHA10	July	m^3/s	38.91	15.97	36.42	42.28
IHA11	August	m^3/s	48.18	19.29	39.97	32.7
IHA12	September	m^3/s	25.75	7.79	21.18	20.34
IHA13	1day-min	m^3/s	2.56	0.3	0.63	1.43
IHA14	3 day-min	m^3/s	3.19	0.24	0.85	1.97
IHA15	7 day-min	m^3/s	3.56	0.29	1.15	2.21
IHA16	30 day-min	m^3/s	7.28	0.84	2.84	3.36
IHA17	90 day-min	m^3/s	10.08	0.91	6.68	5.15
IHA18	1 day-max	m^3/s	435.98	66.1	286.19	260.18
IHA19	3 day-max	m^3/s	329.97	55.33	205.25	192.94
IHA20	7 day-max	m^3/s	219.25	49.42	141.09	133.33
IHA21	30 day-max	m^3/s	81.72	31.73	67.62	50.98
IHA22	90 day-max	m^3/s	51.79	23.8	41.1	24
IHA23	Zero days	days	93	6	44.1	63.79
IHA24	Base flow	/	0.24	0.03	0.05	0.1
IHA25	Date of max	/	276.2	122	274.93	61.4
IHA26	Date of min	/	260	159.6	222.9	117.792
IHA27	Low count	/	12	4	4.55	3.25
IHA28	Low duration	days	16	7.22	22.94	23.62
IHA29	High count	/	13	2	7.69	3.6
IHA30	High duration	days	28.75	6	15.77	18.58
IHA31	Fall rate	/	0.52	0.19	0.63	0.39
IHA32	Rise rate	/	−0.11	−0.18	−0.76	0.62
IHA33	Reversal	/	135	81	121.8	32.62

In Table 1, eco-hydrological indictors are ordered from IHA1 to IHA33. IHA1–IHA12 are the annual mean values of monthly flow from January to December. IHA13–IHA22 are the annual mean values of the maximum or minimum t-day (t = 1, 3, 7, 9, 30, 90) flows. IHA23 and IHA24 are the number of zero flow days and the base flow, respectively. IHA25 and IHA26 are the Julian date of each annual one-day maximum and minimum, respectively; IHA27 and IHA29 are the number of high or low pulses, respectively, within each year (days); IHA28 and IHA30 are the mean duration of high or low pulses, respectively, within each year; IHA31 and IHA32 are the means of all positive and negative differences between consecutive days, respectively; and IHA33 is the total number of reversals [4].

3.3. Pareto-Optimum Solutions

To deal with the trade-off between economic and ecological objectives, VEGA with a population size of 1000 was adopted to solve the model within one year. With the series of daily inflow from

1976–2005, a Pareto-optimal frontier for every year can be obtained. Using three typical years (frequency = 25%, 50%, 75%) as examples, after running 100 generations, the algorithm is stopped; the result of the Pareto-optimal frontiers of the typical years is shown in Figure 5. Each individual solution represents one possible trade-off between the economic and ecological objectives with different inflow scenarios. Here, a frequency of 75% means the annual runoff for the year will be exceeded in 75 years out of 100. In Figure 5, each frontier solution is calculated by a searching direction in the genetic algorithm, and prioritization between the objectives is decided by the decision-maker. For example, for the scenario of frequency being 75%, the solution at point A represents an emphasis on the ecological objective, while the solution at point D represents an emphasis on the economic objective. Solutions at points B and C are the destinations with two sub-population evolution directions, which represent the trade-offs of the two competitive objectives.

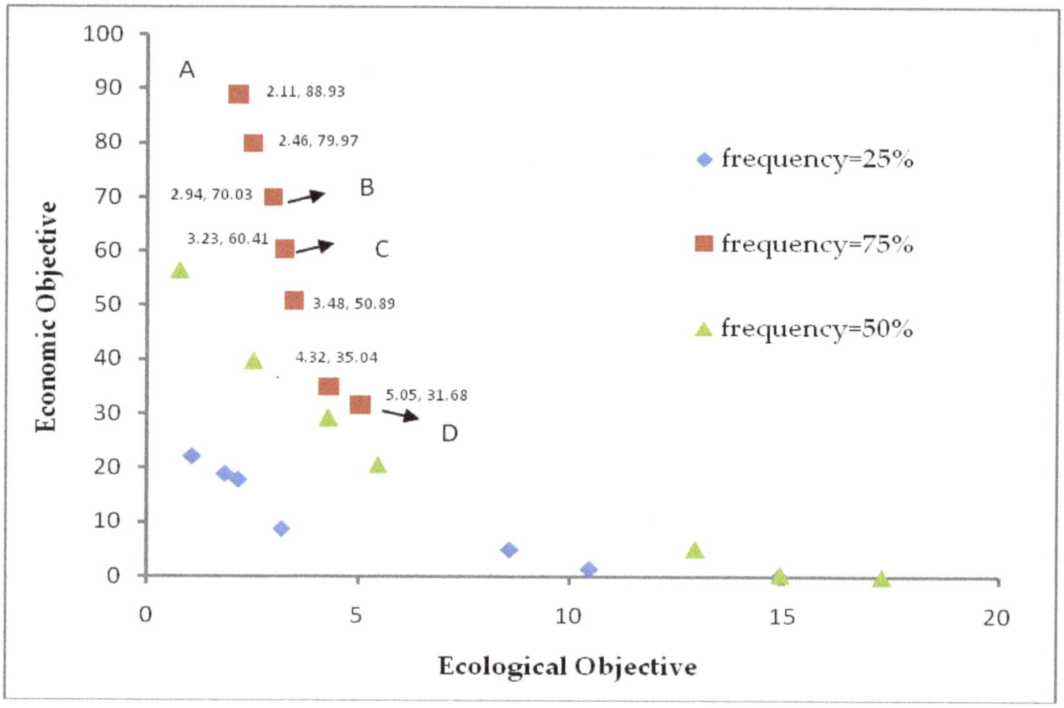

Figure 5. Trade-offs between economic and ecosystem objectives.

The corresponding decision variables of points A, B, C, and D are presented in Table 2.

Table 2. Decision variables of points of A, B, C, and D of the year with 75% frequency.

Pt.	*MinW1*	*MinR1*	*MaxW1*	*MaxR1*	*aE1*	*bE1*	*cE1*	*aS1*	*bS1*	*cS1*	*SL1*	*MinW2*
A	5.14	3.14	8.09	729.09	0.6	0.44	2.89	0.7	−0.95	1.89	102.51	5.13
B	5.9	5.58	8.03	764.83	0.93	−0.35	1.83	0.74	0.99	0.82	102.99	5.33
C	5.48	1.71	8.09	705.23	0.62	−0.8	1.83	0.32	0.05	4.81	102.99	5.66
D	5.35	2.72	8.09	732.35	0.47	0.38	2.32	0.4	−0.07	2.61	102.55	5.4

Pt.	*MinR2*	*MaxW2*	*MaxR2*	*aE2*	*bE2*	*cE2*	*aS2*	*bS2*	*cS2*	*SL2*	*aE3*	*bE3*
A	8.09	8.06	712.42	0.35	0.01	2.5	0.52	−0.95	0.61	97.72	0.2	−0.7
B	6.5	8.09	714.88	0.79	−0.01	2.5	0.63	−0.7	3.09	98.03	0.34	0.83
C	7.6	8.1	776.57	0.82	0	1.08	0.76	0.16	3.69	99.14	0.83	0.1
D	5.53	8.07	737.35	0.63	−0.45	2.55	0.47	0.17	2.45	98.49	0.51	0

Pt.	*cE3*	*aS3*	*bS3*	*cS3*	*SL3*	*aE4*	*bE4*	*cE4*	*aS4*	*bS4*	*cS4*	*SL4*
A	5.06	0.79	0.23	1.03	103.44	0.15	−0.97	3.58	0.05	−0.39	2.19	96.99
B	3.77	0.04	0.38	4.64	102.71	0.71	−0.08	2.77	0.49	−0.05	3.99	98.17
C	1.54	0.66	−0.38	2.65	102.97	0.49	−0.78	2.92	0.77	−0.44	1.26	97.09
D	2.04	0.48	0	2.13	62.4	0.23	−0.18	1.27	0.13	0.22	0.74	97.65

3.4. Monte Carlo Simulation

In the above section, we use the inflow data of a given year to demonstrate the efficiency of the model by generating a set of optimized parameters. However, the obtained optimal parameters for the piecewise-linear multi-objective hedging rule (PMHR) are not applicable for other streamflow conditions. To obtain the optimal PMHR in the context of hydrological uncertainty, a Monte Carlo simulation is applied in this section to deal with different inflow conditions, rather than the inflow of one year.

Based on 29 years of historical hydrological data, a synthetic daily inflow for 100 years was generated, according to the statistical characteristics of the 29-year inflow. To be more specific, the annual mean discharge of these 100 years were obtained by a P-III curve based on samplings from the 29-year historical inflow from 1976–2005. The daily inflow of a synthetic year was decomposed from its annual discharge, according to the historical daily inflow of the year in which the frequency is equal to the synthetic year. The statistical properties of the historical and synthetic flow series is shown in Table 3. Here, the coefficient of variation (C_V) is defined as the ratio of the standard deviation to the mean. It reflects the extent of variability in relation to the mean of inflow. The deviation coefficient (C_S) is measured as the ratio of the difference between the mean and median value of standard deviation. It reflects the skewness of the mean of the inflow.

Table 3. Statistical parameters of the historical inflow series and the synthetic flow series.

	Average Annual Discharge	C_v	C_s
historical inflow series	24.37	0.33	0.83
synthetic flow series	18.70	0.40	0.74

In this paper, the main purposes of synthetic inflow generation are (1) to generate more input scenarios that are close to the historical inflow for the Monte Carlo simulation of the optimal operation model, and (2) to test the effectiveness of the model under the typical frequency of these inflow scenarios. The synthetic flow series can be seen as the input series of a mathematical experiment, while results and discussions are mostly based on the synthetic flow series.

For 100 years of daily inflow data, the optimization model is applied 100 times, and 100 sets of Pareto-optimal frontiers and 100 sets of optimized parameters are calculated correspondingly. In this way, through a Monte Carlo simulation, each set of the optimized parameter forms a possible operation rule for the case reservoir. These 100 groups of parameters represent the optimal operation rules under 100 various inflow scenarios. For each Pareto-optimal frontier, the expected and median values represent the entire probability distribution under different conditions. By using the expected and median values of parameters as the operation rules, optimal economic and ecological water release can be derived under any inflow conditions for decision-makers.

Searching directions used to obtain points B and C in Figure 5 are applied in the model with 100 years' synthetic daily inflow. In this way, 200 groups of optimal decision variables with different searching directions (similar to the method used to obtain points B and C) are obtained. For each searching direction, we calculated the expected and median values of the optimal decision variables ($PMHR_B$ and $PMHR_C$), which represent the optimal operation rules for the case reservoir. Table 4 shows the statistical values of the optimal parameters for 100 years.

Table 4. Statistical Analysis for the Parameters of Counterbalance Solutions.

Parameters		*MinW1*	*MinR1*	*MaxW1*	*MaxR1*	*aE1*	*bE1*	*cE1*	*aS1*	*bS1*	*cS1*	*SL1*	*MinW2*
$PMHR_B$	Expected	5.43	4.84	8.08	736.31	0.55	0.07	2.37	0.58	0.05	3.2	102.69	5.46
	Median	5.44	5.06	8.08	737.19	0.57	0.24	2.52	0.57	0.1	3.41	102.71	5.43
	SD	0.23	2.13	0.02	19.63	0.25	0.47	1.21	0.17	0.49	1.4	0.44	0.23
	Skew	0.26	−0.16	0.09	0.17	−0.18	−0.38	0.3	0.1	0.15	−0.35	−0.64	0.01
$PMHR_C$	Expected	5.44	3.84	8.08	743.23	0.54	0.18	2.66	0.52	0.02	2.85	102.68	5.41
	Median	5.44	4.27	8.08	741.9	0.56	0.28	2.78	0.57	0.1	3	102.67	5.43
	SD	0.25	2.05	0.02	18.96	0.27	0.46	1.16	0.24	0.42	1.26	0.43	0.23
	Skew	0.18	0.05	0.01	0.29	−0.36	−0.53	0.25	−0.46	−0.77	0.04	−0.67	0.05

Parameters		*MinR2*	*MaxW2*	*MaxR2*	*aE2*	*bE2*	*cE2*	*aS2*	*bS2*	*cS2*	*SL2*	*aE3*	*bE3*
$PMHR_B$	Expected	5.21	8.08	739.83	0.55	−0.13	2.69	0.48	0.1	3.22	98.48	0.45	0.02
	Median	5.47	8.08	739.72	0.59	−0.03	2.63	0.44	0.19	3.2	98.55	0.4	−0.01
	SD	2	0.02	19.61	0.24	0.2	1.33	0.24	0.5	1.1	0.35	0.21	0.49
	Skew	−0.32	−0.34	−0.18	−0.25	−1.41	−0.06	−0.01	−0.32	−0.02	−0.43	0.74	0.18
$PMHR_C$	Expected	4.47	8.08	743.41	0.54	−0.27	2.58	0.46	0.33	2.67	98.45	0.45	0.08
	Median	4.33	8.08	742.96	0.53	−0.2	2.49	0.44	0.39	2.72	98.51	0.45	0.07
	SD	2.07	0.03	16.29	0.25	0.3	1.11	0.21	0.41	1.26	0.38	0.21	0.45
	Skew	0.13	0.23	−0.14	−0.17	−0.61	0.11	−0.08	−1.27	0.09	−0.29	0.65	−0.04

Parameters		*cE3*	*aS3*	*bS3*	*cS3*	*SL3*	*aE4*	*bE4*	*cE4*	*aS4*	*bS4*	*cS4*	*SL4*
$PMHR_B$	Expected	2.57	0.5	0.03	2.96	102.71	0.4	−0.36	1.87	0.42	0.04	2	97.68
	Median	2.46	0.52	0.02	3.01	102.06	0.32	−0.24	1.82	0.42	0.11	1.96	97.7
	SD	0.98	0.27	0.48	1.29	0.36	0.28	0.37	1.19	0.27	0.43	1.29	0.48
	Skew	0.63	0.23	−0.02	0.1	−0.83	0.5	−0.36	0.63	0.18	−0.07	0.47	−0.36
$PMHR_C$	Expected	2.76	0.47	0.08	2.82	102.08	0.35	−0.37	2	0.4	0.25	1.93	97.55
	Median	2.65	0.43	0.15	2.99	102.2	0.23	−0.34	1.75	0.37	0.18	1.78	97.65
	SD	0.87	0.24	0.54	1.22	0.36	0.26	0.28	1.34	0.25	0.38	1.31	0.45
	Skew	0.16	0.57	−0.47	0	−0.58	0.72	−0.22	0.55	0.24	−0.4	0.49	−0.53

4. Discussion

To test the effectiveness and the robustness of the proposed model, PMHR$_{B}$ and PMHR$_{C}$ in Table 4 were used to generate the economic and ecological water release by which the corresponding objectives were derived. Using the 100 years of synthetic inflows generated in Section 3.4, the economic and ecological water release processes, as well as the corresponding objectives under different inflow scenarios by the two PMHR rules, were calculated. For comparative study, the economic and ecological water release and their objectives were calculated by the traditional and simplest operating rule SOP, which meant releasing water as close to the delivery targets as possible in order to meet the demand.

4.1. Economic Versus Ecological Objectives under Different Inflow Scenarios

To demonstrate the effectiveness of the HR rules comparing with the SOP rule, the values of economic and ecological objectives optimized by the PMHR$_{B}$, the PMHR$_{C}$, and the SOP are compared in Table 5.

Table 5. Comparison of economic and ecological objectives.

<table>
<tr><th rowspan="2">No.</th><th rowspan="2">Frequency</th><th rowspan="2">Rules</th><th colspan="2">Economic Objective</th><th colspan="2">Ecological Objective</th></tr>
<tr><th>Expected</th><th>Median</th><th>Expected</th><th>Median</th></tr>
<tr><td rowspan="3">1</td><td rowspan="3">95%</td><td>SOP</td><td colspan="2">0.99</td><td colspan="2">238.65</td></tr>
<tr><td>PMHR$_{B}$</td><td>0.57</td><td>0.58</td><td>17.89</td><td>20.38</td></tr>
<tr><td>PMHR$_{C}$</td><td>0.6</td><td>0.64</td><td>88.02</td><td>229.25</td></tr>
<tr><td rowspan="3">2</td><td rowspan="3">90%</td><td>SOP</td><td colspan="2">1.00</td><td colspan="2">165.64</td></tr>
<tr><td>PMHR$_{B}$</td><td>0.89</td><td>0.83</td><td>12.86</td><td>11.97</td></tr>
<tr><td>PMHR$_{C}$</td><td>0.92</td><td>0.95</td><td>43.29</td><td>61.25</td></tr>
<tr><td rowspan="3">3</td><td rowspan="3">75%</td><td>SOP</td><td colspan="2">1.00</td><td colspan="2">148.68</td></tr>
<tr><td>PMHR$_{B}$</td><td>0.90</td><td>0.82</td><td>7.15</td><td>10.1</td></tr>
<tr><td>PMHR$_{C}$</td><td>0.95</td><td>0.97</td><td>13.43</td><td>21.36</td></tr>
<tr><td rowspan="3">4</td><td rowspan="3">50%</td><td>SOP</td><td colspan="2">1.00</td><td colspan="2">114.05</td></tr>
<tr><td>PMHR$_{B}$</td><td>0.99</td><td>0.99</td><td>3.53</td><td>7.23</td></tr>
<tr><td>PMHR$_{C}$</td><td>1.00</td><td>1.00</td><td>4.97</td><td>18.08</td></tr>
<tr><td rowspan="3">5</td><td rowspan="3">25%</td><td>SOP</td><td colspan="2">1.00</td><td colspan="2">193.8</td></tr>
<tr><td>PMHR$_{B}$</td><td>0.99</td><td>0.99</td><td>3.77</td><td>6.37</td></tr>
<tr><td>PMHR$_{C}$</td><td>1.00</td><td>1.00</td><td>2.66</td><td>7.23</td></tr>
<tr><td rowspan="3">6</td><td rowspan="3">10%</td><td>SOP</td><td colspan="2">1.00</td><td colspan="2">137.92</td></tr>
<tr><td>PMHR$_{B}$</td><td>1.00</td><td>0.99</td><td>5.21</td><td>3.81</td></tr>
<tr><td>PMHR$_{C}$</td><td>1.00</td><td>1.00</td><td>28.01</td><td>52.21</td></tr>
</table>

According to Equations (5) and (6), the range of ecological objectives is greater than or equal to 0, and a greater value means a less satisfied ecologic objective. According to Equation (7), the range of economic objectives is from 0 to 1, where a greater value means a higher water supply deficit. With the annual inflow data ranging from a frequency of 10% to 95%, two objective values are compared, as shown in Table 3.

The SOP can satisfy almost all of the economic water demands under a majority of scenarios, but with a greater alteration to the natural flow regime. For the water supply objective (economic objective), the SOP guarantees meeting 100% of the water demands, except in extremely dry years (frequency = 95%). At the same time, the corresponding ecologic objective is greater than 100, which implies a highly altered flow regime. Meanwhile, PMHR$_{B}$ and PMHR$_{C}$ are able to effectively improve the ecological objective by decreasing the ecologic objectives to the range of 2.66–52.21, and satisfy the economic water demands in wet years (frequency $<$ 50%). In dry years, PMHR$_{B}$ and PMHR$_{C}$ have to reduce the water supply for human demand in order to meet the demands of the ecosystem. For a typical dry year (when the inflow frequency is 75%), PMHR$_{B}$ and PMHR$_{C}$ each reduce nearly 10% of the water supply for human demands in order to balance the ecological objective. When it comes to the extreme dry year (frequency > 95%), $PMHR_B$ and $PMHR_C$ each reduce about 50% of human water supply to meet the demands of ecosystem.

4.2. Ecological Release

Under the synthetic inflow conditions, the processes of ecological release in a typical dry year (75% frequency), generated by three rules (SOP, $PMHR_B$ and $PMHR_C$), were compared, as shown in Figures 6 and 7. The processes generated by the expected values of $PMHR_B$ and $PMHR_C$ are shown in Figure 6; processes generated by the median values of $PMHR_B$ and $PMHR_C$ are shown in Figure 7. Compared to the SOP, the ecological release processes generated by $PMHR_B$ and $PMHR_C$ are closer to the natural inflow process. The similarity of the ecological water release process to the inflow process can be quantified by the correlation coefficient. The higher the correlation coefficient is, the closer the release is to the inflow of the reservoir. This means that the smaller hydrologic alteration is made through reservoir operations. Using Figure 6 as an example, the correlation coefficient of the inflow and water release under the SOP is 0.54, while the correlation coefficient of inflow and water release under the $PMHR_B$ and $PMHR_C$ are 0.81 and 0.66, respectively.

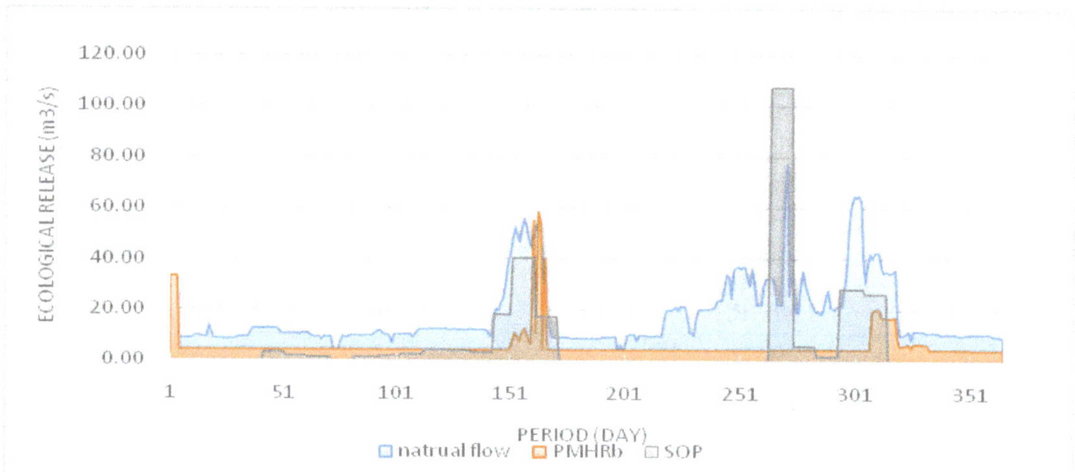

(**a**)

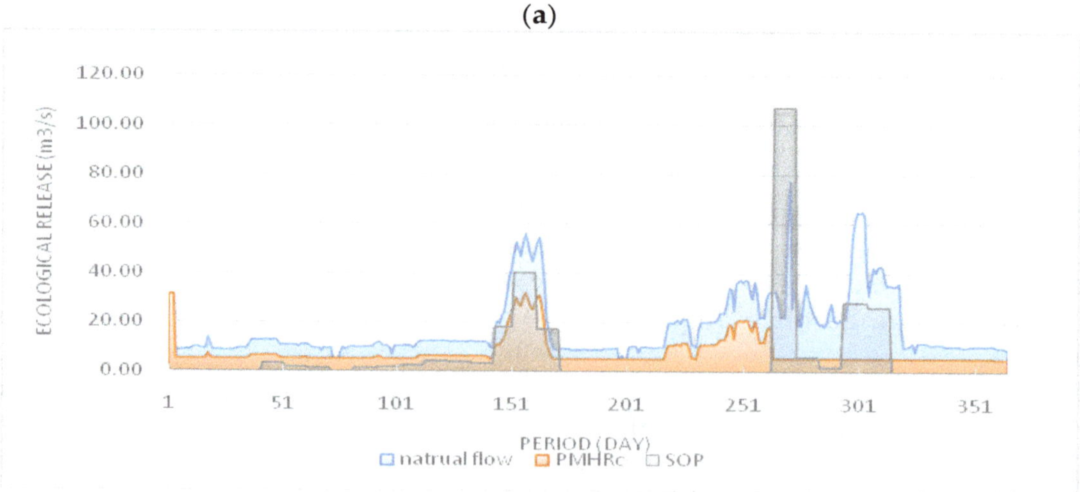

(**b**)

Figure 6. Comparison of release processes between different operating rules, calculated by expectation. (**a**) Comparision of release process of natural flow, $PMHR_B$ and SOP; (**b**) Comparision of release process of natural flow, $PMHR_C$ and SOP.

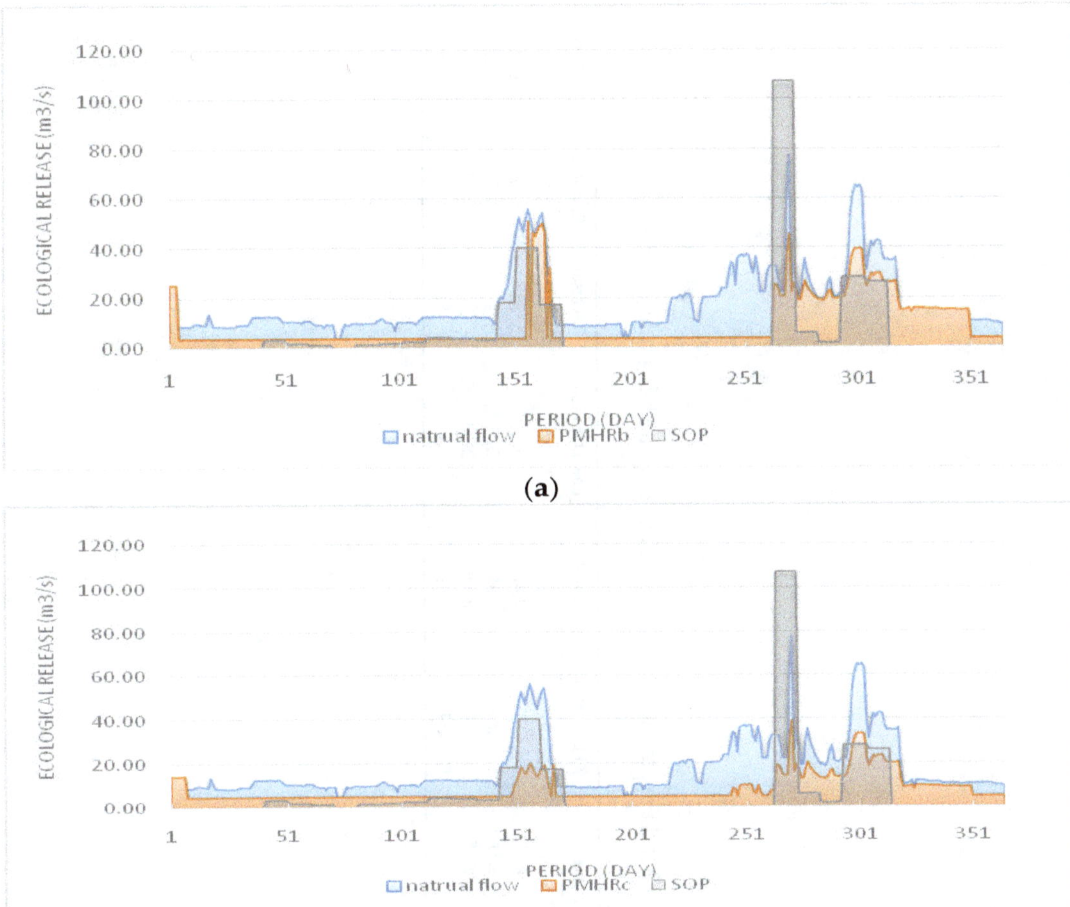

(a)

(b)

Figure 7. Comparison of release processes between different operating rules, calculated by median values. (**a**) Comparision of release process of natural flow, $PMHR_B$ and SOP; (**b**) Comparision of release process of natural flow, $PMHR_C$ and SOP.

4.3. Monte Carlo Simulation Key Indicators Analysis

Figures 6 and 7 demonstrate that reservoir water release alters the flow regime of the river course. This will threaten fish communities and the integrity of river ecosystems downstream. Compared to the operation results under the SOP, the ecological release under the $PMHR_B$ and the $PMHR_C$ can recover the altered flow regime significantly.

The most altered IHA indicators under the SOP are in Table 6, as well as the indicators under $PMHR_B$ and $PMHR_C$. The upper and lower limits (25% and 75% frequency of natural distribution, respectively) of each indicator are also listed for reference. If the value of indicators falls within the range of the upper and lower limits, it means a more acceptable ecological condition for sustainability (as the bold number in Table 6).

Table 6. Comparison of indicator analysis of standard operating policy (SOP) and the piecewise-linear multi-objective hedging rule (PMHR).

	October	November	December	January	April	May	July	August	September	30 Day-Min	90 Day-Min	Low Duration	Reversal	Rise Rate	Fall Rate
SOP	0	1.69	0.92	2.92	0	0	12.6	8.49	0	0	0	74	5	3.37	−0.57
$PMHR_B$	7.03	**4.84**	4.84	4.84	**4.84**	4.84	**18.74**	14.8	6.94	**4.84**	**4.84**	**20.31**	68	**0.2**	**−0.14**
$PMHR_C$	6.65	3.84	3.84	3.84	**3.84**	3.84	**25.03**	**21.04**	**9.53**	**3.84**	**3.84**	**21.5**	47	0.78	**−0.14**
Upper limit	19.81	20.52	20.71	14.59	17.81	23.1	38.91	48.18	25.75	7.28	10.08	28.75	135	0.52	−0.11
Lower limit	8.21	4.26	5.02	6.6	2.16	8.2	15.97	19.29	7.79	0.84	0.91	6	81	0.19	−0.18

4.4. Extended Analysis: Impact of Long-Term Forecast Information

In this section, the impacts of using long-term annual forecast inflows in an operation are discussed preliminarily. We suppose that the annual forecast information, which simply indicates the inflow of the next year and is either larger (wet year) or smaller (dry year) than the average level, is known in advance. Thus, the synthetic inflows used by the Monte Carlo simulation can be divided into two parts: the inflow in wet years and the inflow in dry years. By using the inflow of wet or dry years instead of all the synthetic inflows, the parameters of PMHR can be optimized in two groups: parameters adapted to wet years and parameters adapted to dry years. The expected value of these parameters can be used for wet and dry years, respectively.

Figure 8 shows the environmental flow release using rules in both wet and dry years with long-term forecast information, as well as the release using rules of all possible hydrological years without long-term forecast information. Figure 8a presents the results of a typical wet year (frequency = 25%), in which the correlation coefficient between the water release process and the inflow process is 0.95 with forecast information, but 0.90 without forecast. Figure 8b presents the results of a typical dry year, in which the correlation coefficient is 0.66 with forecast information and 0.61 without forecast information. This indicates that the PMHR combined with forecast information is able to improve the restoration of an ecological flow regime compared to one without the forecast.

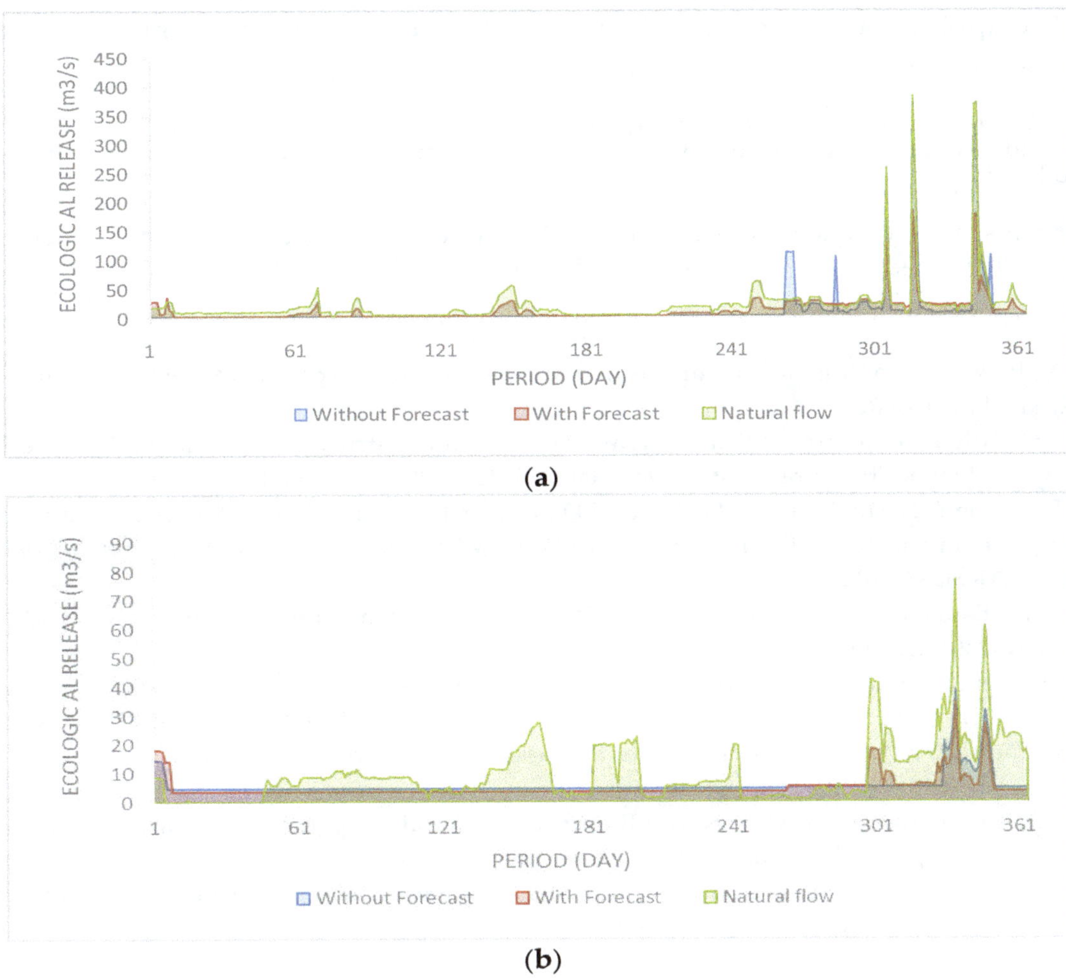

Figure 8. Comparison of ecological releases, between those with and without a long-term forecast. (**a**) Comparison of wet years; (**b**) Comparison of dry years.

5. Conclusions

This study adopts the restoration of natural stream flows as an objective of reservoir operation. An optimization model is proposed by using piecewise-linear multi-objective hedging rules (PMHR) to balance the ecological objective of recovering the natural flow regime and the economic objective of satisfying human consumptive water demand. The proposed model can reflect the daily interval variations of the flow by engaging real-time daily inflow forecasts and storage information into decision-making. The parameters of the PMHR were optimized, and Pareto frontiers were discovered by a vector evaluated genetic algorithm. A Monte Carlo simulation was used to deal with inflow uncertainty during real operation.

The results of Baiguishan Reservoir show the balance between ecological and economic objectives. With the consideration of ecological objectives, the ecological release of a reservoir was restored, reflected by the decrease of IHA to different degrees. To deal with inflow uncertainty, synthetic inflow was generated and used in a Monte Carlo simulation under uncertain hydrological conditions. To test the effectiveness of the model, a set of typical frequency inflows was selected for analysis. The model with PMHR has improved the ecological objective and guaranteed the water supply under most of the hydrological conditions in the case reservoir. The alteration degree of the hydrologic indicators, which has been seriously altered, was recovered into the acceptable range. The impact of involving long-term hydrologic forecast information on improving reservoir operation is also demonstrated by the comparisons of two hydrological conditions (wet and dry years) in the case reservoir. The application of forecast inflow information can obviously improve the scheme of ecological water release.

Author Contributions: The conceptualization and framework of this article are derived from J.Z. Model computation and result analysis are completed by Y.L., supervised by J.Z. and H.Z. The article is modified and reviewed by H.Z.

Acknowledgments: We are grateful to the editors and the three anonymous reviewers for their constructive comments and detailed suggestions, which helped us to substantially improve the paper.

References

1. Zhong, Y.; Power, G. Environmental impacts of hydroelectric projects on fish resources in China. *Regul. River* **1996**, *12*, 81–98. [CrossRef]
2. Joy, M.K.; Death, R.G. Control of fish and cray fish community structure in Taranaki, New Zealand: Dams, Diadromy or Habitat Structure. *Freshw. Biol.* **2001**, *46*, 417–429. [CrossRef]
3. Homa, E.S.; Vogel, R.M.; Smith, M.P.; Apse, C.D.; Huber-Lee, A.; Seiber, J. An optimization approach for balancing human and ecological flow needs. In *World Water and Environmental Resources Congress*; ASCE: Anchorage, AK, USA, 2005.
4. Richter, B.D.; Baumgartner, J.V.; Wigington, R.; Braun, D.P. How much water does a river need? *Freshw. Biol.* **1997**, *37*, 231–249. [CrossRef]
5. Rehn, A.C. Benthic macroinvertebrates as indicators of biological condition below hydropower dams on West Slope Sierra Nevada Streams, California, USA. *River Res. Appl.* **2009**, *25*, 208–228. [CrossRef]
6. Yin, X.A.; Yang, Z.F.; Petts, G.E. A new method to assess the flow regime alterations in riverine ecosystems. *River Res. Appl.* **2015**, *31*, 497–504. [CrossRef]
7. Poff, N.L.R.; Allan, J.D.; Bain, M.B.; Karr, J.R.; Prestegaard, K.L.; Richter, B.D.; Sparks, R.E.; Stromberg, J.C. The natural flow regime. *BioScience* **1997**, *47*, 769–784. [CrossRef]
8. Tharme, R.E. A global perspective on environmental flow assessment: Emerging trends in the development and application of environmental flow methodologies for rivers. *River Res. Appl.* **2003**, *19*, 397–441. [CrossRef]
9. Olden, J.D.; Poff, N.L. Redundancy and the choice of hydrologic indices for characterizing streamflow regimes. *River Res. Appl.* **2003**, *19*, 101–121. [CrossRef]
10. Petts, G.E. Instream flow science for sustainable river management. *J. Am. Water Resour. Assoc.* **2009**, *45*, 1071–1086. [CrossRef]

11. Poff, N.L. Managing for variability to sustain freshwater ecosystems. *J. Am. Water Resour. Plan. Manag.* **2009**, *135*, 1–4. [CrossRef]
12. Bunn, S.E.; Arthington, A.H. Basic principles and ecological consequences of altered flow regimes for aquatic biodiversity. *Environ. Manag.* **2002**, *30*, 492–507. [CrossRef]
13. Bayley, P.B. The Flood Pulse Advantage and the Restoration of River-floodplain Systems. *River Res. Appl.* **1991**, *6*, 75–86. [CrossRef]
14. Sparks, R.E. Need for ecosystem management of large rivers and floodplains. *BioScience* **1995**, *45*, 168–182. [CrossRef]
15. Jager, H.I.; Smith, B.T. Sustainable reservoir operation: Can we generate hydropower and preserve ecosystem values? *River Res. Appl.* **2008**, *24*, 340–352. [CrossRef]
16. Suen, J.P.; Eheart, J.W. Reservoir management to balance ecosystem and human needs: Incorporating the paradigm of the ecological flow regime. *Water Resour. Res.* **2006**, *42*, W03417. [CrossRef]
17. Cardwell, H.; Jager, H.I.; Sale, M.J. Designing instream flows to satisfy fish and human water needs. *J. Am. Water Resour. Plan. Manag.* **1996**, *122*, 356–363. [CrossRef]
18. Ripo, G.C.; Jacobs, J.M.; Good, J.C. An algorithm to integrate ecological indicators with streamflow withdrawals. In *ASCE/EWRI World Water and Environmental Resources Congress*; American Society of Civil Engineers: Philadelphia, PA, USA, 2003.
19. Guo, W.X.; Wang, H.X.; Xu, J.X.; Xia, Z.Q. Ecological operation for Three Gorges Reservoir. *Water Sci. Eng.* **2011**, *4*, 143–156. [CrossRef]
20. Yang, Y.C.E.; Cai, X.M. Reservoir reoperation for fish ecosystem restoration using daily inflows-case study of Lake Shelbyvile. *J. Am. Water Resour. Plan. Manag.* **2011**, *137*, 470–480. [CrossRef]
21. Shiau, J.T.; Wu, F.C. Optimizing environmental flows for multiple reaches affected by a multipurpose reservoir system in Taiwan: Restoring natural flow regimes at multiple temporal scales. *Water Resour. Res.* **2013**, *49*, 565–584. [CrossRef]
22. Cai, W.J.; Zhang, L.L.; Zhu, X.P.; Zhang, A.J.; Yin, J.X.; Wang, H. Optimized reservoir operation to balance human and environmental requirements: A case study for the Three Gorges and Gezhouba Dams, Yangtze River basin, China. *Econ. Inform.* **2013**, *18*, 40–48. [CrossRef]
23. Lane, B.A.; Sandoval-Solis, S.; Porse, E.C. Environmental flows in a human-dominated system: Integrated water management strategies for the Rio Grande/Bravo Basin. *River Res. Appl.* **2015**, *31*, 1053–1065. [CrossRef]
24. Yin, X.A.; Yang, Z.F.; Petts, G.E. Optimizing environmental flows below dams. *River Res. Appl.* **2012**, *28*, 703–716. [CrossRef]
25. Haghighi, A.T.; Klove, B. Development of monthly optimal flow regimes for allocated environmental flow considering natural flow regimes and several surface water protection targets. *Ecol. Eng.* **2015**, *82*, 390–399. [CrossRef]
26. Hui, W.; Brill, E.D.; Ranjithan, R.S.; Sankarasubramanian, A. A framework for incorporating ecological release in single reservoir operation. *Adv. Water Resour.* **2015**, *78*, 9–21. [CrossRef]
27. Yin, X.A.; Yang, Z.F.; Petts, G.E. Reservoir operating rules to sustain environmental flows in regulated rivers. *Water Resour. Res.* **2011**, *47*, W08509. [CrossRef]
28. Zhu, X.P.; Zhang, C.; Yin, J.X.; Zhou, H.C.; Jiang, Y.Z. Optimization of Water Diversion Based on Reservoir Operating rules: Analysis of the Biliu River Reservoir, China. *J. Hydrol. Eng.* **2014**, *19*, 411–421. [CrossRef]
29. Tilmant, A.; Beevers, L.; Muyunda, B. Restoring a flow regime through the coordinated operation of a multireservoir system: The case of the Zambezi River basin. *Water Resour. Res.* **2010**, *46*, W07533. [CrossRef]
30. Chen, Q.; Chen, D.; Li, R.; Ma, J.; Blanckaert, K. Adapting the operation of two cascaded reservoirs for ecological flow requirement of a de-watered river channel due to diversion-type hydropower stations. *Ecol. Model.* **2013**, *252*, 266–272. [CrossRef]
31. Koutsoyiannis, D.; Economou, A. Evaluation of the parameterization-simulation-optimization approach for the control of reservoir systems. *Water Resour. Res.* **2003**, *39*, WR00214. [CrossRef]
32. Hashimoto, T.; Stedinger, J.R.; Loucks, D.P. Reliability, resilience vulnerability criteria for water resources system performance evaluation. *Water Resour. Res.* **1982**, *18*, 14–20. [CrossRef]
33. Drapper, A.J.; Lund, J.R. Optimal hedging and carryover storage value. *J. Am. Water Resour. Plan. Manag.* **2004**, *130*, 83–87. [CrossRef]

34. Zhao, J.S.; Cai, X.M.; Wang, Z.J. Optimality conditions for a two-stage reservoir operation problem. *Water Resour. Res.* **2011**, *47*, W08503. [CrossRef]
35. Loucks, D.P.; Stedinger, J.R.; Haith, D.A. *Water Resource Systems Planning and Analysis*; Prentice-Hall: Englewood Cliffs, NJ, USA, 1981.
36. Tejada-Guibert, J.A.; Johnson, S.A.; Stedinger, J.R. Comparison of two approaches for implementing multireservoir operating policies derived using stochastic dynamic programming. *Water Resour. Res.* **1993**, *29*, 3969–3980. [CrossRef]
37. Karamouz, M.; Houck, M.H.; Delleur, J.W. Optimization and simulation of multiple reservoir system. *J. Am. Water Resour. Plan. Manag.* **1992**, *118*, 71–81. [CrossRef]
38. Tu, M.Y.; Hsu, N.S.; Tsai, T.C.; Yeh, W.W.G. Optimization of Hedging Rules for Reservoir Operations. *J. Am. Water Resour. Plan. Manag.* **2008**, *134*, 3–13. [CrossRef]
39. Taghian, M.; Dan, R.; Haghighi, A.; Madsen, H. Optimization of conventional rule curves coupled with hedging rules for reservoir operation. *J. Am. Water Resour. Plan. Manag.* **2014**, *140*, 693–698. [CrossRef]
40. Shiau, J.T. Analytical optimal hedging with explicit incorporation of reservoir release and carryover storage targets. *Water Resour. Res.* **2011**, *47*, 238–247. [CrossRef]
41. Gao, Y.X.; Vogel, R.M.; Kroll, C.N.; Poff, N.L.; Olden, J.D. Development of representative indicators of hydrologic alteration. *J. Hydrol.* **2009**, *374*, 136–147. [CrossRef]
42. Zhao, T.; Zhao, J.S.; Yang, D.W.; Wang, H. Generalized martingale model of the uncertainty evolution of streamflow forecast. *Adv. Water Resour.* **2013**, *57*, 41–45. [CrossRef]
43. Schaffer, J. Multiple objective optimization with vector evaluated genetic algorithms. In Proceedings of the 1st International Conference on Genetic Algorithms, Pittsburgh, PA, USA, 24–26 July 1985; Psychology Press: Hillsdale, NJ, USA, 1985; pp. 93–100.

11

Quantification of Seasonal Precipitation over the Upper Chao Phraya River Basin in the Past Fifty Years Based on Monsoon and El Niño/Southern Oscillation Related Climate Indices

Tsuyoshi Kinouchi [1,*], Gakuji Yamamoto [1], Atchara Komsai [1] and Winai Liengcharernsit [2]

1 Department of Transdisciplinary Science and Engineering, School of Environment and Society, Tokyo Institute of Technology, Yokohama 226-8503, Japan; yamamoto.g.ab@m.titech.ac.jp (G.Y.); komsai.a.aa@m.titech.ac.jp (A.K.)

2 Department of Environmental Engineering, Faculty of Engineering, Kasetsart University, Bangkok 10900, Thailand; fengwnl@ku.ac.th

* Correspondence: kinouchi.t.ab@m.titech.ac.jp

Abstract: For better water resources management, we proposed a method to estimate basin-scale seasonal rainfall over selected areas of the Chao Phraya River Basin, Thailand, from existing climate indices that represent variations in the Asian summer monsoon, the El Niño/Southern Oscillation, and sea surface temperatures (SST) in the Pacific Ocean. The basin-scale seasonal rainfall between 1965 and 2015 was calculated for the upper Ping River Basin (PRB) and the upper Nan River Basin (NRB) from a gridded rainfall dataset and rainfall data collected at several gauging stations. The corresponding climate indices, i.e., the Equatorial-Southern Oscillation Index (EQ-SOI), Indian Monsoon Index (IMI), and SST-related indices, were examined to quantify seasonal rainfall. Based on variations in the rainfall anomaly and each climate index, we found that IMI is the primary variable that can explain variations in seasonal rainfall when EQ-SOI is negative. Through a multiple regression analysis, we found that EQ-SOI and two SST-related indices, i.e., Pacific Decadal Oscillation Index (PDO) and SST anomalies in the tropical western Pacific (SST_{NW}), can quantify the seasonal rainfall for years with positive EQ-SOI. The seasonal rainfall calculated for 1975 to 2015 based on the proposed method was highly correlated with the observed rainfall, with correlation coefficients of 0.8 and 0.86 for PRB and NRB, respectively. These results suggest that the existing indices are useful for quantifying basin-scale seasonal rainfall, provided a proper classification and combination of the climate indices are introduced. The developed method could forecast seasonal rainfall over the target basins if well-forecasted climate indices are provided with sufficient leading time.

Keywords: seasonal rainfall; upper Chao Phraya River Basin; El Niño/Southern Oscillation; Indian Monsoon; sea surface temperatures

---

1. Introduction

Recently, many regions have endured significant impacts from extreme floods and droughts, including Southeastern Asian countries [1–4]; disastrous floods and droughts are expected to occur more frequently due to the changing climate [2,5–7]. To mitigate the impacts of extreme hydro-meteorological events, well-prepared water resource management practices are required, for which the quantification of rainfall and resulting runoff is key information.

In Thailand, a devastating flood occurred in the Chao Phraya River Basin in 2011 [8,9], resulting in a change in policy to focus more on flood mitigation. However, the basin suffered a serious drought during 2015 and 2016 due to the reduced water storage of the major reservoirs, i.e., the Bhumibol and Sirikit Dam reservoirs during the preceding years and the limited amount of rainfall in the wet season of 2015. To sustain socio-economic activities in the downstream areas of the Chao Phraya River Basin, sufficient water storage in these reservoirs is required before the dry season while enough empty volume should be maintained for flood-control during the rainy season. Therefore, water release from the reservoirs needs to be well-managed by forecasting the areal rainfall and its resulting inflows to the reservoirs with sufficient leading time.

Different approaches have been adopted to quantify or forecast rainfall in Thailand using climate indices. Most approaches are based on a simple linear correlation method [10,11], and a limited number of works have considered multiple indices using a linear multiple-regression model and a non-parametric method [12]. The linear regression approach is conventional, but not suitable if the rainfall is non-linearly related to climate indices. The non-parametric method is advantageous in this situation, although its implementation is complicated.

Furthermore, previous studies focused on rainfall at a limited number of sites with shorter duration (for example, [12]); however, for water resources management purposes, rainfall must be estimated to represent the entire rainy season at regional or basin-scales, rather than the smaller area represented by individual gauging stations.

In this study, we attempt to develop a method that can relate seasonal rainfall over the two major upstream sub-basins of the Chao Phraya River Basin draining to the Bhumibol and Sirikit Dam reservoirs, i.e., the upper Ping River Basin (PRB) and the upper Nan River Basin (NRB), to multiple climate indices that are predictable prior to the rainy season. The Bhumibol and Sirikit Dam reservoirs are the first and second-largest reservoirs in Thailand and provide water for various economic and environmental purposes [13]. The areal-averaged rainfall data of the rainy seasons between 1965 and 2015 were used to determine a relationship between the total rainfall and selected indices, i.e., the Southern Oscillation index (Equatorial SOI, hereafter EQ-SOI), Indian Monsoon index (IMI), El Niño/Southern Oscillation index (SSTs), and the Pacific Decadal Oscillation index (PDO).

2. Study Area and Data Analysis

2.1. Study Area

Two major sub-basins of the Chao Phraya River Basin (CPRB) draining to the Bhumibol and Sirikit Dam reservoirs, i.e., the PRB and NRB, were selected as study sites, respectively (Figure 1). The PRB and NRB cover 26,300 km^2 and 11,950 km^2, respectively. The Bhumibol and Sirikit Dam reservoirs discharge a significant amount of water for domestic and industrial use, as well as for irrigation and environmental purposes [13,14]. Although domestic usage receives priority in the water allocation of the Chao Phraya River Basin, a large area of farmland requires a large amount of water for irrigation, especially during the dry season, and depends on the two major dams and other medium- to small-scale reservoirs. Therefore, it is highly important to know the volume of water available in these reservoirs before the dry season arrives.

The climate of the study area is characterized by distinct rainy and dry seasons (Figure 2). During the rainy season, the amount of precipitation and its areal extent in the CPRB are caused by the southwest monsoons, together with typhoons, monsoon troughs, and depressions from the South China Sea. The mean monthly rainfall over the study area is large during May and October (PRB) or September (NRB), which corresponds to the southwest monsoon periods [15] (Figure 2).

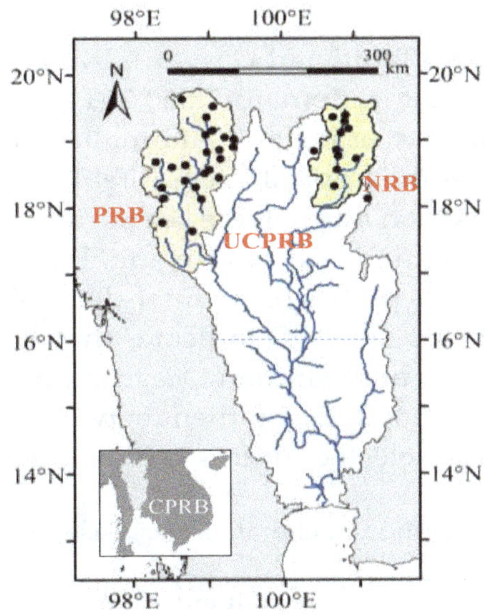

Figure 1. Location of the upper Ping River Basin (PRB) and the upper Nan River Basin (NRB). The inset shows the area of the Chao Phraya River Basin (CPRB). The Upper Chao Phraya River Basin (UCPRB) is defined to cover the area north of 16° N. The black dots indicate the locations of rain gauges used to estimate areal rainfall since 2001.

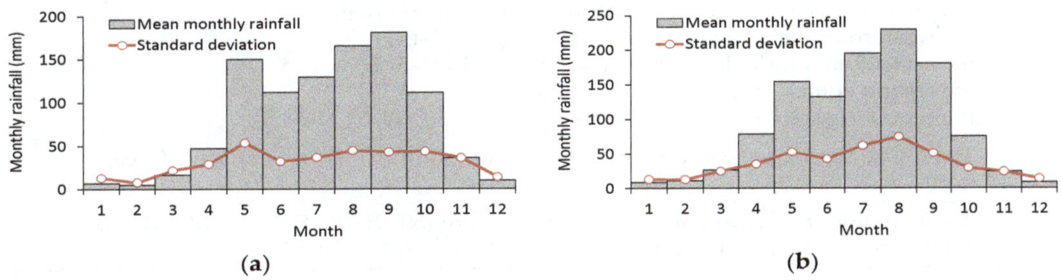

Figure 2. Mean monthly areal-averaged rainfall for the period from 1975 to 2015 over (**a**) PRB and (**b**) NRB.

2.2. Data Source

To determine the relationship between seasonal rainfall and regional climate conditions, we used the areal rainfall obtained from gridded and gauged rainfalls, and indices representing the regional climate conditions, such as the monsoon and El Niño/Southern Oscillation (ENSO). The area-averaged rainfall during the rainy season was calculated for the period between April and October utilizing a gridded rainfall dataset (APHRODITE, [16]) and rain-gauge data collected by the Royal Irrigation Department of Thailand (RID) at multiple sites located in the PRB and NRB (Figure 1). Rainfall in April is included in the rainy season as the onset of the summer rainy season begins earlier in the inland Indochina in late April to May [17], and the monthly rainfall in April is comparable to that of October for NRB (Figure 2). As the APHRODITE dataset only contains data until 2007, the areal rainfall during 2008–2015 was estimated from the original relationship between the areal-mean monthly rainfalls obtained from the APHRODITE dataset and those calculated by the Thiessen Polygon applied to the rain-gauge data of RID for the period from 1980 to 2000 (see the supplement for detailed method information to calculate the areal rainfall).

The climate indices used in the analysis include the EQ-SOI, IMI, SST anomaly over selected domains in the Pacific Ocean, PDO, and the Indian Ocean Dipole Mode index (DMI). The definition of each index is listed in Table 1. EQ-SOI is superior to SOI in representing the strength of trade winds over the tropical Pacific as one of the domains for calculating EQ-SOI (SLP.2 in Figure 3) is closer to locations in the Southern Hemisphere (Tahiti and Darwin). IMI is the difference between the 850-hPa zonal wind in domain U850.1 and that in domain U850.2 (Figure 3) [18]. A larger IMI indicates a stronger summer monsoon over the Bay of Bengal. Anomalies in sea surface temperature (SST) are defined for the selected domains over the tropical Pacific (Table 1). If EQ-SOI is positive, which likely corresponds to stronger trade winds than normal, the sea surface temperature in these domains may affect the supply of atmospheric vapor in the leeward region. PDO is an anomaly pattern of SST in the Pacific Ocean (north of 20° N) that shifts phases on an inter-decadal time scale that usually covers over 10 years [19]. A positive (negative) value of PDO indicates that SST is lower (higher) than normal in the northwestern Pacific. The intensity of the Indian Ocean dipole mode (IOD) is represented by the Dipole Mode Index (DMI) as an anomalous SST gradient between two selected domains (Table 1) [20]. The domain for calculating the finally selected indices is illustrated in Figure 3.

Table 1. Description of climate indices used in this study.

Index	Definition	Data Sources
EQ-SOI	The difference between the area-averaged sea level pressure in an area of the eastern equatorial Pacific (80° W–130° W, 5° N–5° S; indicated as SLP.1 in Figure 3) and an area over Indonesia (90° E–140° E, 5° N–5° S; SLP.2 in Figure 3)	[21]
IMI	The difference between the 850-hPa zonal wind in the domain of 40° E–80° E, 5° N–15° N (indicated as U850.1 in Figure 3) and that in the domain of 70° E–90° E, 20° N–30° N (U850.2 in Figure 3)	[22]
SST	Anomalies in sea surface temperature for the domain over	[23]
NINO.WEST	the western tropical Pacific (130° E–150° E, 15° N–EQ)	
NINO.3	the eastern equatorial Pacific (150° W–90° W, 5° N–5° S)	
NINO.4	the central equatorial Pacific (160° E–150° W, 5° N–5° S)	
NINO.34	the central/eastern equatorial Pacific (170° W–120° W, 5° N–5° S)	
PDO	The leading principal component of the mean monthly SST in the Pacific Ocean north of 20° N [19]	[24]
DMI	An anomalous SST gradient between the western (50° E–70° E, 10° S–10° N) and southeastern equatorial Indian Ocean (90° E–110° E, 10° S–0° N)	[25]

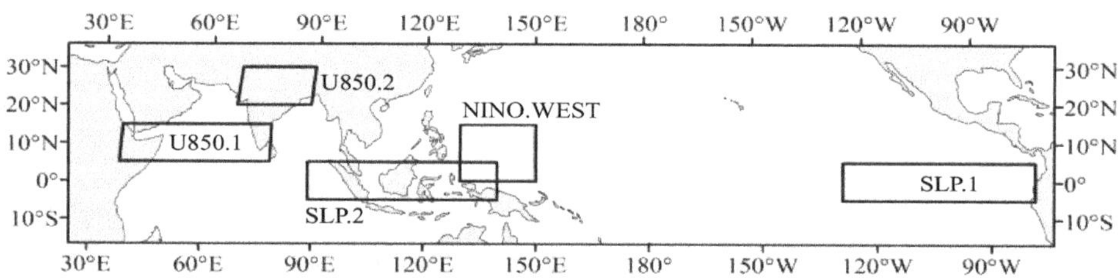

Figure 3. Domains related with each climate index finally employed in the proposed method (the domain related with PDO is not shown).

The EQ-SOI and IMI data were obtained from the NOAA National Weather Service [21] and the International Pacific Research Center (IPRC), University of Hawaii [22], respectively. The SST indices over NINO.WEST, NINO.3, NINO.4, and NINO.34 were obtained from the Japan Meteorological Agency [23]. The PDO and DMI data were obtained from the NOAA Earth System Research Laboratory (ESRL) [24] and the Japan Agency for Marine-Earth Science and Technology (JAMSTEC) [25], respectively.

2.3. Overview of Data Analysis

To quantify the seasonal rainfall from existing climate indices, we assume that a functional relationship between the total rainfall from April to October and the selected climate indices averaged over the same period can be applied, which may take the form of Equation (1):

$$\mathrm{TR} = f(\mathrm{EQ} - \mathrm{SOI}, \mathrm{IMI}, \mathrm{PDO}, \mathrm{DMI}, \mathrm{SST}) \tag{1}$$

where TR (mm) is the areal-averaged seasonal rainfall during April and October, and SST represents the SST anomaly over a certain domain in the Pacific Ocean.

Based on our preliminary check for coherent variations in TR and each index, we found that the influential climate factors differed depending on EQ-SOI; thus, we classified the data into two categories with respect to EQ-SOI. When EQ-SOI is negative, we simplified Equation (1) to utilize a single independent variable (IMI), however, when EQ-SOI is positive, multiple regression analysis was used to determine the exact form of Equation (1). We obtained the functional relationship by parameter optimization (when EQ-SOI is negative) and the multiple regression analysis (when EQ-SOI is positive) based on the data for 1975 to 2015, and verified it for data between 1965 and 1974.

3. Results

3.1. Characteristics of Rainfall and Corresponding Climate Indices

Areal rainfall data indicates that over 90% of the annual rainfall occurs during April and October almost every year, with the largest mean monthly rainfall occurring in September (PRB) or August (NRB) (Figure 2). In the NRB, the mean monthly rainfall in April is larger than that in October. The anomaly of seasonal rainfall from April to October ranges from −291 to 367 mm for PRB and from −315 to 444 mm for NRB, with mean rainfall of 890 mm and 1073 mm, and standard deviations of 124 mm and 165 mm for PRB and NRB, respectively (Figure 4).

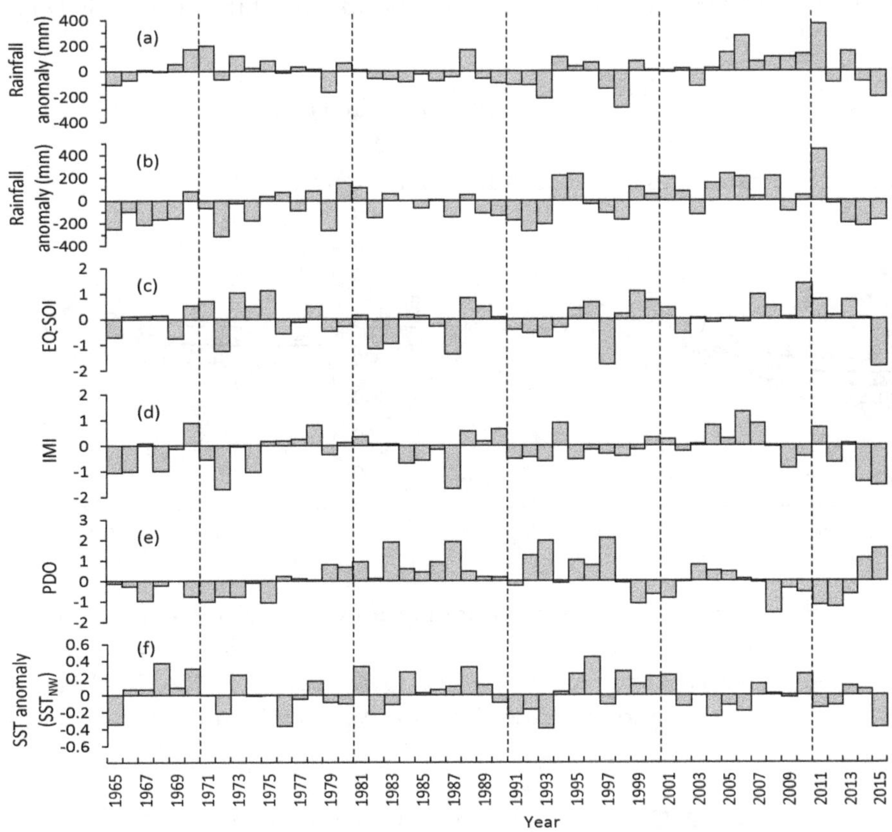

Figure 4. Anomalies of the seasonal rainfall and finally selected climate indices for the period between 1965 and 2015. (**a**) Rainfall anomaly in PRB; (**b**) Rainfall anomaly in NRB; (**c**) EQ-SOI; (**d**) IMI; (**e**) PDO; and (**f**) SST_{NW} (SST anomaly over NINO.WEST). Each decade is separated by dashed lines. The rainfall anomaly is calculated as the deviation from the mean rainfall for the period from 1975 to 2015.

The areal rainfall over PRB tends to be smaller and less variable than that over NRB, which could be due to the effect of the Dawna Range located on the western side of PRB and the limited influence of the easterly wind. The rainfall anomalies for PRB and NRB tended to be positively large between 2004 and 2011 (excluding some years), while negative anomalies were predominant during 1981–1993 and since 2012 (Figure 4). The average of each climate index between April and October is plotted in Figure 4. In years with negative EQ-SOI and IMI (1979, 1991, 1992, 1993, 1997, and 2015), it is likely that the rainfall anomalies are also negative. A positive rainfall anomaly was found in years with positively large IMI (1994 and 2006) and negative EQ-SOI. PDO was large and negative between 2008 and 2013, when the rainfall anomaly is mostly positive. EQ-SOI was negatively correlated with PDO on inter-annual and decadal time scales. Coherent variation is observed between EQ-SOI and the SST anomaly over NINO.WEST (SST_{NW}) in both the negative and positive phases. However, the influence of SST_{NW} on rainfall anomalies is not clear from Figure 4.

3.2. Relationship between Seasonal Rainfall and Climate Indices

Based on the coherent variation of the rainfall anomaly and each climate index described in the previous section, we classified the rainfall dataset and climate indices into two groups depending on EQ-SOI (the threshold value was slightly larger than 0). Seasonal rainfall is primarily related to IMI when EQ-SOI is negative and approaches the lower (upper) bound when IMI is negatively (positively) large (Figure 5). This suggests that, when EQ-SOI is negative, the effect of the monsoon on rainfall over the Indochina Peninsula is predominant due to the lower influence of the weaker surface trade winds. A nonlinear equation based on a sigmoid function (Equation (2)) was applied to the relationship between IMI and seasonal rainfall TR (mm) by calibrating the parameters in Equation (2) to minimize the root mean square error (RMSE) between the actual and calculated rainfalls (Figure 5):

$$\mathrm{TR} = \frac{a}{1 + \exp(-\lambda \times \mathrm{IMI})} + b \tag{2}$$

where a, b, and λ are the parameters calibrated for each basin (Table 2).

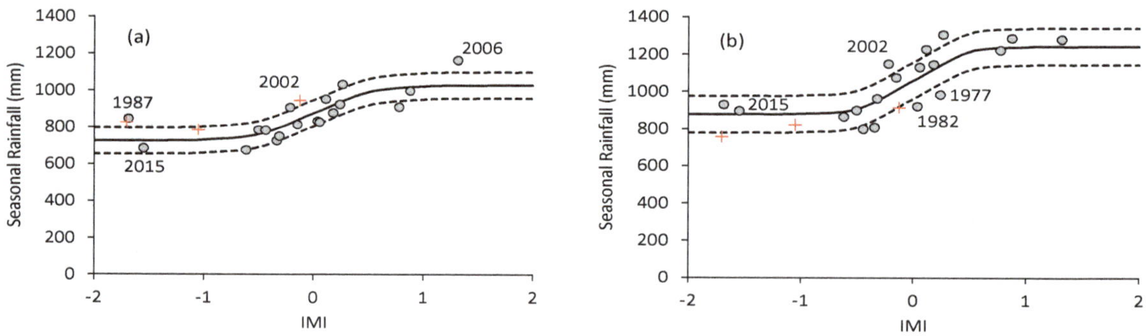

Figure 5. The relation between the seasonal rainfall and IMI for (**a**) PRB and (**b**) NRB. Two dashed lines are separated from the solid line (Equation (1)) by the standard deviation of the error between the observed and estimated seasonal rainfalls. The gray circles are the data used for calibrating the parameters, and the cross marks indicate data for 1965, 1969, and 1972, which were not used for calibration.

Table 2. Calibrated parameters used in Equation (1).

Basin	a (mm)	b (mm)	λ
PRB	302.4	723.0	4.107
NRB	395.1	870.6	4.734

The most recent drought year, 2015, is plotted close to the curve of Equation (2) (Figure 5). The difference between the estimated (solid line) and observed rainfall in some years is large (Figure 6). Although we checked the influence of other factors (EQ-SOI, PDO, and other SST-related indices), none of them were likely to be responsible for this difference. Data for years before 1975 are also plotted in Figure 5, indicating that the seasonal rainfall for these years can also be represented by Equation (2) despite the limited number of gauging stations used to obtain the areal rainfall. Furthermore, it was found that Equation (2) is applicable to the upper Chao Phraya River Basin (UCPRB, Figure 1, the basin area is 109,000 km^2) with the correlation coefficient $r = 0.77$ and RMSE = 83.1 mm for calibration period from 1976 to 2006 (sample number = 17).

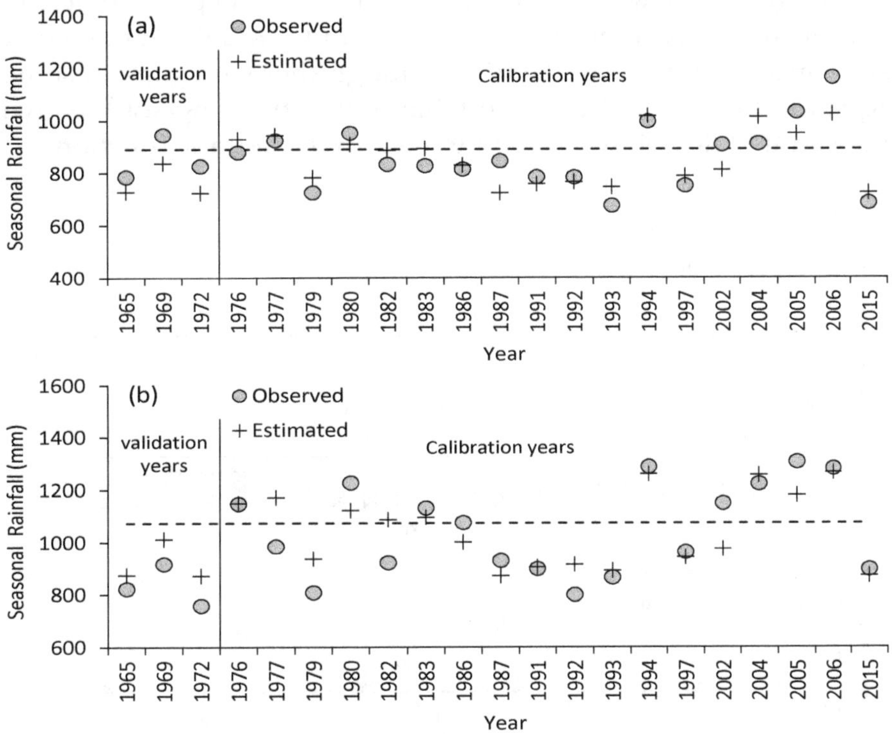

Figure 6. Comparison of observed and estimated seasonal rainfall for (**a**) PRB and (**b**) NRB for calibration and validation years with EQ-SOI < 0.02. Equation (2) with calibrated parameters (Table 2) was applied. The dashed line indicates the mean rainfall for the period from 1975 to 2015.

When EQ-SOI is positive, i.e., when the surface trade winds are larger than normal, seasonal rainfall was not significantly related to any one of the tested indices; however, it was positively correlated to EQ-SOI (Figure 9a). Since the arrival and passage of typhoons and tropical cyclones are major causes of increased rainfall, it is likely that the trade winds and SSTs in the Pacific Ocean contribute to these phenomena, with additional influence on seasonal rainfall from the Indian Ocean and monsoon [26]. In this study, we applied an equation that considers interactive and non-linear relationships between the seasonal rainfall and ENSO and SST-related indices for the Pacific Ocean using quadratic functions (Equation (3)). In addition, we tested the inclusion of IMI and DMI as additional components in Equation (6) to consider the influence of the Indian Ocean and monsoon, respectively; however, there was no significant improvement for rainfall estimation:

$$\mathrm{TR} = (g_1 \times \mathrm{PDO}^2 + g_2 \times \mathrm{PDO} + g_3) \times \mathrm{EQ} - \mathrm{SOI} + d_0 \quad (3)$$

$$g_1 = d_1 \times \mathrm{SST}^2 + d_2 \times \mathrm{SST} + d_3 \quad (4)$$

$$g_2 = d_4 \times \mathrm{SST}^2 + d_5 \times \mathrm{SST} + d_6 \quad (5)$$

$$g_3 = d_7 \times \text{SST}^2 + d_8 \times \text{SST} + d_9 \quad (6)$$

where SST is any one of the four SST anomalies over the domain of NINO.WEST, NINO.3, NINO.4, and NINO.34, and d_0–d_9 are the coefficients of the multiple regression analysis. Equation (3) was evaluated with each SST index by the RMSE, multiple correlation coefficient, the similarity of coefficients d_1–d_9 between PRB and NRB, and the consistency with the perceived physical influences, which will be discussed in the following sections.

On average, comparatively high correlation coefficients and small RMSE were obtained for PRB and NRB by applying Equation (3) with the SST anomaly over NINO.WEST (SST_{NW}). Figure 7 compares the observed and estimated seasonal rainfall during the parameter setting period by the multiple regression analysis. A favorable agreement is found throughout the period, including flood years (1995 and 2011) and drought years (2003), although there are gaps in some years. Thus, we determined that Equation (3) with SST_{NW} is suitable within the range of the given climate indices examined, despite the limited flexibility caused by the form (quadratic function).

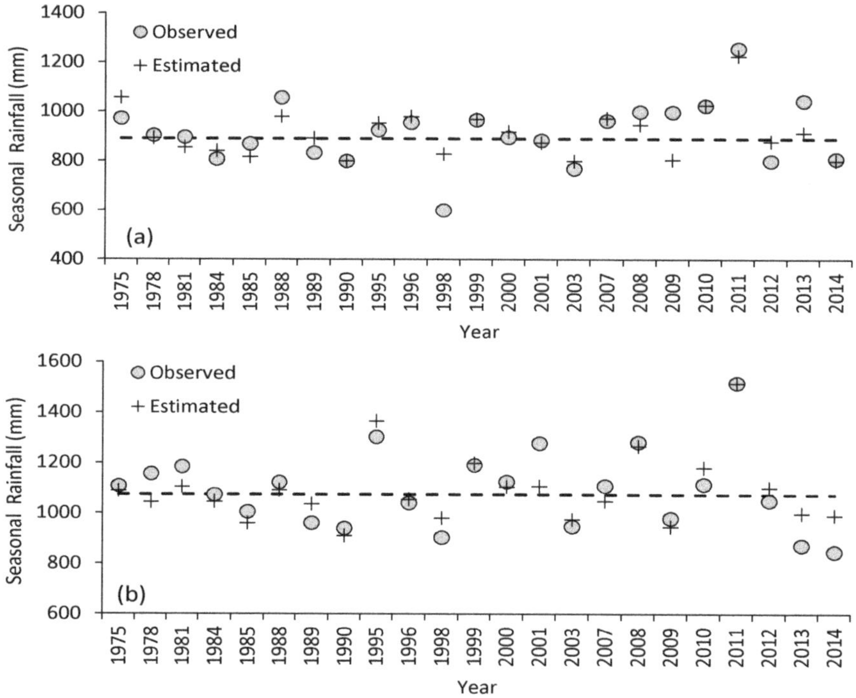

Figure 7. Comparison of observed and estimated seasonal rainfall for (**a**) PRB and (**b**) NRB for years with EQ-SOI > 0.02. Equation (3) with NINO.WEST was applied. The dashed line indicates the mean rainfall for the period from 1975 to 2015.

We validated this method by comparing the observed seasonal rainfall during 1966–1974 (when EQ-SOI is positive) with those estimated using Equation (3), resulting in RMSE of 91.0 and 101.4 mm for PRB and NRB, respectively (Figure 8). These RMSEs are larger than those for the calibration period (during 1975–2014) probably due to the limited number of gauging stations used for APHRODITE products during the period. In addition, we applied multiple regression analysis using Equation (3) to UCPRB (Figure 1), which resulted in satisfactory agreement for both the period of regression analysis between 1975 and 2007 (sample number = 16, $r = 0.86$, RMSE = 44.6 mm) and the period of validation between 1966 and 1974 (sample number = 6, $r = 0.92$, RMSE = 40.2 mm). Table 3 summarizes the results of the proposed Equations (2) and (3) applied to all years between 1975 and 2015. The correlation coefficients were greatly improved from those in previous studies (such as [11]), although the domain and duration differed.

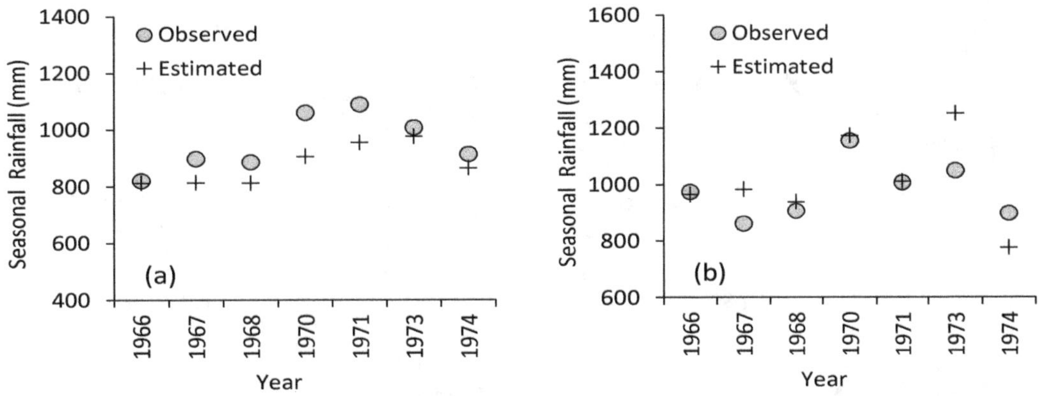

Figure 8. Same as Figure 7 but for validation years between 1966 and 1974 for (**a**) PRB and (**b**) NRB.

Table 3. Correlation coefficient r and RMSE for the period from 1975 to 2015 for years of negative EQ-SOI, positive EQ-SOI and all years.

Basin		EQ-SOI < 0.0	EQ-SOI > 0.0	All Years
PRB	r	0.83	0.79	0.81
	RMSE (mm)	69.0	78.4	74.4
NRB	r	0.82	0.89	0.86
	RMSE (mm)	95.4	72.1	83.1

4. Discussion

The seasonal rainfall data and three climate indices (EQ-SOI, PDO, and SST_{NW}) used in Equation (3) are plotted in Figure 9 to visualize the influence of these indices during years with positive EQ-SOI. Figure 9a shows that EQ-SOI is positively correlated with seasonal rainfall for both PRB and NRB at a significance level less than 5%, while there is no clear difference depending on the PDO phase (Figure 9a). Therefore, it is reasonable to include EQ-SOI as a multiplier of the quadratic function in Equation (3).

In PRB and NRB, the seasonal rainfall during the cool PDO phase (PDO < 0) is 220 mm and 150 mm larger than that in the warm phase (PDO > 0), respectively, under negative SST_{NW} conditions (Figure 9b). However, for years with a positive SST_{NW}, the differences in seasonal rainfall between the negative and positive PDO phases are much smaller for both basins (below 35 mm). There was a negative correlation between PDO and seasonal rainfall in PRB for all plots ($r = -0.42$, $p = 0.02$), while there was no clear correlation for NRB ($r = -0.26$, $p = 0.16$), and seasonal rainfall increases with positive-phase PDO when EQ-SOI exceeds 0.03 (Figure 9b, right). Similar characteristics were observed by Sen Roy and Sen Roy [27], who concluded that monsoon precipitation in eastern Myanmar is positively related to PDO during its warm phase, and negatively correlated during the cold phase. This is considered in the quadratic function.

Seasonal rainfall in both the PRB and NRB is positively correlated with SST_{NW}, with a p-value below 5%, when PDO is positive (Figure 9c), while seasonal rainfall negatively related to SST_{NW} when PDO is negative. A previous study found that the positive (negative) phase of PDO is associated with deficit (excess) rainfall over India [28]. The same phase characteristics were found between annual rainfall in Thailand and an ENSO-related index (MEI) and PDO [29]. Our results indicate that seasonal rainfall cannot be solely related to a single index, and multiple indices (including EQ-SOI) need to be considered when quantifying rainfall.

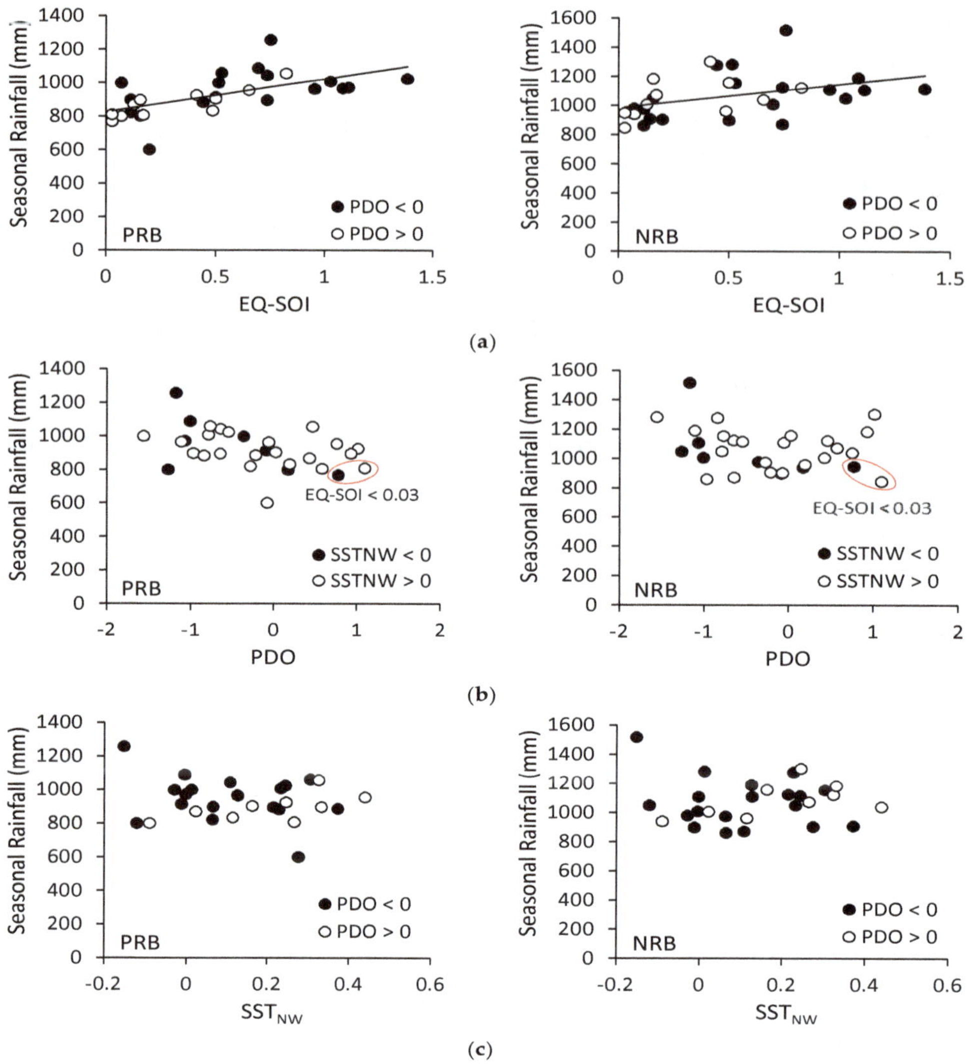

Figure 9. Relationship between seasonal rainfall and (**a**) EQ-SOI; (**b**) PDO and (**c**) SST_{NW} for (left) PRB and (right) NRB between 1965 and 2015 (when EQ-SOI is positive).

To highlight the influences of PDO and SST_{NW}, we define the normalized rainfall anomaly (Equation (7)) by rearranging Equation (3):

$$\frac{\mathrm{TR} - d_0}{\mathrm{EQ} - \mathrm{SOI}} = g_1 \times \mathrm{PDO}^2 + g_2 \times \mathrm{PDO} + g_3 \tag{7}$$

where g_1, g_2, and g_3 are given by Equations (4)–(6) with calibrated parameters. Figure 10 shows the relationships between the normalized rainfall anomaly and PDO and SST_{NW}. Each curve in Figure 10 was drawn using Equation (3), and the data plots were limited for years between 1975 and 2015 with EQ-SOI larger than 0.2 for clear indication. Most of the data plots in each class of SST_{NW} (Figure 10a) or PDO (Figure 10b) are consistent with the corresponding curves with similar SST_{NW} or PDO value, respectively, although there are some outliers.

The lines in Figure 10a indicate that PDO increases the normalized rainfall anomaly when it is largely negative or positive for the same or similar SST_{NW}. Therefore, we confirmed that the sensitivity of Equation (3) is consistent with the recognized influence of PDO, i.e., abundant precipitation tends to occur during La Niña years and the PDO cool phase [29]. Furthermore, the results in Figure 10a suggest that the seasonal rainfall in PRB and NRB can increase with both negative and positive PDO as there was no clear relationship between EQ-SOI and PDO when EQ-SOI is positive, which is consistent with the data plots in Figure 9b (especially for NRB) as well as previous analysis [27].

Unlike PDO, the influence of SST_{NW} on the normalized rainfall anomaly appears to be more complicated (Figure 10b). For NRB (Figure 10b right), SST_{NW} positively affects the normalized rainfall anomaly when PDO is positive and SST_{NW} is within a certain positive range, while the normalized rainfall anomaly is likely to decrease as SST_{NW} becomes larger (Figure 10b). These characteristics may be applied to seasonal rainfall itself as there is no certain correlation between EQ-SOI and SST_{NW} when EQ-SOI is positive. When PDO is negative, local minima are found for the normalized rainfall anomaly, as indicated by the circle plots and corresponding curve (PDO = −1.0) for NRB in Figure 10b. Similar characteristics were found for PRB, although the variation in the normalized rainfall anomaly is smaller than that for NRB as there was less seasonal rainfall with lower variation over NRB. Although the mechanistic reason for these characteristics is unclear, the normalized rainfall anomaly could be amplified when PDO and SST_{NW} are in the same phase. Hoell and Funk [30] reported that, for La Niña events (when EQ-SOI tends to be positive), precipitation is enhanced over the western Pacific and extended to the Indian Ocean during a strong western Pacific SST gradient (WPG), which is defined as the standardized difference between area-averaged SST over the central Pacific Ocean and the domain similar to NINO.WEST. This suggests that SST over NINO.WEST, combined with PDO, could affect precipitation over the Indochina Peninsula through the SST gradient in the western Pacific.

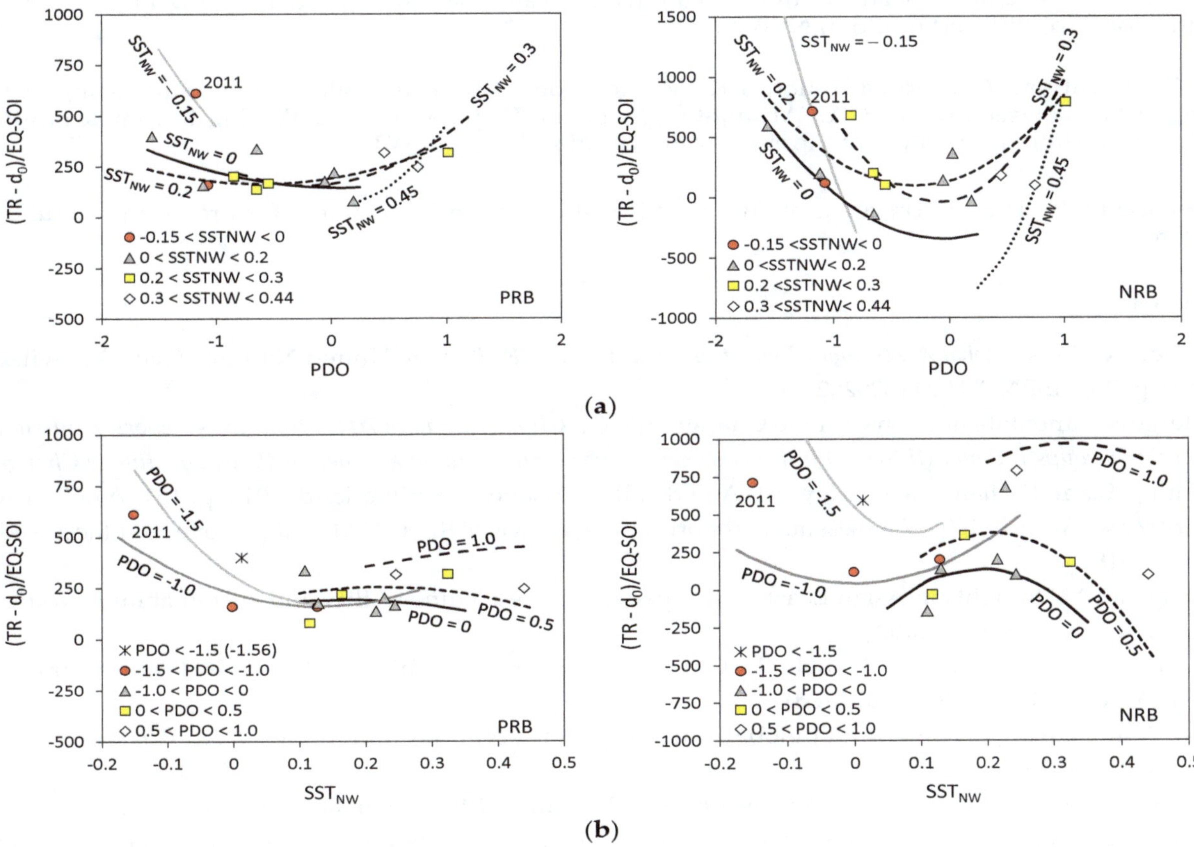

Figure 10. Relationship between seasonal rainfall and (**a**) PDO and (**b**) SST_{NW} for (left) PRB and (right) NRB. The vertical axis indicates the seasonal rainfall anomaly divided by EQ-SOI. The solid and dashed curves in each panel are given by Equation (3). Each symbol represents the data when EQ-SOI is positive.

5. Conclusions

The seasonal rainfall over the 50 years during 1965–2015 over the upper Ping River Basin (PRB) and upper Nan River Basin (NRB) in Thailand were analyzed to elucidate a quantitative relationship with existing climate indices. We showed that, when the Equatorial Southern Oscillation index (EQ-SOI) is negative, seasonal rainfall is predominantly related to the Indian Monsoon index (IMI) using a sigmoid function; however, when EQ-SOI is positive, seasonal rainfall is estimated by the nonlinear function using the El Niño/Southern Oscillation index (SST_{NW}), Pacific Decadal Oscillation index (PDO), and EQ-SOI. The areal rainfall calculated for the period between 1975 and 2015 from the proposed functions exhibits a high correlation with the observed seasonal rainfall, with correlation coefficients of 0.8 and 0.86 for PRB and NRB, respectively.

Although our proposed method exhibited a relatively high performance for estimating seasonal rainfall, further verification and improvement are necessary by updating the rainfall data and climate indices. Furthermore, to apply the method in practical water resource management, such as dam reservoir operation and the assessment of climate change impacts [31], the use of the forecasted or projected results of selected climate indices should be examined to evaluate the accuracy and uncertainty of seasonal rainfall resulting from the proposed method.

Supplementary Materials
Figure S1: Relationship between the areal-mean monthly rainfalls obtained from the APHRODITE dataset (Aphrodite-based) and those calculated by the Thiessen Polygon applied to the rain-gauge data of RID (RID-based) for the period from 1980 to 2000 for (left) PRB and (right) NRB; Figure S2. The number of gauging stations used to calculate the RID-based areal-mean monthly rainfall (red line) and that used to generate the APHRODITE dataset (black line) for (top) PRB and (bottom) NRB.

Author Contributions: Conceptualization, T.K.; Methodology, T.K.; Data collection and processing, A.K., G.Y. and W.L.; Data Analysis, T.K. and G.Y.; Writing-Original Draft Preparation, T.K.; Writing-Review & Editing, T.K. and W.L.; Visualization, T.K. and G.Y.; Project Administration, T.K. and W.L.

Acknowledgments: The authors are grateful to the Royal Irrigation Department for providing useful data for this research.

References

1. United Nations. *Global Assessment Report on Disaster Risk Reduction*; United Nations: Geneva, Switzerland, 2009; p. 207. ISBN 9789211320282.
2. Intergovernmental Panel on Climate Change (IPCC). *Climate Change 2014: Synthesis Report. Contribution of Working Groups I, II and III to the Fifth Assessment Report of the Intergovernmental Panel on Climate Change*; Core Writing Team, Pachauri, R.K., Meyer, L.A., Eds.; IPCC: Geneva, Switzerland, 2014; p. 151. Available online: https://www.ipcc.ch/pdf/assessment-report/ar5/syr/SYR_AR5_FINAL_full_wcover.pdf (accessed on 29 April 2018).
3. Miyan, M.A. Droughts in Asian Least Developed Countries: Vulnerability and sustainability. *Weather Clim. Extr.* **2015**, *7*, 8–23. [CrossRef]
4. Tanoue, M.; Hirabayashi, Y.; Ikeuchi, H. Global-scale river flood vulnerability in the last 50 years. *Sci. Rep.* **2016**, *6*, 36021. [CrossRef] [PubMed]
5. Dai, A. Increasing drought under global warming in observations and models. *Nat. Clim. Chang.* **2013**, *3*, 52–58. [CrossRef]
6. Cai, W.; Borlace, S.; Lengaigne, M.; van Rensch, P.; Collins, M.; Vecchi, G.; Timmermann, A.; Santoso, A.; McPhaden, M.J.; Wu, L.; et al. Increasing frequency of extreme El Niño events due to greenhouse warming. *Nat. Clim. Chang.* **2014**, *4*, 111–116. [CrossRef]
7. Cai, W.; Wang, G.; Santoso, A.; McPhaden, M.J.; Wu, L.; Jin, F.-F.; Timmermann, A.; Collins, M.; Vecchi, G.; Lengaigne, M.; et al. Increased frequency of extreme La Niña events under greenhouse warming. *Nat. Clim. Chang.* **2015**, *5*, 132–137. [CrossRef]
8. Komori, D.; Nakamura, S.; Kiguchi, M.; Nishijima, A.; Yamazaki, D.; Suzuki, S.; Kawasaki, A.; Oki, K.; Oki, T. Characteristics of the 2011 Chao Phraya River flood in central Thailand. *Hydrol. Res. Lett.* **2012**, *6*, 41–46. [CrossRef]

9. Promchote, P.; Wang, S.S.-Y.; Johnson, P.G. The 2011 Great Flood in Thailand: Climate Diagnostics and Implications from Climate Change. *J. Clim.* **2016**, *29*, 367–379. [CrossRef]
10. Ueangsawat, K.; Jintrawet, A. The Impacts of ENSO Phases on the Variation of Rainfall and Stream Flow in the Upper Ping River Basin, Northern Thailand. *Environ. Nat. Resour. J.* **2013**, *11*, 97–119.
11. Tsai, C.; Behera, S.K.; Waseda, T. Indo-China Monsoon Indices. *Sci. Rep.* **2015**, *5*, 8107. [CrossRef] [PubMed]
12. Singhrattna, N.; Rajagopalan, B.; Clark, M.; Krishna Kumar, K. Seasonal forecasting of Thailand summer monsoon rainfall. *Int. J. Climatol.* **2005**, *25*, 649–664. [CrossRef]
13. Divakar, L.; Babel, M.S.; Perret, S.R.; Das Gupta, A. Optimal allocation of bulk water supplies to competing use sectors based on economic criterion—An application to the Chao Phraya River Basin, Thailand. *J. Hydrol.* **2011**, *401*, 22–35. [CrossRef]
14. Takeda, M.; Laphimsing, A.; Putthividhya, A. Dry season water allocation in the Chao Phraya River basin, Thailand. *Int. J. Water Resour. Dev.* **2015**, *32*, 321–338. [CrossRef]
15. Kripalani, R.H.; Singh, S.V.; Panchawagh, N.; Brikshavana, M. Variability of the summer monsoon rainfall over Thailand—Comparison with features over India. *Int. J. Climatol.* **1995**, *15*, 657–672. [CrossRef]
16. Yatagai, A.; Kamiguchi, K.; Arakawa, O.; Hamada, A.; Yasutomi, N.; Kitoh, A. APHRODITE: Constructing a Long-term Daily Gridded Precipitation Dataset for Asia based on a Dense Network of Rain Gauges. *Bull. Am. Meteorol. Soc.* **2012**, *93*, 1401–1415. [CrossRef]
17. Matsumoto, J. Seasonal transition of summer rainy season over Indochina and adjacent monsoon region. *Adv. Atmos. Sci.* **1997**, *14*, 231–245. [CrossRef]
18. Wang, B.; Fan, Z. Choice of South Asian summer monsoon indices. *Bull. Am. Meteorol. Soc.* **1999**, *80*, 629–638. [CrossRef]
19. Mantua, N.J.; Hare, S.R.; Zhang, Y.; Wallace, J.M.; Francis, R.C. A Pacific decadal climate oscillation with impacts on salmon. *Bull. Am. Meteorol. Soc.* **1997**, *78*, 1069–1079. [CrossRef]
20. Saji, N.H.; Goswami, B.N.; Vinayachandran, P.N.; Yamagata, T. A dipole mode in the tropical Indian Ocean. *Nature* **1999**, *401*, 360–362. [CrossRef] [PubMed]
21. NOAA National Weather Service Climate Prediction Center. Available online: http://www.cpc.ncep.noaa.gov/data/indices/ (accessed on 29 April 2018).
22. University of Hawaii, International Pacific Research Center (IPRC). Available online: http://apdrc.soest.hawaii.edu/projects/monsoon/index.html (accessed on 29 April 2018).
23. Japan Meteorological Agency. Available online: http://ds.data.jma.go.jp/tcc/tcc/products/elnino/index/ (accessed on 29 April 2018).
24. NOAA ESRL Physical Sciences Division. Available online: http://www.esrl.noaa.gov/psd/gcos_wgsp/Timeseries/PDO/ (accessed on 29 April 2018).
25. JAMSTEC. Available online: http://www.jamstec.go.jp/frsgc/research/d1/iod/iod/dipole_mode_index.html (accessed on 29 April 2018).
26. Meyers, G.; McIntosh, P.; Pigot, L.; Pook, M. The years of El Niño, La Niña, and interactions with the tropical Indian Ocean. *J. Clim.* **2007**, *20*, 2872–2880. [CrossRef]
27. Sen Roy, S.; Sen Roy, N. Influence of Pacific decadal oscillation and El Niño Southern oscillation on the summer monsoon precipitation in Myanmar. *Int. J. Climatol.* **2011**, *31*, 14–21. [CrossRef]
28. Krishnamurthy, L.; Krishnamurthy, V. Influence of PDO on South Asian summer monsoon and monsoon–ENSO relation. *Clim. Dyn.* **2014**, *42*, 2397–2410. [CrossRef]
29. Limsakul, A.; Singhruck, P. Long-term trends and variability of total and extreme precipitation in Thailand. *Atmos. Res.* **2016**, *169*, 301–317. [CrossRef]
30. Hoell, A.; Funk, C. The ENSO-Related West Pacific Sea Surface Temperature Gradient. *J. Clim.* **2013**, *26*, 9545–9562. [CrossRef]
31. Singhrattna, N.; Babel, M.S. Changes in summer monsoon rainfall in the Upper Chao Phraya River Basin, Thailand. *Clim. Res.* **2011**, *49*, 155–168. [CrossRef]

12

Evaluating Regime Change of Sediment Transport in the Jingjiang River Reach, Yangtze River

Li He [1,*], Dong Chen [1,2], Shiyan Zhang [1], Meng Liu [1,2] and Guanglei Duan [3]

[1] Key Laboratory of Water Cycle and Related Land Surface Processes, Institute of Geographic Sciences and Natural Resources Research, Chinese Academy of Sciences, Beijing 100101, China; dchen@igsnrr.ac.cn (D.C.); zhangshy@igsnrr.ac.cn (S.Z.); lium.17b@igsnrr.ac.cn (M.L.)

[2] College of Resources and Environment, University of Chinese Academy of Sciences, Beijing 100049, China

[3] Jingjiang Hydrology and Water Resources Surveying Bureau, Yangtze River Water Resources Commission, Wuhan 434020, China; jjduangl@cjh.com.cn

* Correspondence: heli@igsnrr.ac.cn

Abstract: The sediment regime in the Jingjiang river reach of the middle Yangtze River has been significantly changed from quasi-equilibrium to unsaturated since the impoundment of the Three Gorges Dam (TGD). Vertical profiles of suspended sediment concentration (SSC) and sediment flux can be adopted to evaluate the sediment regime at the local and reach scale, respectively. However, the connection between the vertical concentration profiles and the hydrologic conditions of the sub-saturated channel has rarely been examined based on field data. Thus, vertical concentration data at three hydrological stations in the reach (Zhicheng, Shashi, and Jianli) are collected. Analyses show that the near-bed concentration (within 10% of water depth from the riverbed) may reach up to 15 times that of the vertical average concentration. By comparing the fractions of the suspended sediment and bed material before and after TGD operation, the geomorphic condition under which the distinct large near-bed concentrations occur have been examined. Based on daily discharge-sediment hydrographs, the reach scale sediment regime and availability of sediment sources are analyzed. In total, remarkable large near-bed concentrations may respond to the combination of wide grading suspended particles and bed material. Finally, several future challenges caused by the anomalous vertical concentration profiles in the unsaturated reach are discussed. This indicates that more detailed measurements or new measuring technologies may help us to provide accurate measurements, while a fractional dispersion equation may help us in describing. The present study aims to gain new insights into regime change of sediment suspension in the river reaches downstream of a very large reservoir.

Keywords: sediment regime; suspended sediment concentration; vertical profiles of concentration; the Jingjiang River Reach; the Yangtze River

---

1. Introduction

Characteristics of the sediment regime are one of the most important factors that control the processes in the transportation and sedimentation zones of a fluvial system [1]. Changes in the sediment regime cause not only varied fluvial evolution processes but also the inner physical processes of sediment suspension. Using a reach scale, the sediment regime of a river reach can be expressed as the suspended sediment load (sediment flux). Regarding the physical processes of suspension on a local scale, they may be related to the vertical profiles of suspended sediment concentration (SSC), which are essential for estimating sediment flux, and also numerical simulation [2]. Thus, a large number of researchers have paid, and are paying, attention to the vertical concentration profiles.

During the early stage of physical or analytical analyses, many studies have investigated steady and equilibrium sediment transport [3]. For equilibrium sediment transport, traditional advection-diffusion equation (ADE) theory and the improvements (including sediment turbulent diffusion coefficient, vertical flow velocity distribution, and suspension index) are widely used [4,5]. For calculating profiles of SSC, numerical models produce good results for uniform sediments in equilibrium situations [6,7]. However, this is not the case for non-uniform sediments in equilibrium situations, not to mention non-uniform sediments in non-equilibrium situations [6,7]. For instance, Nicholas et al. [8] compared observed flow widths with results from a theoretical model developed for non-equilibrium (aggradation) conditions. Analysis results illustrated that models derived for equilibrium conditions may have limited utility in non-equilibrium situations, despite their widespread use to date [8].

In the non-equilibrium condition, the deviations of the vertical profiles of SSC from the local equilibrium state are different from those of the equilibrium state [9,10]. For hydraulic engineers, the vertical concentration profiles are essential for estimating erosion and deposition using the sediment balanced method, and also for estimating sediment-associated contaminants [11,12]. For geomorphologies, non-equilibrium suspended sediment transport (SST) is one of the key processes driving the channel evolution by erosion or deposition [13]. Thus, various approaches have been employed in non-uniform sediments in non-equilibrium sediment transport. However, most of these analyses are based on flume data or theoretical analysis [14,15]. The connection between the hydrologic condition of the non-equilibrium channel and the vertical concentration profiles has rarely been examined based on field data [16].

The regime of sediment transport in the Jingjiang river reach (JJRR, about 102 km downstream of the Three Gorges Dam (TGD)) has been significantly changed from quasi-equilibrium to sub-saturation, as the impoundment of the TGD. More and more researchers are focusing on the changed sediment regime of the sub-saturated channel in the reach scale. However, field observation has revealed frequent occurrences of remarkable large concentrations in the near-bed zone by extending the vertical measuring to the near-bed zone (i.e., with a distance of 0.1 m from the riverbed). Thus, it is also necessary to analyze the changed sediment regime in the local scale. The contribution of the remarkable, large, near-bed concentration on the sediment flux in the unsaturated channel has been analyzed [17]. Here, this paper aims to evaluate the changed sediment regime of the JJRR from both a local-scale view (vertical concentration profiles) and a reach-scale view (sediment flux). It also helps one to understand the connection between the concentration profiles (local-scale) and the hydrologic conditions of the sub-saturated channel.

Firstly, the vertical profiles of SSC are analyzed by their characteristics, hydrodynamic conditions, and geomorphic conditions to exhibit the changed sediment regime at the local scale. Secondly, the temporal variation of sediment flux is analyzed with the cumulative anomaly method to exhibit the changed sediment regime at the reach scale. Thirdly, the relationship between SSC and discharge is analyzed with the method of the sediment rating curve (SRC), which is used to investigate the conditions for the occurrence of the remarkable large near-bed concentration. Finally, the challenges that may be caused by the remarkable, large, near-bed concentration are discussed.

2. Study Reach, Data, and Methodology

2.1. The Study Reach

The length of the Yangtze River (YR) is 6.3×10^3 km, and the drainage area is 1.8×10^6 km^2 (Figure 1). The TGD, located at the end of the upper reach of the YR, controls a drainage area of 1.0×10^6 km^2. The dam is 185 m high, and storage capacity of the reservoir is 3.93×10^4 Hm^3. The main purposes of the project are flood control, power generation, and navigation.

The JJRR (between Zhicheng and Chenglingji stations) is about 60 km downstream from the Yichang hydrological station. There are five hydrological stations in the JJRR: Zhicheng, Shashi (Jing 45), Xinchang (Jing 84), Jianli, and Chenglingji (Figure 1b). Moreover, there are three distributary channels on the reach through which YR delivers water and sediment into Dongting Lake, Songzikou, Taipingkou, and Ouchikou. The river reach can be divided by Ouchikou into the upper and lower sub-reaches (Figure 1b). The lengths of these two sub-reaches are 172 km and 176 km, respectively.

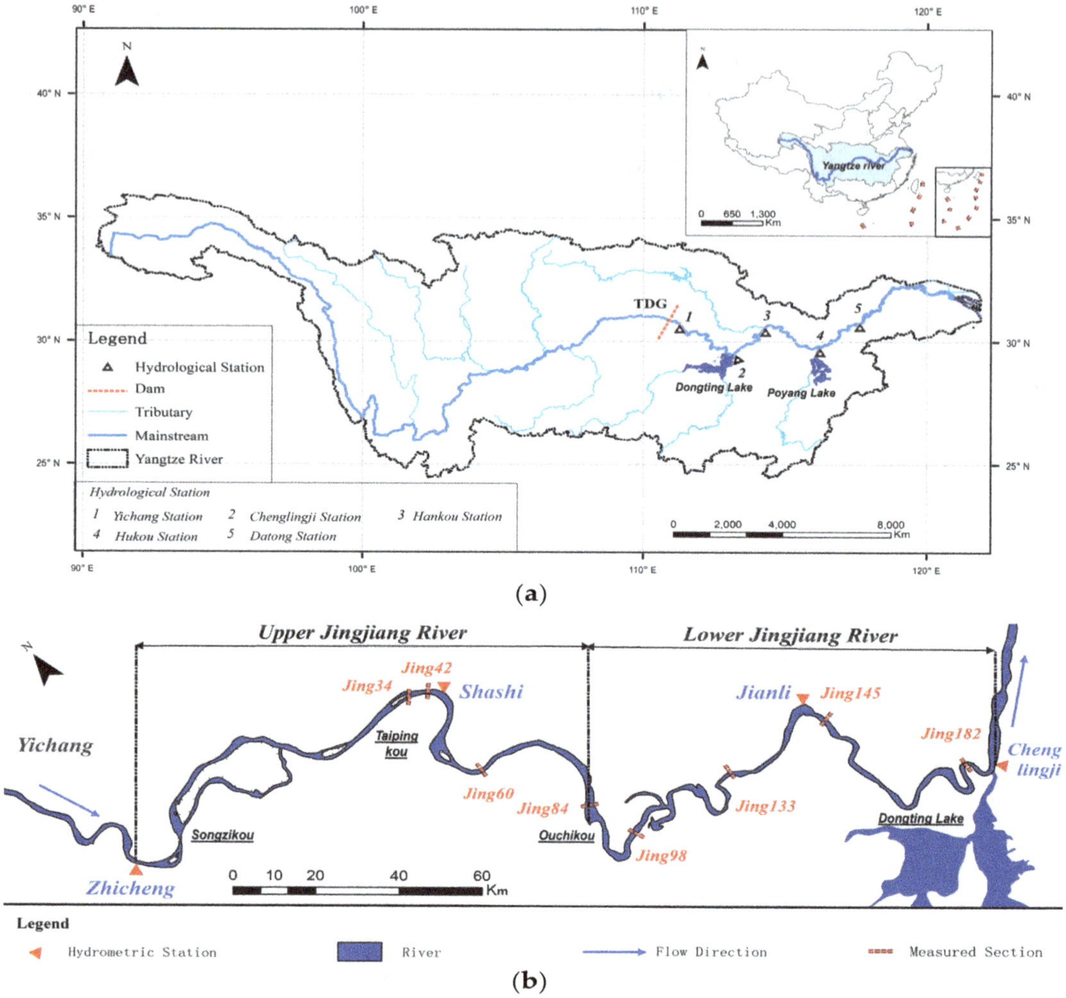

Figure 1. Sketch of the YR (**a**) and JJRR (**b**); (**b**) is drawn from [17].

2.2. Data

Both daily data and field surveyed data were used in the present study.

Daily discharge and suspended sediment load (SSL) are measured and published by the Yangtze River Water Resources Commission (YRWRC). Data at three stations (Zhicheng, Shashi, and Jianli) in 1987–2014 were obtained from yearbooks. Distances of these three stations downstream from the Yichang station are 60 km, 157 km, and 300 km, respectively.

Field surveyed data were provided by the Jingjiang Hydrology and Water Resources Surveying Bureau (JHWRSB), YRWRC. Data include vertical concentration, sediment gradation, and corresponding velocities. Table 1 manifests the years of measurement and the numbers of vertical profiles. Profiles measured before and after 2010 are different. Before 2010, there are five measuring points in each vertical profile. The normalized depths are approximately 1.0H, 0.8H, 0.4H, 0.2H, and 0.1H (H is the water depth of the vertical line, in meters), respectively. After 2010, there are seven measuring points in each vertical profile. The normalized depths of the upper five points are approximately 1.0H, 0.8H, 0.4H, 0.2H, and 0.1H, respectively. The other two points were measured in the near-bed region with constant distances from the riverbed of 0.5 m and 0.1 m.

For profiles measured at Shashi station, measured flow depths range from 2.09 m to 19.7 m, and the normalized depths of these two near-bed points are approximately (0.02–0.2)H. For profiles measured at Jianli station, measured flow depths range from 2.13 m to 21.2 m, and the normalized depths of these two near-bed points are approximately (0.0047–0.047)H.

Table 1. Statistics of data measured at the JJRR (following [17]).

Stations	Observation Year	No. of Verticals
Zhicheng	1996, 1998, 2002	55
Shashi	1996, 1998, 2010, 2011, 2012, 2013	130
Jianli	1986, 1998, 2002, 2010, 2011, 2012, 2013	180

Equipment adopted in the field survey includes global navigation satellite system (GNSS, antenna, and receiver), total station, vessel-mounted acoustic Doppler current profilers (ADCP), digital level, GPS, and laser particle size analyzer (LPSA) [17]. All of this equipment has been strictly examined by professional surveyors. According to the technical manual by JHWRSB, several technical standards and criteria are illustrated, as follows:

(1) Water depth is measured with ADCP, and verified with fish lead.
(2) ADCP is applied to measure the flow twice. The deviation between each measured discharge and the average discharge should be less than $\pm 5\%$. Otherwise, the data should be re-measured.
(3) Vertical lines for measuring the velocity are positioned with a real-time kinematic (RTK) GNSS.
(4) Suspended sediments are sampled with samplers in the field, and its grain-size distributions are analyzed by the sieving method. All the sampling procedures should be finished within dozens of minutes, and sieving analyses should be finished following the manual operation. For all the five points measured before 2010 and the upper five points measured after 2010, the sampler shown in Figure 2a is adopted. For the two near-bed points measured after 2010, bottom-touched automatic-closing samplers shown in Figure 2b,c are adopted. With this kind of sampler (Figure 2b,c), the distances of the two near-bed points from the riverbed can be ensured. However, the potential disturbance on the riverbed when settling the sampler may still be questionable. Therefore, except a slowed down settling velocity of the sampler, the near-bed concentrations may be double-checked after sampling and grain-size analyses. The assumption is that the measured concentration is reliable only if the vertical concentration profiles of particles finer than 0.062 mm have no abrupt changing point in the near-bed region [17]. Thus, bed materials can be distinguished from suspended sediments, which may be caused by disturbance during the sampling processes.

(a)

Figure 2. *Cont.*

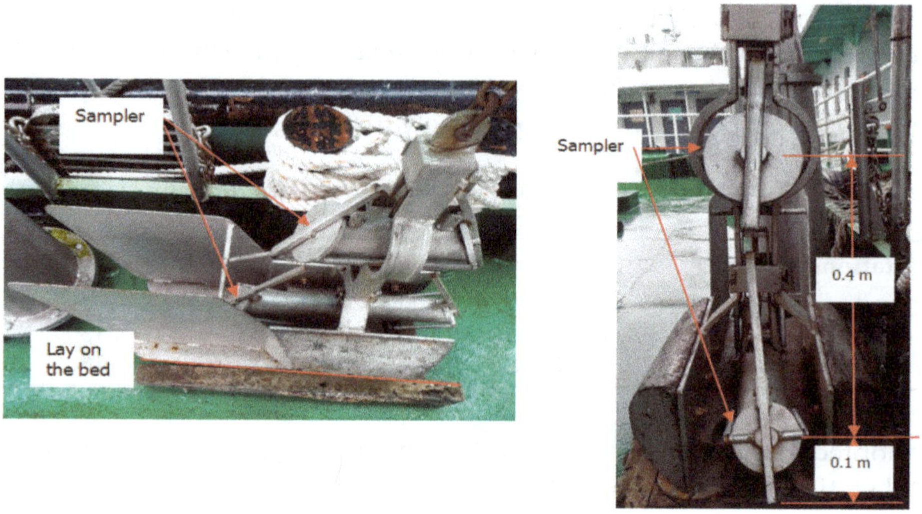

(b) (c)

Figure 2. Samplers for suspended sediment in the field survey; (**a**) is the sampler with fish lead, and (**b**,**c**) are the samplers for sampling suspended sediment at the two near-bed points (different camera angles).

2.3. Methodology

Methods of cumulative anomaly and SRCs are applied to investigate the changed sediment regime of un-saturated channel in reach scales. ADEs and the sediment-balance method are adopted to evaluate new challenges for the non-saturated channel.

2.3.1. Cumulative Anomaly

The cumulative anomaly has been widely applied to detect the trends and stages of time series [18]. For time series, the cumulative anomaly (X_t) for data point x_i can be expressed as:

$$X_t = \sum_{i=1}^{t}(x_i - x_m) \qquad 1 \leq t \leq n \qquad x_m = \frac{1}{n}\sum_{i=1}^{n} x_i \tag{1}$$

in which x_i is the value of the time series; x_m is the mean value of the series x_i; and n is the length of the time series.

The mass curve of cumulative anomaly values describes the change process of a particular parameter by comparing it with the mean value of that parameter over the whole study period [18]. For instance, if the observed data of a given year is greater than the overall mean, the anomaly values will be positive and the mass curve will rise.

2.3.2. SRCs

The SRC is defined as the statistical relationship between SSC (kg·m^{-3}) and discharge. SRCs are normally used to describe the flow-sediment relationships in river systems for various purposes [19]. The method is often applied to estimate the SSC or SSL in ungauged regions [20].

The power function of SRC is usually expressed as one of the following two formats:

$$\mathrm{SSC} = eQ^f \tag{2}$$

$$\log(\mathrm{SSC}) = \log(e) + \mathrm{f}\log(Q) \tag{3}$$

in which Q is the discharge (m^3·s^{-1}), and e and f are the sediment rating coefficient and exponent, respectively. Linear regression, i.e., Equation (3), is often used to derive the values of the rating coefficient e and exponent f from empirical data (e.g., [21]). The e-parameter contains information on the conversion of Q into SSC, namely, the erosion severity index [22]. A high value of the e-parameter indicates easily-erodible materials and high loads of transported materials [20]. The exponent f-coefficient corresponds to the erosive power of the river and the influence on the sediment supplement. High values of the f-coefficient indicate a slight increase in discharge would significantly enhance the erosive power of the river [20]. Moreover, the f-coefficient is also related to climate, channel morphology, the grain-size distribution of sediments, and erodibility within the river basin [22].

2.3.3. ADEs

Three equations are adopted to evaluate the challenge in simulating the vertical concentration profiles in the unsaturated channel: the Rouse Equation, the Han Equation, and the fADE. These three equations may represent typical concentration profiles in equilibrium transport, non-equilibrium transport, and non-Fickion suspension in the non-equilibrium condition, respectively.

By assuming Fick's first law for sediment diffusion in turbulence, the diffusion theory of sediment suspension can be expressed with traditional ADE model. The equation for equilibrium sediment transport can be expressed as:

$$\omega S + \varepsilon_{sy} \frac{\partial S}{\partial y} = 0 \tag{4}$$

in which S is the sediment concentration (W·L^{-3}), y is the vertical coordinate (L), ω is the sediment settling velocity (L·T^{-1}), and ε_{sy} is the sediment turbulent diffusion coefficient in the y direction (L·T^{-1}).

By replacing ε_{sy} with the fluid eddy viscosity and assuming $\varepsilon_m = \kappa u_*(1 - y/H)y$ based on the Karman-Prandtl logarithmic velocity profile, Rouse et al. [4] obtained the analytical vertical concentration profile:

$$\frac{S}{S_a} = \left[\frac{\frac{H}{y} - 1}{\frac{H}{a} - 1}\right]^{\frac{\omega}{\kappa u_*}} \tag{5}$$

in which S_a is a reference concentration at a given height above the river bed (W·L^{-3}), α is the reference height (L), H is the flow depth (L), κ is von Karman's constant, and u_* is the shear velocity (L·T^{-1}). The frictional flow velocity can be calculated as $u_* = \sqrt{\tau_b/\rho}$, in which τ_b is the near-bed shear stress and ρ is the water density. The Rouse equation, Equation (5), has been widely used for decades. However, its drawbacks are obvious: the concentration is calculated as zero at the water surface and infinity at the river bed. These drawbacks are caused by implicit assumptions, equilibrium sediment transport, and Fick's first law. Extensive efforts have been put forward to improve the equation [10,23–25].

The assumption of equilibrium sediment transport limits its application in the field in which the sediment regime is far from equilibrium [26]. Among others, Han et al. [9] have modified Equation (4) to meet the non-equilibrium condition:

$$\omega S + \varepsilon_{sy} \frac{\partial S}{\partial y} = q_s \tag{6}$$

in which q_s is the net flux due to the imbalance between downward sediment settling and upward turbulent dispersion. Han et al. [9] obtained the solution of Equation (6):

$$\frac{S}{S_a} = \left[\frac{\frac{H}{y} - 1}{\frac{H}{a} - 1}\right]^{\frac{c\omega}{\kappa u_*}} \tag{7}$$

in which c is non-equilibrium coefficient defined as a function of the degree of saturation $c = 1 - f\left(\frac{\bar{S}}{\bar{S}_*}\right)$. $\bar{S}$ is the depth-averaged sediment concentration (kg·m^{-3}), and $\bar{S}_*$ is the depth-averaged sediment capacity. If the saturation degree $\frac{\bar{S}}{\bar{S}_*} = 1$, $c = 1$; if $\frac{\bar{S}}{\bar{S}_*} < 1$, $c > 1$, which means the flow is unsaturated with sediment; and if $\frac{\bar{S}}{S_*} > 1$, $c < 1$, which indicates an super-saturated flow. The von Karman's constant κ is usually given a value of 0.408 for an unstratified flow.

According to Fick's First Law, particles may make "local" jumps, as most particle jumps induced by turbulence are constrained to a small distance (Δy) in a given Δt according to the Central Limit Theorem (CLT), and Δy is characterized by the length of Representative Elementary Volume (REV) [27,28]. However, modern observations of coherent structures in turbulence have proven that high-speed currents may intermittently sweep the bed, carry much sediment, and eject directly into the upper part of the water column [27,29]. The non-Fickion suspension indicates that particles may make many "nonlocal" jumps in burst-like suspension events [27]. The non-Fickion suspension in non-equilibrium sediment suspension can be described by a fractional advection-dispersion equation (fADE) [27]:

$$\omega S + \bar{\varepsilon}_{sy}\frac{\partial^a S}{\partial^a y} = q_s \tag{8}$$

in which α is the order of the fractional derivative ($0 < \alpha \leq 1$); $\bar{\varepsilon}_{sy}$ ($L^{\alpha}\cdot T^{-1}$) is the depth-averaged diffusivity expressed as follows:

$$\bar{\varepsilon}_{sy} = \frac{\int_{0.01H}^{H}\kappa u_*(1 - y/h)ydy}{0.99H} = 0.175\kappa u_* H \tag{9}$$

We assume:

$$q_s = -c_0 s \tag{10}$$

in which c_0 is called the non-equilibrium coefficient, and c_0 is a function of (S/S_*):

$$c_0 = f(S - S_*) \tag{11}$$

in which S_* is the sediment transport capacity of flow; if $S = S_*$, then $c_0 = q_s = 0$, which means the flow reaches equilibrium; if $S < S_*$, then $c_0 > 0$ and $q_s < 0$, which means the flow is unsaturated with sediment; and if $S > S_*$, then $c_0 < 0$ and $q_s > 0$, which means flow is supper-saturated with sediment.

The analytical solution of Equation (8) is obtained:

$$\frac{S}{S_a} = E_a\left[-\frac{\omega + c_0}{\bar{\varepsilon}_{sy}}(y - a)^{\alpha}\right] \tag{12}$$

in which $E_a()$ is called Mittag-Leffler function ($y \leq a \leq H$), which can be calculated by series expansion or the MATLAB open-source code.

2.3.4. Contribution of Near-Bed Concentrations on SST Rate

The SST rate can be estimated with measured vertical concentration profiles, and the deposition/erosion amount of the channel can be estimated by the sediment balance method. Changed vertical concentration profiles in the unsaturated channel lead to new challenges in field surveying and estimating the SST rate. The field survey should be conducted according to its new characteristic (as extended measuring of near-bed zone by JHWRSB). The effect of the changed vertical concentration profiles on the SST rate can be estimated by evaluating the contribution of two near-bed concentrations on the SST rate:

$$R = \frac{Q_{s(7)} - Q_{s(5)}}{Q_{s(5)}} \times 100\% \tag{13}$$

in which $Q_{s(7)}$ and $Q_{s(5)}$ are cross-sectional SST rates by seven points and, artificially, by five points, respectively. The cross-sectional sediment flux can be estimated with all measured vertical profiles. The delta-shaped area between the left bank and the first vertical line from the left bank is also considered [17]. The delta-shaped area between the right bank and the last vertical line from the left bank is not considered due to the difficulty of identifying the right bank [17].

3. Change of the Sediment Regime

Vertical concentration profiles and SST rates are analyzed to evaluate the sediment regime of local and reach scales, respectively. Then, sediment rating curves are analyzed to investigate the availability of sediment for suspension from the riverbed.

3.1. Vertical Profiles of SSC

3.1.1. Remarkable Large Concentrations in the Near-Bed Zone

e depth-averaged concentration S_{avg} (kg·m^{-3}) is calculated as $S_{avg} = \frac{1}{H}\int S_i dy$, in which S_i is the measured concentration at each point in a vertical line (kg·m^{-3}) and y is the vertical axis of the Cartesian coordinate system. The coordinate origin is located on the riverbed. To present the deviation of the near-bed concentration from the average, the relative concentration is defined as S_a/S_{avg}, in which S_a is the measured concentration at the reference height. The reference height is the minimum distance of measuring points from the river bed. The reference height a equals 0.4–0.5 m for profiles measured before 2010, while the value is 0.1 m for profiles after 2010.

Figure 3 shows the typical vertical profiles measured at Shashi station before and after the impoundment of TGD. The relative concentrations (S_a/S_{avg}) of these four vertical profiles are 2.47, 4.1, 2.10, and 1.52, respectively. Thus, remarkable large concentrations can be observed in the near-bed zone. The remarkable large near-bed concentration is defined by comparison to the vertical average concentration. This can be observed in vertical concentration profiles with large or small measured near-bed concentrations (3.01 kg·m^{-3} for Figure 3a and 0.041 kg·m^{-3} for Figure 3b).

The statistic characteristics of hundreds of vertical profiles are summarized in Table 2. Before TGD operation, the maximum relative concentrations (S_a/S_{avg}) are 2.76 (Zhicheng), 2.30 (Shashi), and 3.62 (Jianli), respectively. After TGD operation, the maximum relative concentrations (S_a/S_{avg}) are 15.05 (Shashi), and 14.45 (Jianli), respectively. The average relative concentrations before TGD are 1.35 (Shashi) and 1.57 (Jianli), respectively. These values jump to 3.22 (Shashi) and 4.89 (Jianli) after TGD, respectively. Thus, both the average and maximum values of relative concentrations (S_a/S_{avg}) increase significantly after the impoundment of the TGD.

Relatively high concentrations near the bed region have also been pronounced in the estuary of YR, Qiantangjiang River, and the Taizhou sea area [30–33]. The near-bed concentration of the S-type vertical distribution is also relatively large [34]. However, the relative concentrations are relatively small (shown in Table 2). The maximum relative concentration (S_a/S_{avg}) is approximately 4.3, and the minimum relative height of measured points from river bed is approximately 0.02. In total, the tailing phenomena in the unsaturated channel are much more apparent.

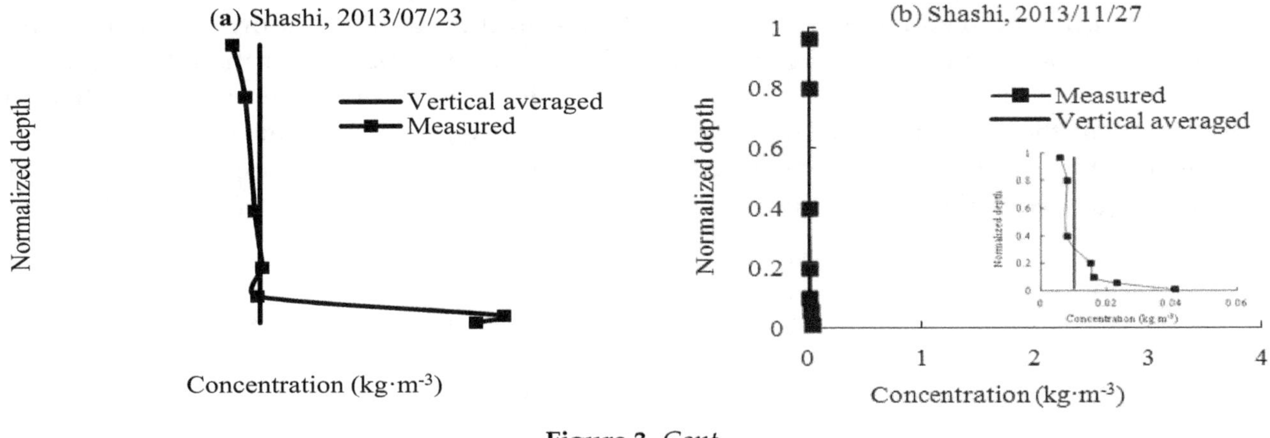

Figure 3. *Cont.*

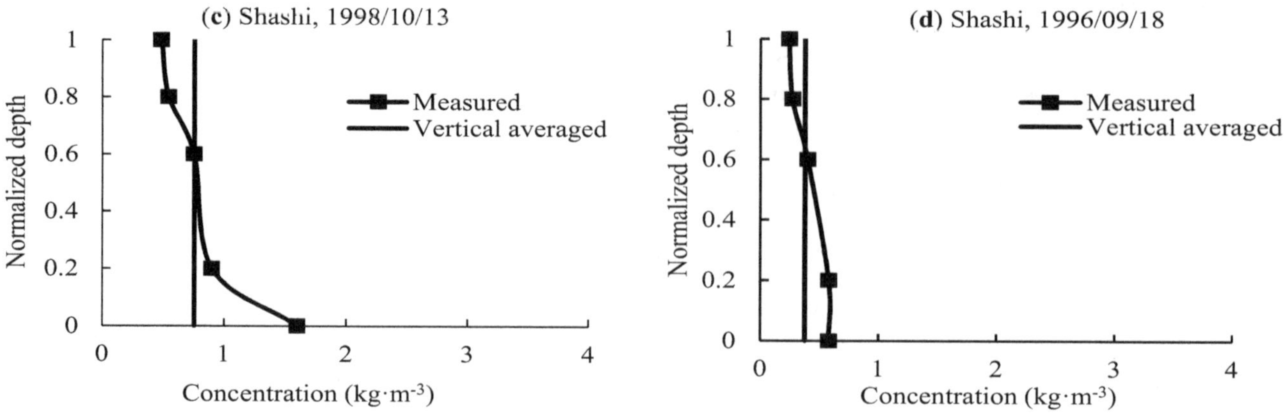

Figure 3. Typical measured vertical profiles of SSC and vertical averaged concentration in the JJRR, measured at Shashi hydrological station before (**c**,**d**) and after 2010 (**a**,**b**).

The coarsening of suspended sediment can be revealed by mean diameters of vertical profiles (shown in Table 2). For vertical lines measured at Jianli station, the average mean diameters after TGD range from 0.0087 mm to about 0.22 mm, while the values before TGD may range from 0.0037 mm to 0.07 mm. For vertical lines measured at Shashi station, the average mean diameters after TGD change from 0.0058–0.027 mm to 0.0098–0.224 mm. Thus, the coarsening process of suspended sediment after TGD operation is apparent, especially at Shashi station.

Both of the near-bed SSC and sediment flux are connected with hydrodynamic conditions [35]. Thus, hydrodynamic parameters (depth, velocity, and vertical average concentration) are also summarized in Table 2. The values of ω/u_* vary dramatically (shown in Table 2). Remarkable large values of ω/u_* indicate that relatively coarser particles may be suspended with a small bed shear velocity. Tang et al. [32] pointed out that the near-bed concentration may be more apparent when $\omega/\kappa u_*) > 5$. Therefore, larger values of $\log(\omega/u_*)$ in non-equilibrium channels contribute to remarkable large concentrations in the near-bed zone.

Normally, concentration at the water surface should be smaller than that at the riverbed. This means that the concentration gradient (estimated with $d(S_i - S_{i-1})/d(y_i - y_{i-1})$) should be negative. However, the occurrence of vertical concentration profiles with positive gradients increases (Table 3). For data measured before TGD, about 20% (Zhicheng, Shashi, and Jianli stations) vertical concentration profiles are characterized with positive gradients. For data measured after TGD, about 28–47% (Shashi and Jianli) of vertical concentration profiles are characterized with positive gradients (Table 3). The differential equation for vertical SSC distribution of non-uniform particles indicates that the concentration of finer particles group near the surface may be greater than that near the bed with wide grading non-uniformed suspended sediment [36]. Moreover, concentration gradients may be positive when the diameter of coarser particles is about eight times that of finer particles [36]. This indicates wide grading sediment particles in the non-equilibrium channel contribute to the remarkable large near-bed concentration.

Table 2. Statistical characteristics of vertical profiles.

River Reach or Hydrological Station		S_a/S_{avg} Maximum Value	S_a/S_{avg} Average Value	Minimum Relative Depth (Reference Height a)	Water Depth (m)	(ω/u_*)	Mean Diameter (mm)	Vertical Averaged Concentration ($kg \cdot m^{-3}$)	Velocity ($m \cdot s^{-1}$)	Source
Post-reservoir	Shashi	15.05	3.22	0.0051 (0.1 m from river bed)	2.09 ∣ 19.70	0.16 ∣ 108.89	0.0098 ∣ 0.224	0.0052 ∣ 1.29	0.42 ∣ 2.12	Field survey
	Jianli	14.45	4.89	0.0047 (0.1 m from river bed)	2.13 ∣ 21.2	0.13 ∣ 76.15	0.0087 ∣ 0.220	0.015 ∣ 1.26	0.27 ∣ 2.69	
Pre-reservoir	Zhicheng	2.76	1.33	0.04	1.69 ∣ 52.88	∣	0.0052 ∣ 0.0134	0.55 ∣ 1.76	∣	
	Shashi	2.30	1.35	0.018 (0.4 m from river bed)	4.3 ∣ 22.6	0.08 ∣ 1.34	0.0058 ∣ 0.0265	0.188 ∣ 1.99	0.08 ∣ 1.34	
	Jianli	3.62	1.57	0.019 (0.4 m from river bed)	3 ∣ 21.3	0.024 ∣ 12.12	0.0037 ∣ 0.0699	0.26 ∣ 1.71	0.17 ∣ 2.62	
Changjiang Estuary		4.33		0.044	∣	∣	0.009	0.11 ∣ 0.83	0.24 ∣ 2.04	[30]
Oujiangkou and Taizhou sea area		2.38		0.02	∣	∣	∣	∣	∣	[31]
Wuhan and Jiujiang River Reach		1.32		0.19	10 ∣ 25	0.007 ∣ 0.036	0.003 ∣ 0.01	0.01 ∣ 0.1	∣	[32]
Silty Coast		2.15		0.045 (0.012 m from river bed)	0.14 ∣ 1.95	∣	0.12 ∣ 0.33	1.50 ∣ 5.30	∣	[33]

Table 3. Characteristics of vertical profiles with positive and negative gradations at Shashi and Jianli hydrological stations.

Stations	Post-Reservoir				Pre-Reservoir	
	Negative		Positive		Negative	Positive
	No. of Verticals	D_{50} (mm)	No. of Verticals	D_{50} (mm)	No. of Verticals	No. of Verticals
Shashi	50	0.01–0.06	45	0.03–0.07	29	7
Jianli	76	0.06–0.10	29	0.06–0.10	60	15

3.1.2. Coarsening of Suspended and Bed Materials

Relatively large concentrations in the near-bed zone in estuary areas can be observed during tide periods and non-flood seasons [30]. Thus, the hydrologic condition (suspended sediment and bed materials) under which the remarkable large concentrations in the near-bed zone may occur is also analyzed.

With the operation of TGD, both the suspended and bed materials become much coarser, except for the dramatic decrease of the annual suspended sediment load (SSL) [37]. The temporal variation of suspended sediment size in the downstream channel may exhibit some complex processes [38]. This means that the saturation recovery process of fine sediments that responds to changing hydraulic conditions is different from that of the coarser one. Additionally, under non-uniform and non-equilibrium conditions, suspended sediment dynamics require a relatively large distance or time to approach equilibrium [13]. Thus, the variation trend of non-uniform sediment varies in different river reaches.

The mean diameter of SSC measured at Yichang station decreases from 0.009 mm (before impoundment) to 0.005 mm (2003–2008) [37]. Size distributions of suspended sediment and bed materials in the JJRR are shown in Figures 4 and 5. For Zhicheng station, the proportion of grouped particles with diameters of 0.008–0.062 mm increases, while the proportions of finer (<0.008 mm) and coarser (0.062–0.5 mm) groups decreases. For Shashi station, the non-uniform sediment becomes a kind of wide grading (as shown in Figure 4b), as the proportion of grouped particles with diameters finer than 0.008 mm decreases and the proportion of grouped particles with diameters coarser than 0.062 mm increases. The mean diameter measured at Jianli station decreases from 0.009 mm (before impoundment) to 0.045 mm (2003–2008) [37]. This indicates that the coarser particles (>0.062 mm) around Shashi station have recovered by bed erosion. When flow transports the reach around Jianli station, the change of coarser particles (>0.062 mm) is much more apparent. Thus, wide grading suspended sediment at Shashi and Jianli stations may contribute to the remarkable large concentration in the near-bed zone.

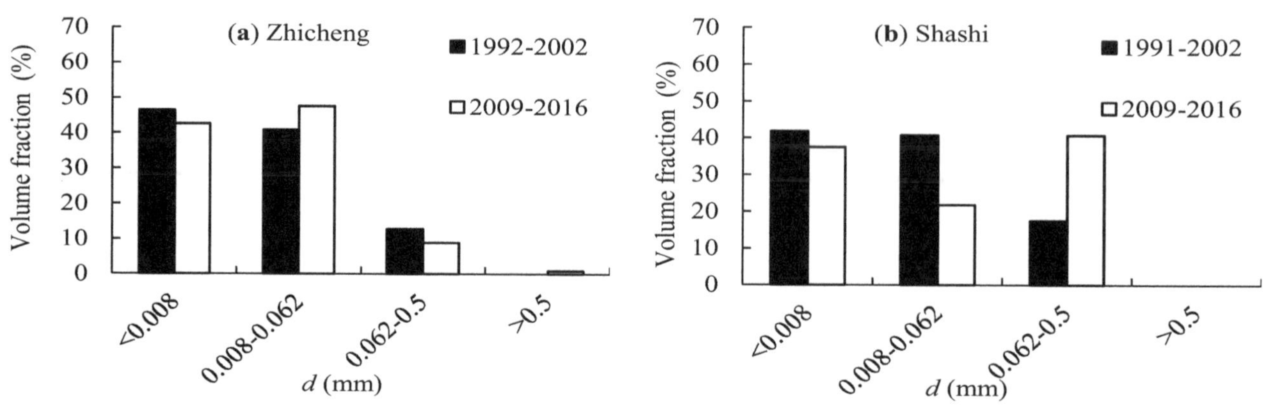

Figure 4. *Cont.*

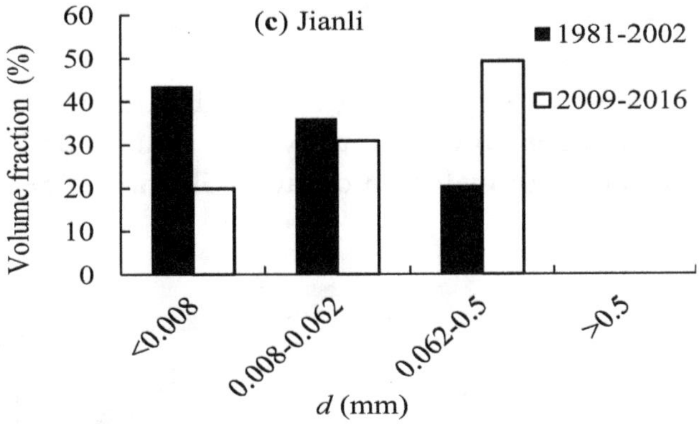

Figure 4. Volume fractions of suspended sediment measured at Zhicheng (**a**), Shashi (**b**), and Jianli (**c**) stations.

The variations of volume fractions of bed materials indicate a remarkable coarsening process in the JJRR, especially the river reach of Zhicheng–Shashi (Figure 5). Zhang [30] pointed out that concentrations in the near-bed zone are more apparent with larger values of relative particle size (D_i/D_m, particle-size over averaged particle size) of non-uniform sediment. Thus, the lifting and suspension of wide grading bed materials may also contribute to the remarkable large concentration in the near-bed region of the non-equilibrium channel.

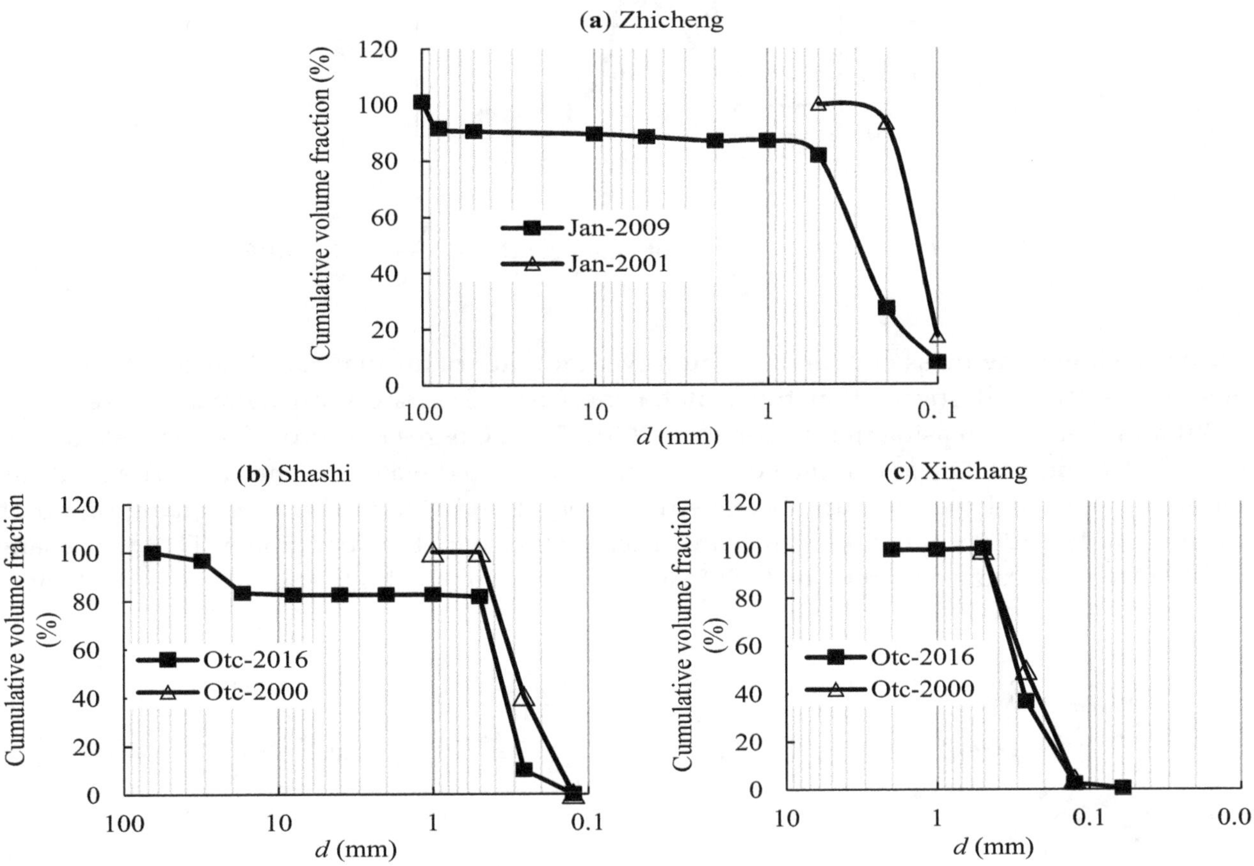

Figure 5. Cumulative volume fractions of bed materials measured at Zhichegn (**a**), Shashi (**b**), and Xinchang (**c**) stations.

3.2. Temporal Change of SSL

At the reach scale, the changed sediment regime can be described with temporal and spatial variations of SSL. The temporal and spatial variations of water runoff have also been analyzed to make a comparison. The temporal variation of water runoff and SSL show a decreasing trend, especially the SSL (Figure 6).

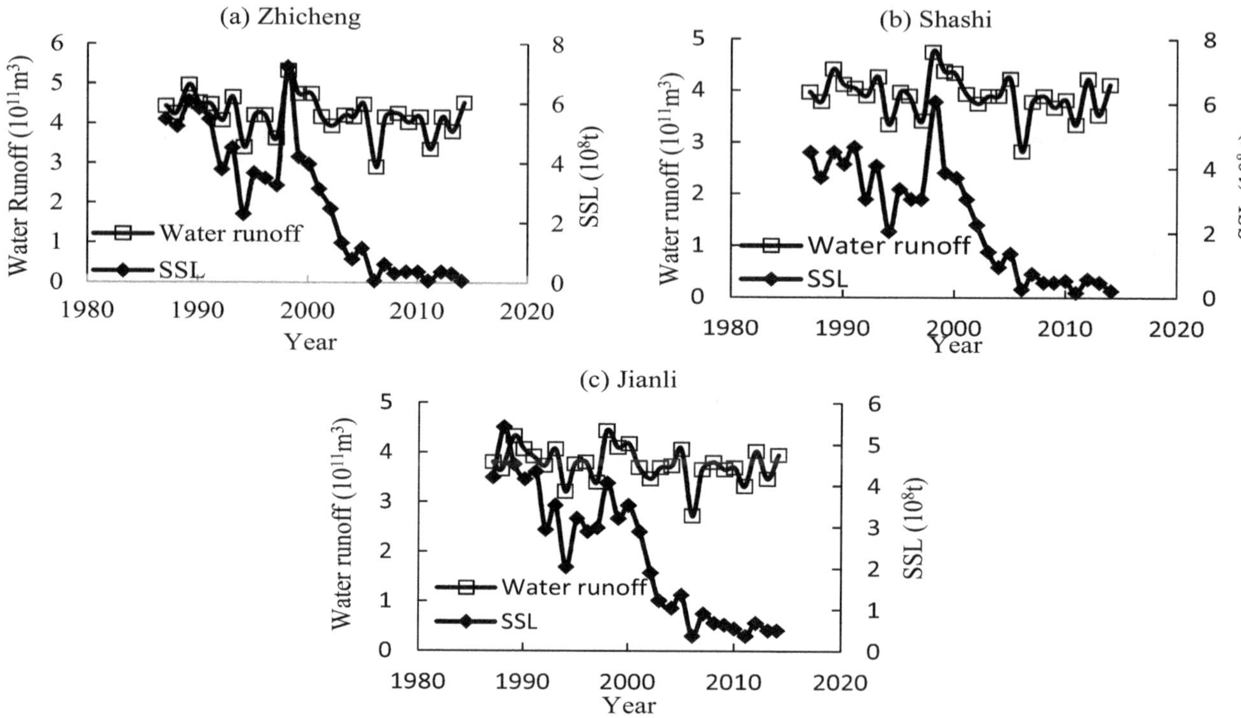

Figure 6. Temporal variation of water runoff and SSL in the JJRR.

Figure 7 shows the mass curves of cumulative anomaly of annual runoff and SSL at the three stations (1987–2014). It shows that the water runoff and SSL show an increasing trend before 2000–2001 and then a decreasing trend after 2000–2001. Three Gorges Reservoir (TGR) was launched to construct in 1994, and TGD was completed and firstly impounded water in June 2003. The TGR started to operate regularly in June, 2006 and became fully operation in 2008. The other changing points of water runoff are 1993, 2005, and 2010, which are in accordance with the operation of TDR. The changing point of 1997 may be caused by the great 1998 flood, which was the highest in recorded history.

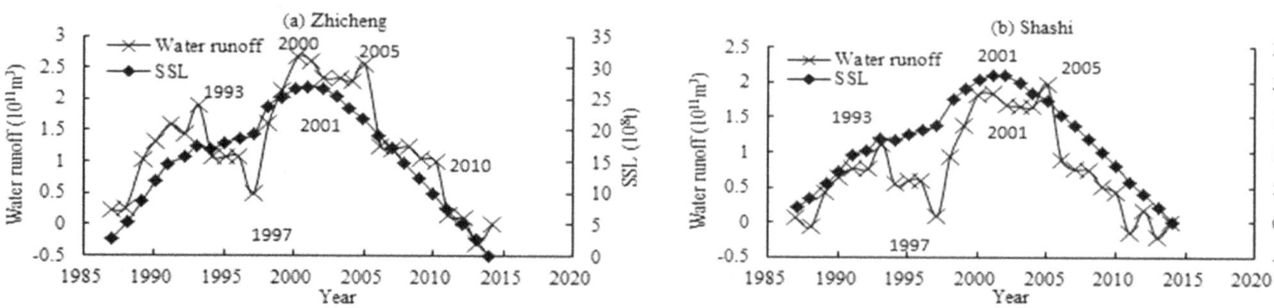

Figure 7. *Cont.*

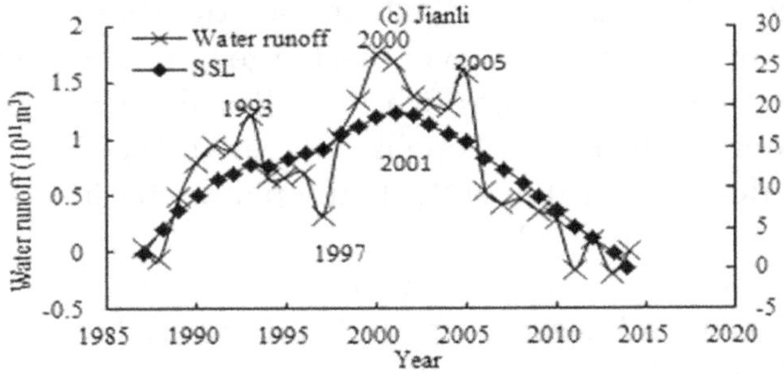

Figure 7. Mass curves of anomalies of runoff and SSL at Zhicheng (**a**), Shashi (**b**), and Jianli (**c**) hydrological stations (1987–2014).

Thus, after 2005, both the runoff and SSL are less than the long-term average value.

3.3. Availability of Sediment Sources

The availability of weathered sediments in the unsaturated channel can be revealed by rating parameters. Thus, the SSC–Q rating curves with data measured before and after TGD operation at the JJRR is adopted to investigate the hydrological condition.

According to data measured in 2013, values of the f-parameter are 2.27 (Shashi) and 1.67 (Jianli), respectively, and values of the log(e)-parameter are −7.70 (Shashi) and −10.70 (Jianli), respectively (Figure 8). The long-term average values by Yang et al. [39] are illustrated as data before dam operation (Figure 8). The long-term average values of the log(e)-parameter and the f-parameter for Xinchang station are −1.92 and 1.17, respectively, and the values for Jianli station are −0.97 and 0.96, respectively [39]. Historical data adopted in the analysis of Yang et al. [39] were measured at four stations: Yichang (1950–1985), Xinchang (1956–1984), Jianli (1954–1986), and Chenglingji (1951–1982). According to Figures 6 and 7, parameters by Yang et al. [39] may represent the hydrological conditions during the period before the changing point. Thus, with the operation of TGD, the values of the f-parameter increase while the values of the log(e)-parameter decrease.

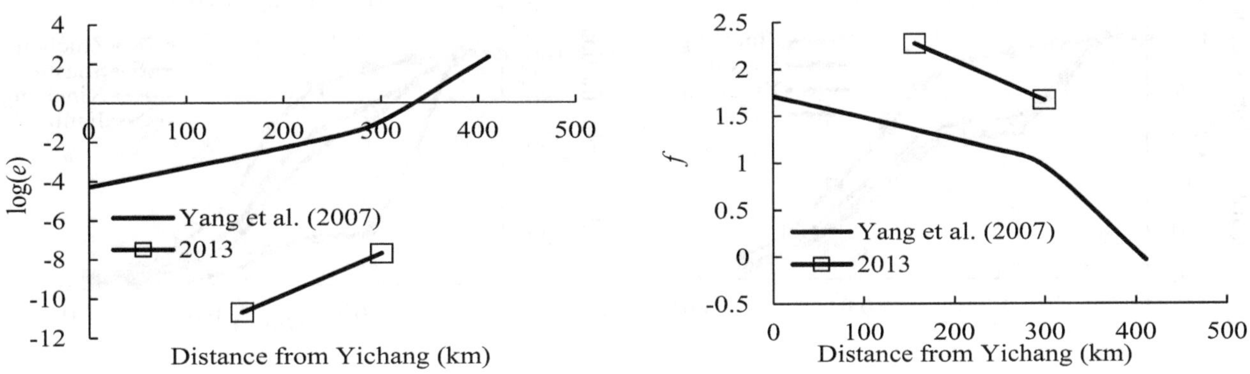

(**a**) (**b**)

Figure 8. Coefficients and exponents of sediment rating curves (**a**) for log(e) and (**b**) for f; solid lines are data drawn from Yang et al. [39].

Instantaneous SST rates are not only a function of the transport capacity of a river, but also of sediment availability [22]. For sediment rating parameters estimated through regression analysis, the e-coefficient is an erosion severity index in the river channel and is associated with the availability of weathered sediments in the basin area [22]. A high e-value indicates that this area is characterized with easily erodible materials and high loads of materials transported by runoffs [20]. The exponent f-coefficient corresponds to the erosive power of the river and the influence on the sediment supply from the entire basin surface. High values indicate a considerable increase in erosive power and sediment-carrying capacity with an increase in river discharge [20]. Thus, a slight increase in discharge enhances the erosive power of the river significantly. For the JJRR, the increased value of the f-parameter indicates a considerable increase of erosive power and sediment-carrying capacity with an increase in river discharge. A decreased value of the log(e)-parameter indicates a decreased availability of suspended sediments in the river channel.

The saturation recovery process in the JJRR can be described with Figures 4 and 5, while the changed interactions between suspended sediment and bed material can be explained by the changed values of rating curves' coefficients (Figure 8).

For river reaches around Zhicheng station, coarsening of bed material is obvious. As to its limited distance from the TGD dam (60 km), the saturation recovery process of suspended sediment is limited.

For river reaches around Shashi station, coarsening of bed material is also obvious. Finer-grouped suspended sediments (particles with diameter < 0.008 mm) are transported from the upper river reach instead of local bed materials. Coarser suspended particles (with diameter of 0.062–0.05 mm) have recovered, which are eroded from local bed materials as an increased value of the sediment rating exponent (f-parameter).

For river reaches around Jianli station, the coarsening of the bed material is limited. The coarsening of suspended sediment is obvious, especially after flooding (Figure 9). Coarser suspended particles (with diameter of 0.062–0.05 mm) may come from the erosion of the local river bed and transportation from upper reaches by an increased value of the f-parameter.

The variability in the relationship between the sediment concentration and water discharge (namely hysteretic patterns) has also been used to explain the variability of sediment sources from one flood to another [40]. Whitaker et al. [41] compared the estimated sediment yield for an un-sampled flood event, a single-event surrogate SSC–Q rating curve, or a long-term SSC–Q rating curve. Analysis revealed that the long-term SSC–Q rating curve estimates approximately 37% higher than the single-event SSC–Q estimates, which indicates a high degree of uncertainty. Analysis shows a high variability in the rating curve between similar floods [41]. This is the reason for erosion during the flood season.

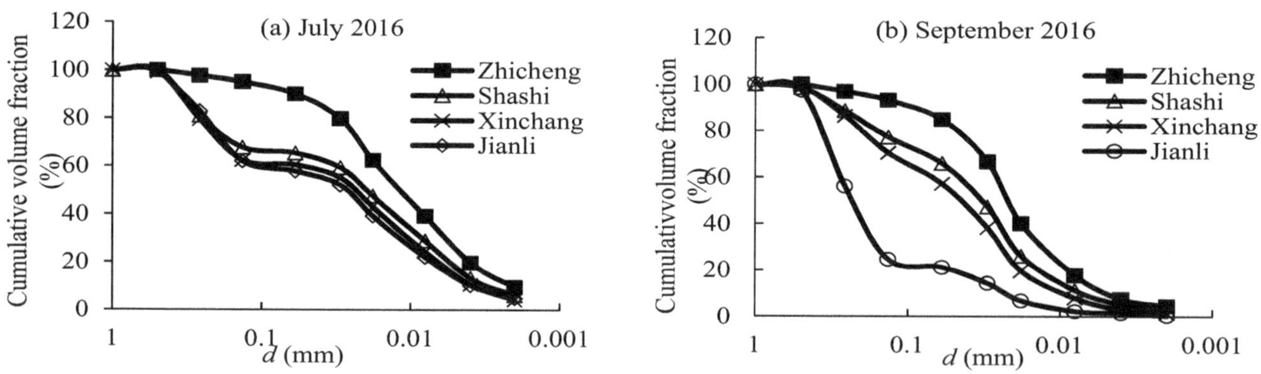

Figure 9. Cumulative volume fractions of suspended sediment at 2016: (**a**) measured in July 2016 and (**b**) data measured in September 2016.

4. Future Challenges

4.1. Accurate Measuring in Field Surveys

Accurate measurement of vertical concentration profiles is important for investigating the vertical distribution of SSC and the accurate estimation of sediment transport. For the field survey in the JJRR, the two near-bed points are measured with constant distances of 0.5 m and 0.1 m from the bed, respectively. However, the small distances may be doubtful due to difficulties in distinguishing the bed material and suspended sediment, especially when the bed is disturbed during settling sampling.

Equipment and criteria adopted in the field survey have been described in detail in the Methodology section, and their availability and methods of improvement with regard to accuracy in the future are discussed.

The availabilities of ADCP and GNSS have been certified by field surveys conducted by other researchers. The error of hydrologic measurement with ADCP is $\pm 1\%$, and the random uncertainty is 2–5% for measuring cross-sections in the middle and lower YR channel [42]. The integrated system (GPS and ADCP) delivers positioning with <15 cm accuracy vertically and <30 cm horizontally [42]. For GNSS, the differences between coordinates of the ground control points estimated using RTK–GNSS, and those computed from the classical topographic measurements during processing of the photogrammetric blocks have been analyzed by Pagliari et al. [43]. It was pointed out that an RTK–GNSS survey may be sufficient to reach the requested tolerance [43]. GNSS receivers have been adopted in field surveys in recent years, as their main advantages are of being faster, cheaper, and easier to use during the surveying phase than classical topographic survey [43,44]. For sediment grain-size analysis, the results by LPSA and sieving methods are comparable when particles are spherical [45]. Sieving methods for coarser particles are more acceptable than for finer ones [45]. Thus, the sieving method adopted by JHWRSB for fine suspended particles needs further verification.

In total, various efforts have being applied to measure the SSC profiles in field surveying, including methods and instrumentation [46–48]. Some new field instrumentation, including magnetic tracers, drone-based or plane-based topography surveys, and video analysis by spectral cameras can also be used to measure sediment transport and morphology evolution in future field surveys [49]. Except for these new instruments, already existing instruments can also be used with more details. For instance, ADCPs are increasingly used to measure the three-dimensional velocity distribution and sediment transport by measuring the Doppler shift in the backscattered signals from an array of acoustic beams [49,50]. According to the backscatter intensity of ADCP beams, SSC can also be estimated with inversion [51].

4.2. Describing Vertical Concentration Profiles

The vertical distribution of SSC needs to be described accurately for an accurate simulation of channel migration, i.e., a one-dimensional simulation for the transport of sediment mixture in non-equilibrium conditions [52]. Thus, finding a proper way to describe the vertical concentration profiles of the unsaturated channel is one of the key problems for hydraulic researchers [53].

Measured sediment profiles are compared with calculations by three equations, i.e., the Rouse Equation (Equation (5)), the Han Equation (Equation (7)), and the new fADE (Equation (12)). Figure 10 indicates the Rouse Equation is not applicable in the non-equilibrium reach. Compared to the Rouse Equation, the Han Equation for non-equilibrium, sediment-laden flow performs better in the upper part of a vertical but underestimates the concentration near the river bed, i.e., within 10% of the water depth. The fADE is able to characterize the dynamics of the non-equilibrium sediment suspension in "starving" river reaches.

No doubt, Equation (7) for non-equilibrium conditions can offer a better description of sediment profiles in river reaches with unsaturated or oversaturated flows. The inaccurate prediction is partly due to our poor understanding of the mechanism of sediment suspension, i.e., the spatially-stochastic transport behavior in sediment dispersion. For the changed sediment regime of the JJRR, more efforts are required to improve the description of vertical profiles of SSC, especially the physical processes. Additionally, "nonlocal" jumps in burst-like suspension events indicate that bursting eddies may sweep particles from the water bottom and directly eject them to the upper part of water bodies, which may lead to ecological-related problems. The ecology consequences include water quality of the riverine ecosystems, as nutrients and pollutants are often closely connected with sediments [49]. The three outlets (Songzikou, Taipingkou, and Ouchikou) have a special setting of the surface water system between YR and Dongting Lake, and the three diversion channels could interact directly with the lake. Thus, it may also have a profound impact on the biological connection between the main stream and the downstream lakes of Poyang and Dongting.

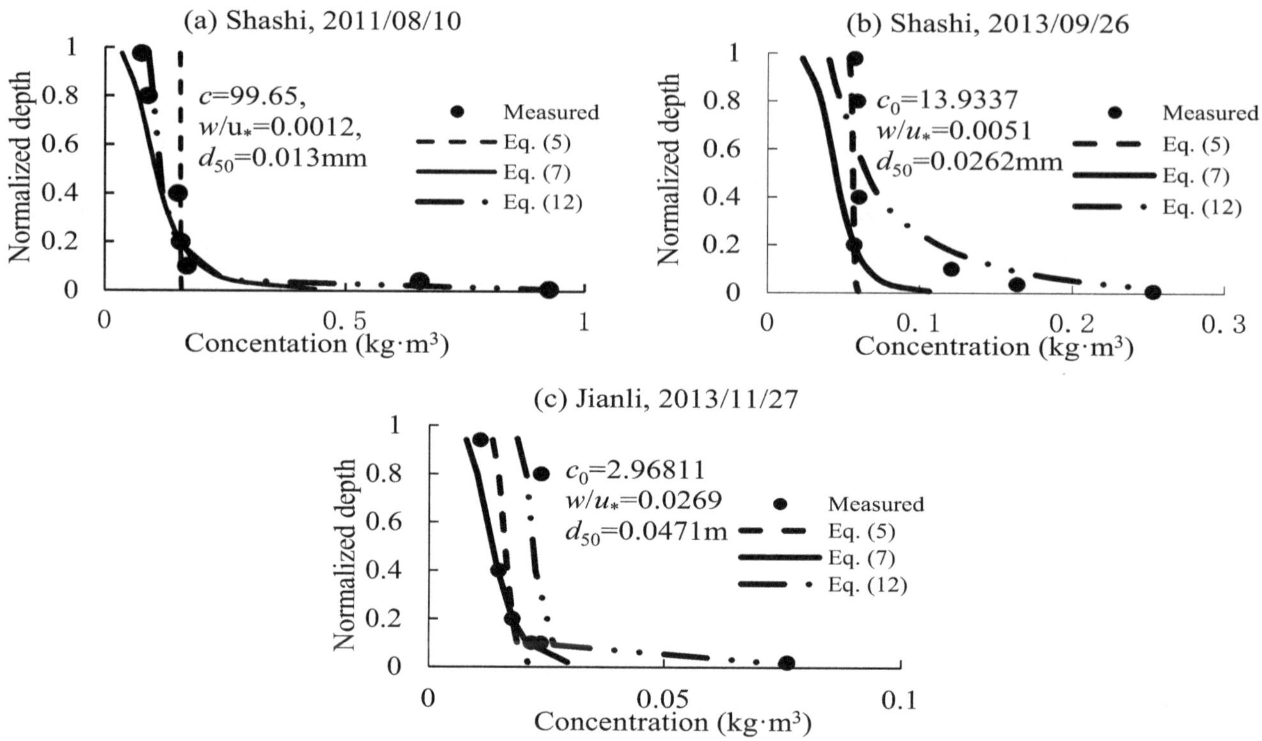

Figure 10. Comparison between measured and calculated vertical profiles of suspended sediment concentration. Data in (**a**) were measured on 10 August 2011 at Shashi station, data in (**b**) were measured on 26 September 2013 at Shashi station, and data in (**c**) were measured on 27 November 2017 at Jianli station.

4.3. Estimating SST Rate

Measuring and describing the vertical SSC profiles accurately is also imperative to the quantification of the SST rate [2]. If near-bed concentrations are not measured or simulated correctly, the estimation of the SST rate may also be influenced.

Figure 11 shows the estimated contributions by two near-bed points. Both of these two extremely large ratios (larger than 60%) occurred at Jianli station in 1996. For these two sections, the estimated sectional average velocities are different from measured data with a large deviation. This indicates that the assumption for Equation (13) is not adoptable in these two cross-sections. Thus, these two sections are excluded from discussion. Figure 11 shows that, if the near-bed concentrations are not included, the sediment flux may be underestimated by approximately 23.5% (Zhicheng), 9.35% (Shashi), and 18.68% (Jianli), respectively. Thus, for the unsaturated JJRR, sediment concentration within 10% of the water depth from the riverbed cannot be ignored. Otherwise, it will lead to underestimated results when calculating the erosion/deposition of the JJRR by the sediment-transport balance method and volume method. Additionally, an underestimated SST rate at Zhicheng station may also lead to overestimated sedimentation in the TGR, which is important for the water storage capacity of the reservoir. The reservoir's functions of water supply, energy production, navigation, flood control, and increased maintenance costs are correlated with the reservoir's water storage capacity [49]. Thus, underestimated sedimentation at Zhicheng station may also affect the reservoir's waterway systems, hydraulic schemes, intakes and outlets, and flow regulation [49].

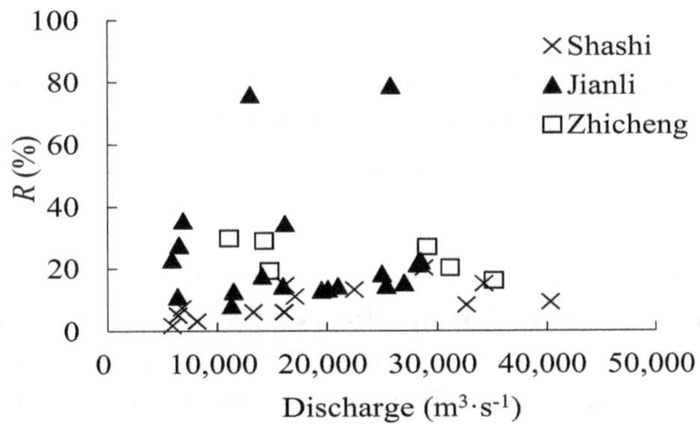

Figure 11. Contribution of the near-bed load on sediment flux.

Fortunately, the influence of near-bed concentration on estimating sediment flux can also be verified by comparing the estimated deposition with the sediment-balance method and the geometric method.

Based on the hydrology data and geometry data measured in 2002–2008, contributing factors that may cause deviation during estimating deposition were distinguished [54,55]. Yuan et al. [53] pronounced that the contribution ratio by near-bed concretion varies in different hydrology years and river reaches (Table 4). For a certain hydrological station, the contribution ratios may also vary with the inflow condition. In total, they partially verified the reliability of measured vertical concentration profiles by comparing contribution ratios.

When suspension index $A = \omega/(\beta k u_*) = 2.0$, suspended sediment may concentrate in areas with a distance of 0.2 times of the water depth from the riverbed, in which β is a coefficient for non-uniform sediment [30]. The near-bed concentration may affect the estimated sediment load only if $A > 0.15$ [30]. If $A = 2.0$, the ratio of near-bed transportation over the total transportation is approximately 30% [30]. The values of ω/u_* with data measured after TGD in Table 2 are much larger than the criteria by Zhang [30], and a large contribution of the near-bed transportation over the total transportation in Figure 11 can be certified. The ratio of the near-bed transportation over the total transportation is approximately 25% when $D_i/D_m > 3$, in which D_i/D_m is the particle size the over average size [30]. The wide grading distribution of SSC after TGD (as shown in Figures 4 and 5) also indicates a large contribution to the total transportation. Thus, the potential doubt that refers to difficulties in distinguishing bed material and suspended sediment can partially be certified. Moreover, the impact of a remarkable large near-bed concentration must be given more attention in the future, especially at river reaches around Shashi station.

Table 4. Contribution by various factors on sediment flux [1].

River Reach	Total Sediment Flux [2] (10^4 m^3)	Contribution by Near-Bed Concentration		Contribution by Other Factors	
		Sediment Flux (10^4 m^3)	Ratio (%)	Sediment Flux (10^4 m^3)	Ratio (%)
Yichang–Zhicheng	10,275	1267	12.33	2047	19.92
Zhicheng–Shashi	11,242	980	8.72	4317	38.40
Shashi–Jianli	12,840	2303	17.94	1494	11.64

[1] Data drawn from Yuan et al. [53] and Duan et al. [54]; [2] total sediment flux is estimated with the geometry method.

In total, fluvial and dynamic characteristics of distinct, large concentrations in near-bed regions may affect sediment flux, erosion/deposition estimation, sediment rating curve, and the channel geometry followed. This brings river engineers a lot of new problems, i.e., degradation of navigation, and dramatic bank erosion [56]. Additionally, the estimation of erosion in the JJRR has a profound impact on the hydrologic connection between the middle YR and Dongting Lake [57].

5. Conclusions

Sediment regime of the JJRR has changed dramatically with the operation of TGD. Based on daily and vertical concentration data at the three hydrological stations (Zhicheng, Shashi, and Jianli) in the JJRR, the sediment regime and its potential challenges are analyzed in this study. The conclusions can be summarized as follows:

(1) In the unsaturated JJRR, measurements have revealed anomalous vertical profiles of SSC, as the near-bed concentrations normalized with the vertical average concentration are dramatically larger than that of the pre-equilibrium channel. The near-bed concentration (within 10% of the water depth from the river bed) may reach up to 15 times that of the vertical average concentration in the non-equilibrium channel.

(2) In the unsaturated JJRR, the combination of wide grading suspended sediment and coarsened bed materials in non-equilibrium channel contribute to a remarkably large concentration in the near-bed zone. For the river reach around Shashi station, the remarkably large near-bed concentration is more apparent.

(3) More detailed measurements or new measuring technologies may assist in the accurate measurement of vertical concentration profiles. A fractional dispersion equation may help to provide accurate descriptions. The outcomes can provide useful information for modeling the morphologic change of the JJRR.

Acknowledgments: The financial support from National Key R&D Program of China (no. 2016YFC0402303); and the National Natural Science Foundation of China (Nos. 51579230, 51779242, 51509234, 41330751 and 51479179).

Author Contributions: L.H. and D.C. conceived and designed the paper, S.Z. analyzed the data, M.L. contributed to figures, G.D. provided materials of vertical concentration, and L.H. wrote the paper.

References

1. Schumm, S.A. *The Fluvial System*; Wiley: New York, NY, USA, 1977.
2. McLean, S.R. Depth-integrated suspended load calculations. *J. Hydraul. Eng.* **1991**, *117*, 1440–1458. [CrossRef]
3. Papanicolaou, A.N.; Elhakeem, M.; Krallis, G.; Prakash, S.; Edinger, J. Sediment transport modelling review-current and future developments. *J. Hydraul. Eng. ASCE* **2008**, *134*, 1–14. [CrossRef]
4. Rouse, H. Modern conceptions of the mechanics of fluid turbulence. *Trans. Am. Soc. Civ. Eng.* **1938**, *102*, 463–543.
5. Hunt, J.N. On the turbulent transport of sediment in open channel. *Proc. R. Soc. Lond. A* **1954**, *224*, 1158. [CrossRef]
6. Aragon, J.A.G.; Salazar, S.S.; Reyes, P.M.; Delgado, C.D. Concentration profiles of suspended sediments calculation in non-uniform sediments, nonequilibrium situations. *Ing. Hidraul. Mex.* **2000**, *15*, 29–35.
7. Maleki, F.S.; Khan, A.A. 1-D coupled non-equilibrium sediment transport modelling for unsteady flows in the discontinuous Galerkin framework. *J. Hydrodyn.* **2016**, *28*, 534–543. [CrossRef]
8. Nicholad, A.P.; Clarke, L.; Quine, T.A. A numerical modelling and experimental study of flow width dynamics on alluvial fans. *Earth Surf. Process. Landf.* **2009**, *34*, 1985–1993. [CrossRef]
9. Han, Q.W.; Chen, X.J.; Xue, X.C. On the vertical distribution of sediment concentration in non-equilibrium transportation. *Adv. Water Sci.* **2010**, *21*, 512–523.
10. Zhao, L.J.; Wu, G.Y.; Wang, J.Y. Review and prospect of vertical distribution by non-equilibrium sediment concentration theory. *J. Hydroelectr. Eng.* **2015**, *34*, 63–69.
11. Regüés, D.; Nadal-Romero, E. Uncertainty in the evaluation of sediment yield form badland areas: Suspended sediment transport estimated in the Araguás catchment (central Spanish Pyrenees). *Catena* **2013**, *106*, 93–100. [CrossRef]
12. Zhu, H.W.; Cheng, P.D.; Li, W.; Chen, J.H.; Pang, Y.; Zhang, D.W. Empirical model for estimating vertical concentration profiles of re-suspended, sediment-associated contaminants. *Acta Mech. Sin.* **2017**, *33*, 846–854. [CrossRef]

13. Cheng, C.; Huang, H.M.; Liu, C.Y.; Jiang, W.M. Challenges to the representation of suspended sediment transfer using a depth-averaged flux. *Earth Surf. Process. Landf.* **2016**, *41*, 1337–1357. [CrossRef]
14. Wu, W. Depth-averaged two-dimensional numerical modelling of unsteady flow and nonuniform sediment transport in open channels. *J. Hydraul. Eng.* **2004**, *130*, 1013–1024. [CrossRef]
15. Tayfur, G.; Singh, V.P. Transport capacity models for unsteady and non-equilibrium sediment transport in alluvial channels. *Comput. Electron. Agric.* **2012**, *86*, 26–33. [CrossRef]
16. Singh, A.K.; Kothyari, U.C.; Raju, K.G.R. Rapidly varying transient flows in alluvial rivers. *J. Hydraul. Res.* **2004**, *42*, 473–486. [CrossRef]
17. He, L. Quantifying the effects of near-bed concentration on the sediment flux after the operation of the Three Gorges Dam, Yangtze River. *Water* **2017**, *9*, 986. [CrossRef]
18. He, Y.; Wang, F.; Tian, P.; Mu, X.M.; Gao, P.; Zhao, G.J.; Wu, Y.P. Impact assessment of human activities on runoff and sediment of Beiluo River in the Yellow River Based on Paired Years of Similar Climate. *Pol. J. Environ. Stud.* **2016**, *25*, 121–135. [CrossRef]
19. Wang, H.J.; Yang, Z.S.; Wang, Y.; Saito, Y.; Liu, J.P. Reconstruction of sediment flux from the Changjiang (Yangtze River) to the sea since the 1860s. *J. Hydrol.* **2008**, *349*, 318–332. [CrossRef]
20. Iadanza, C.; Napolitano, F. Sediment transport time series in the Tiber River. *Phys. Chem. Earth* **2006**, *31*, 1212–1227. [CrossRef]
21. Desilets, S.L.E.; Nijssen, B.; Ekwurzel, B.; Ferre, T.P.A. Post-wildfire changes in suspended sediment rating curves: Sabino Canyon, Arizona. *Hydrol. Process.* **2007**, *21*, 1413–1423. [CrossRef]
22. Asselman, N.E.M. Fitting and interpretation of sediment rating curves. *J. Hydrol.* **2000**, *234*, 228–248. [CrossRef]
23. Huang, S.H.; Sun, Z.L.; Xu, D.; Xia, S.S. Vertical distribution of sediment concentration. *J. Zhejiang Univ. Sci. A* **2008**, *9*, 1560–1566. [CrossRef]
24. Wilson, K.C. Rapid increase in suspended load at high bed shear. *J. Hydraul. Eng.* **2005**, *131*, 46–51. [CrossRef]
25. Brown, G.L. Approximate profile for nonequilibrium suspended sediment. *J. Hydraul. Eng.* **2008**, *134*, 1010–1014. [CrossRef]
26. Jobson, H.E.; Serge, W.W. Predicting concentration profiles in open channels. *J. Hydraul. Div.* **1970**, *96*, 1983–1996.
27. Chen, D.; Sun, H.; Zhang, Y. Fractional dispersion equation for sediment suspension. *J. Hydrol.* **2013**, *491*, 13–22. [CrossRef]
28. Nie, S.Q.; Sun, H.G.; Zhang, Y.; Chen, D.; Chen, W.; Chen, L.; Schaefer, S. Vertical distribution of suspended sediment under steady flow: Existing theories and fractional derivative model. *Discret. Dyn. Nat. Soc.* 2017. [CrossRef]
29. Noguchi, K.; Nezu, I. Particle-turbulence interaction and local particle concentration in sediment-laden open-channel flows. *J. Hydro-Environ. Res.* **2009**, *3*, 54–68. [CrossRef]
30. Sun, Z.L.; Zhang, C.F.; Du, L.H.; Xu, D. Transport rate of nonuniform suspended load. *Shuili Xuebao* **2016**, *47*, 501–508.
31. Dong, X.T.; Li, R.J.; Fu, G.C.; Zhang, H.C. Relationship between vertical distribution of velocity and sediment concentration of sediment-laden flow. *J. Hohai Univ. Natl. Sci.* **2015**, *43*, 371–376. [CrossRef]
32. Tang, M.L.; Cao, H.Q.; Li, Q.Y.; Zhai, W.L. Vertical distribution of suspended sediment in middle reaches of Yangtze River. *Yangtze River* **2017**, *48*, 6–11.
33. Xia, Y.F.; Xu, H.; Chen, Z.; Wu, D.W.; Zhang, S.Z. Experimental study on suspended sediment concentration and its vertical distribution under spilling breaking wave actions in silty coast. *China Ocean Eng.* **2011**, *25*, 565–575. [CrossRef]
34. Feng, Q.; Xiao, Q.L. An new S-type vertical distribution of suspended sediment concentration. *J. Sedim. Res.* **2015**, *1*, 19–24.
35. Parsons, A.J.; Cooper, J.; Wainwright, J. What is suspended sediment? *Earth Surf. Process. Landf.* **2015**, *40*, 1417–1420. [CrossRef]
36. Zhang, X.F.; Tan, G.M. Characteristics of vertical concentration distribution of non-uniform particles. *Shuli Xuebao* **1992**, *10*, 48–52.
37. Han, J.Q.; Sun, Z.H.; Li, Y.T.; Tang, J.W. Changed and caused of lower water level in Yichang–Chenglingji reach after impounding of Three Gorges Reservoir. *Eng. J. Wuhan Univ.* **2011**, *44*, 685–695.
38. Xu, J.X. Complex behaviour of suspended sediment grain size downstream from a reservoir: An example from the Hanjiang River, China. *Hydrol. Sci. J.* **1996**, *41*, 837–849.

39. Yang, G.F.; Chen, Z.Y.; Yu, F.; Wang, Z.; Zhao, Y.; Wang, Z. Sediment rating parameters and their implications: Yangtze River, China. *Geomorphology* **2007**, *85*, 166–175. [CrossRef]
40. Seeger, M.; Errea, M.-P.; Begueria, S.; Arnaez, J.; Marti, C.; Garcia-Ruiz, J.M. Catchment soil moisture and rainfall characteristics as determinant factors for discharge/suspended sediment hysteretic loops in a small headwater catchment in the Spanish Pyrenees. *J. Hydrol.* **2004**, *288*, 299–311. [CrossRef]
41. Whitaker, A.C.; Sato, H.; Sugiyama, H. Changing suspended sediment dynamics due to extreme flood events in a small pluvial-nival system in northern Japan. *Sedim. Dyn. Chang. Environ.* **2008**, *325*, 192–199.
42. Zhang, Q.; Shi, Y.F.; Xiong, M. Geometric properties of river cross sections and associated hydrodynamic implications in Wuhan-Jiujiang river reach, the Yangtze River. *J. Geogr. Sci.* **2009**, *19*, 58–66. [CrossRef]
43. Pagliari, D.; Rossi, L.; Passoni, D.; Pinto, L.; De Michele, C.; Avanzi, F. Measuring the volume of flushed sediments in a reservoir using multi-temporal images acquired with UAS, Geomatics. *Nat. Hazards Risk* **2017**, *8*, 150–166. [CrossRef]
44. Bandini, F.; Jakobsen, J.; Olesen, D.; Reyna-Gutierrez, J.A.; Bauer-Gottwein, P. Measuring water level in rivers and lakes from lightweight unmanned aerial vehicles. *J. Hydrol.* **2017**, *548*, 237–250. [CrossRef]
45. Li, W.K.; Wu, Y.X.; Huang, Z.M.; Fan, R.; Lv, J.F. Measurement results comparison between Laser Particle Analyzer and Sieving method in particle size distribution. *Zhongguo Fenti Jishu* **2007**, *13*, 10–13.
46. Pedocchi, F.; Garcia, M.H. Acoustic measurement of suspended sediment concentration profiles in an oscillatory boundary layer. *Cont. Shelf Res.* **2012**, *46*, 87–95. [CrossRef]
47. Yu, Q.; Flemming, B.W.; Gao, S. Tide-induced vertical suspended sediment concentration profiles: Phase lag and amplitude attenuation. *Ocean Dyn.* **2011**, *61*, 403–410. [CrossRef]
48. Pitarch, J.; Odermatt, D.; Kawak, M.; Wuest, A. Retrieval of vertical particle concentration profiles by optical remote sensing: A model study. *Opt. Express* **2014**, *22*, 947–959. [CrossRef] [PubMed]
49. Schleiss, A.; Franca, M.J.; Juez, C.; Cesare, G. Reservoir sedimentation. *J. Hydraul. Res.* **2016**, *54*, 595–614. [CrossRef]
50. Togneri, M.; Lewis, M.; Neill, S.; Masters, I. Comparison of ADCP observations and 3D model simulations of turbulence at a tidal energy site. *Renew. Energy* **2017**, *114*, 273–282. [CrossRef]
51. Jin, W.F.; Liang, C.J.; Zhou, B.F.; Li, J.D. Measurement and analysis of high suspended sediment concentration in Jintang channel with vessel-mounted ADCP. *J. Mar. Sci.* **2009**, *27*, 31–39.
52. Merkhali, S.P.; Ehteshami, M.; Sadrnejad, S.A. Assessment quality of a nonuniform suspended sediment transport model under unsteady flow condition (case study: Aras River). *Water Environ. J. Promot. Sustain. Solut.* **2015**, *29*, 189–498. [CrossRef]
53. Eggenhuisen, J.T.; McCaffrey, W.D. The vertical turbulence structure of experimental turbidity currents encountering basal obstructions: Implications for vertical suspended sediment distribution in non-equilibrium currents. *Sedimentology* **2012**, *59*, 1101–1120. [CrossRef]
54. Yuan, Y.; Zhang, X.F.; Duan, G.L. Modifying the channel erosion and deposition amount of Yichang–Jianli reach calculated by sediment discharge method. *J. Hydroelectr. Eng.* **2014**, *33*, 163–169.
55. Duan, G.L.; Peng, Y.B.; Guo, M.J. Comparative analysis on riverbed erosion and deposition amount calculated by different methods. *J. Yangtze River Sci. Res. Inst.* **2014**, *31*, 108–118.
56. Xia, J.Q.; Deng, S.S.; Lu, J.Y.; Xu, Q.X.; Zong, Q.L.; Tan, G.M. Dynamic channel adjustments in the Jingjiang reach of the middle Yangtze River. *Sci. Rep.* **2016**, *6*, 22802. [CrossRef] [PubMed]
57. Li, Y.Y.; Yang, G.S.; Li, B.; Wan, R.R.; Duan, W.L.; He, Z. Quantifying the effects of channel change on the discharge diversion of Jingjiang Three outlets after the operation of the Three Gorges Dam. *Hydrol. Res.* **2016**, *47*, 161–174. [CrossRef]

Permissions

All chapters in this book were first published in MDPI; hereby published with permission under the Creative Commons Attribution License or equivalent. Every chapter published in this book has been scrutinized by our experts. Their significance has been extensively debated. The topics covered herein carry significant findings which will fuel the growth of the discipline. They may even be implemented as practical applications or may be referred to as a beginning point for another development.

The contributors of this book come from diverse backgrounds, making this book a truly international effort. This book will bring forth new frontiers with its revolutionizing research information and detailed analysis of the nascent developments around the world.

We would like to thank all the contributing authors for lending their expertise to make the book truly unique. They have played a crucial role in the development of this book. Without their invaluable contributions this book wouldn't have been possible. They have made vital efforts to compile up to date information on the varied aspects of this subject to make this book a valuable addition to the collection of many professionals and students.

This book was conceptualized with the vision of imparting up-to-date information and advanced data in this field. To ensure the same, a matchless editorial board was set up. Every individual on the board went through rigorous rounds of assessment to prove their worth. After which they invested a large part of their time researching and compiling the most relevant data for our readers.

The editorial board has been involved in producing this book since its inception. They have spent rigorous hours researching and exploring the diverse topics which have resulted in the successful publishing of this book. They have passed on their knowledge of decades through this book. To expedite this challenging task, the publisher supported the team at every step. A small team of assistant editors was also appointed to further simplify the editing procedure and attain best results for the readers.

Apart from the editorial board, the designing team has also invested a significant amount of their time in understanding the subject and creating the most relevant covers. They scrutinized every image to scout for the most suitable representation of the subject and create an appropriate cover for the book.

The publishing team has been an ardent support to the editorial, designing and production team. Their endless efforts to recruit the best for this project, has resulted in the accomplishment of this book. They are a veteran in the field of academics and their pool of knowledge is as vast as their experience in printing. Their expertise and guidance has proved useful at every step. Their uncompromising quality standards have made this book an exceptional effort. Their encouragement from time to time has been an inspiration for everyone.

The publisher and the editorial board hope that this book will prove to be a valuable piece of knowledge for researchers, students, practitioners and scholars across the globe.

List of Contributors

André R. Fonseca
Centre for the Research and Technology of Agro-Environmental and Biological Sciences, CITAB, 5000-801 Vila Real, Portugal

João A. Santos
Centre for the Research and Technology of Agro-Environmental and Biological Sciences, CITAB, 5000-801 Vila Real, Portugal
Departamento de Física, Universidade de Trás-os-Montes e Alto Douro, UTAD, 5000-801 Vila Real, Portugal

Simone G.P. Varandas and Rui M.V. Cortes
Centre for the Research and Technology of Agro-Environmental and Biological Sciences, CITAB, 5000-801 Vila Real, Portugal
Departamento de Ciências Florestais e Arquitetura Paisagista, Universidade de Trás-os-Montes e Alto Douro, UTAD, 5000-801 Vila Real, Portugal

Sandra M. Monteiro and Edna Cabecinha
Centre for the Research and Technology of Agro-Environmental and Biological Sciences, CITAB, 5000-801 Vila Real, Portugal
Departamento de Biologia e Ambiente, Universidade de Trás-os-Montes e Alto Douro, UTAD, 5000-801 Vila Real, Portugal

José L. Martinho
Departamento de Geologia, Universidade de Trás-os-Montes e Alto Douro, UTAD, 5000-801 Vila Real, Portugal
Geosciences Center, University of Coimbra, 3030-790 Coimbra, Portugal

Chih-Chiang Wei
Department of Marine Environmental Informatics & Center of Excellence for Ocean Engineering, National Taiwan Ocean University, Keelung 20224, Taiwan

Paweł Marcinkowski and Mateusz Grygoruk
Department of Hydraulic Engineering, Faculty of Civil and Environmental Engineering, Warsaw University of Life Sciences, Nowoursynowska 166 Street, 02-787 Warsaw, Poland

Changming Ji, Hongjie Yu, Jiajie Wu and Xiaoran Yan
School of Renewable Energy, North China Electric Power University, Beijing 102206, China

Rui Li
Overseas Exam Service Center, Shijiazhuang Information Engineering Vocational College, Shijiazhuang 050035, China

Fen Zhao and Chunhui Li
Ministry of Education Key Lab of Water and Sand Science, School of Environment, Beijing Normal University, Beijing 100875, China

Libin Chen
College of Resources and Environment, Yangtze University, Wuhan 430100, China

Yuan Zhang
Sate Key Laboratory of Environmental Criteria and Risk Assessment, Chinese Research Academy of Environmental Sciences, Beijing 100012, China

Shengtian Yang, Changsen Zhao and Hezhen Lou
College of Water Sciences, Beijing Normal University, Beijing 100875, China

Juan Bai
State Key Laboratory of Remote Sensing Science, Faculty of Geographical Science, Beijing Normal University, Beijing Key Laboratory for Remote Sensing of Environment and Digital Cities, Beijing 100875, China
College of Geographical Science, Shanxi Normal University, Linfen 041004, China

Zhiwei Wang
Key Laboratory of Radiometric Calibration and Validation for Environmental Satellites, National Satellite Meteorological Center, China Meteorological Administration, Beijing 100081, China

Yabing Guan, Yichi Zhang, Chunbin Zhang and Xinyi Yu
State Key Laboratory of Remote Sensing Science, Faculty of Geographical Science, Beijing Normal University, Beijing Key Laboratory for Remote Sensing of Environment and Digital Cities, Beijing 100875, China

Chao Zhou and Yi Ding
Changjiang Institute of Survey, Planning, Design and Research, Wuhan 430010, China

Na Sun, Lu Chen, Jianzhong Zhou, Gang Zha, Guanglei Luo, Ling Dai and Xin Yang
School of Hydropower and Information Engineering, Huazhong University of Science and Technology, Wuhan 430074, China

Mikyoung Choi
International Water Resources Research Institute, Chungnam National University, Daejeon 34134, Korea

Yasuhiro Takemon
Disaster Prevention Research Institute, Kyoto University, Kyoto 611-0011, Japan

Kinko Ikeda
ASIA AIR SURVEY CO., LTD., Osaka 530-6029, Japan

Kwansue Jung
Department of Civil Engineering, Chungnam National University, Daejeon 34134, Korea

Ching-Nuo Chen
International Master Program of Soil and Water Engineering, National Pingtung University of Science and Technology, Pingtung 91201, Taiwan

Chih-Heng Tsai
Department of Recreation and Healthcare Management, Chia Nan University of Pharmacy and Science, Tainan 71710, Taiwan

Yueyi Liu and Hang Zheng
School of Environment and Civil Engineering, Dongguan University of Technology, Dongguan 523808, China

Jianshi Zhao
State Key Laboratory of Hydro-Science and Engineering, Department of Hydraulic Engineering, Tsinghua University, Beijing 100086, China

Tsuyoshi Kinouchi, Gakuji Yamamoto and Atchara Komsai
Department of Transdisciplinary Science and Engineering, School of Environment and Society, Tokyo Institute of Technology, Yokohama 226-8503, Japan

Winai Liengcharernsit
Department of Environmental Engineering, Faculty of Engineering, Kasetsart University, Bangkok 10900, Thailand

Li He and Shiyan Zhang
Key Laboratory of Water Cycle and Related Land Surface Processes, Institute of Geographic Sciences and Natural Resources Research, Chinese Academy of Sciences, Beijing 100101, China

Dong Chen and Meng Liu
Key Laboratory of Water Cycle and Related Land Surface Processes, Institute of Geographic Sciences and Natural Resources Research, Chinese Academy of Sciences, Beijing 100101, China
College of Resources and Environment, University of Chinese Academy of Sciences, Beijing 100049, China

Guanglei Duan
Jingjiang Hydrology and Water Resources Surveying Bureau, Yangtze River Water Resources Commission, Wuhan 434020, China

Index

www.ingramcontent.com/pod-product-compliance
Lightning Source LLC
LaVergne TN
LVHW060519210926
848485LV00005B/1108